Communications in Computer and Information Science 2956

Series Editors

Rationale

The CCIS series is devoted to the publication of proceedings of computer science conferences. Its aim is to efficiently disseminate original research results in informatics in printed and electronic form. While the focus is on publication of peer-reviewed full papers presenting mature work, inclusion of reviewed short papers reporting on work in progress is welcome, too. Besides globally relevant meetings with internationally representative program committees guaranteeing a strict peer-reviewing and paper selection process, conferences run by societies or of high regional or national relevance are also considered for publication.

Topics

The topical scope of CCIS spans the entire spectrum of informatics ranging from foundational topics in the theory of computing to information and communications science and technology and a broad variety of interdisciplinary application fields.

Information for Volume Editors and Authors

Publication in CCIS is free of charge. No royalties are paid, however, we offer registered conference participants temporary free access to the online version of the conference proceedings on SpringerLink (http://link.springer.com) by means of an http referrer from the conference website and/or a number of complimentary printed copies, as specified in the official acceptance email of the event.

CCIS proceedings can be published in time for distribution at conferences or as post-proceedings, and delivered in the form of printed books and/or electronically as USBs and/or e-content licenses for accessing proceedings at SpringerLink. Furthermore, CCIS proceedings are included in the CCIS electronic book series hosted in the SpringerLink digital library at http://link.springer.com/bookseries/7899. Conferences publishing in CCIS are allowed to use our online conference service (Meteor) for managing the whole proceedings lifecycle (from submission and reviewing to preparing for publication) free of charge.

Publication process

The language of publication is exclusively English. Authors publishing in CCIS have to sign the Springer CCIS copyright transfer form, however, they are free to use their material published in CCIS for substantially changed, more elaborate subsequent publications elsewhere. For the preparation of the camera-ready papers/files, authors have to strictly adhere to the Springer CCIS Authors' Instructions and are strongly encouraged to use the CCIS LaTeX style files or templates.

Abstracting/Indexing

CCIS is abstracted/indexed in DBLP, Google Scholar, EI-Compendex, Mathematical Reviews, SCImago, Scopus. CCIS volumes are also submitted for the inclusion in ISI Proceedings.

How to start

To start the evaluation of your proposal for inclusion in the CCIS series, please send an e-mail to ccis@springer.com

Nurul Adilla Mohd Subha ·
Norhaliza Abdul Wahab · Pei Ling Leow ·
Herman Wahid · Siti Maherah Hussin ·
Zaharuddin Mohamed
Editors

Systems Modelling and Simulation

Second International Symposium, SMS 2026
Putrajaya, Malaysia, May 4–5, 2026
Proceedings

Editors
Nurul Adilla Mohd Subha
Universiti Teknologi Malaysia
Johor Bahru, Malaysia

Pei Ling Leow
Universiti Teknologi Malaysia
Johor Bahru, Malaysia

Siti Maherah Hussin
Universiti Teknologi Malaysia
Kuala Lumpur, Malaysia

Norhaliza Abdul Wahab
Universiti Teknologi Malaysia
Johor Bahru, Malaysia

Herman Wahid
Universiti Teknologi Malaysia
Johor Bahru, Malaysia

Zaharuddin Mohamed
Universiti Teknologi Malaysia
Johor Bahru, Malaysia

ISSN 1865-0929 ISSN 1865-0937 (electronic)
Communications in Computer and Information Science
ISBN 978-981-92-0305-5 ISBN 978-981-92-0306-2 (eBook)
https://doi.org/10.1007/978-981-92-0306-2

This Springer imprint is published by the registered company Springer Nature Singapore Pte Ltd.
The registered company address is: 152 Beach Road, #21-01/04 Gateway East, Singapore 189721, Singapore

Preface

The 2nd International Symposium on Systems Modelling and Simulation (SMS 2026) was held in Putrajaya, Kuala Lumpur, Malaysia on 4–5th May 2026. This symposium was organized by the Malaysian Simulation Society.

This symposium provides a platform for scientists, academicians, and professionals from around the world to present and discuss their work and new emerging simulation technologies. SMS 2026 focused on important topics in Modelling and Simulation that were related to current and emerging technologies. In particular, applications of modelling and simulation to advanced systems such as robotics, smart manufacturing, intelligent systems and machine learning were of interest. Moreover, SMS 2026 encouraged articles that apply theoretical work to practical applications.

The papers presented at SMS 2026 have been compiled in this volume of Communications in Computer and Information Science (CCIS). SMS 2026 specifically focused on three main tracks, namely: Emerging Trends in Engineering and Computing, Modelling and Simulation for Digital Transformation, and Intelligent System and Applications. Following a single-blind review of 53 submissions in which submissions received three reviews each on average, 23 papers were selected for oral presentation.

As editorial members of the SMS 2026 symposium, we would like to express our gratitude to all the authors who chose this symposium as a venue for their publications. We are also very grateful for the support and tremendous effort of the Program and Organizing Committee members for soliciting and selecting research papers with a balance of high quality and new ideas and applications. Thank you from us.

March 2026

Nurul Adilla Mohd Subha
Norhaliza Abdul Wahab
Pei Ling Leow
Herman Wahid
Siti Maherah Hussin
Zaharuddin Mohamed

Organization

Conference Chair

Herman Wahid	Universiti Teknologi Malaysia, Malaysia

Conference Co-chair

Nurul Adilla Mohd Subha	Universiti Teknologi Malaysia, Malaysia

Program Chair

Norhaliza Abdul Wahab	Universiti Teknologi Malaysia, Malaysia

Program Co-chairs

Gary Tan	National University of Singapore, Singapore
Liang Li	Ritsumeikan University, Japan
Mohd Fua'ad Rahmat	Universiti Teknologi Malaysia, Malaysia
Pei Ling Leow	Universiti Teknologi Malaysia, Malaysia
Xiao Song	Beihang University, China
Zaharuddin Mohamed	Universiti Teknologi Malaysia, Malaysia

Publication Chairs

Nurul Adilla Mohd Subha	Universiti Teknologi Malaysia, Malaysia
Siti Maherah Hussin	Universiti Teknologi Malaysia, Malaysia

Finance Chair

Fazilah Hassan	Universiti Teknologi Malaysia, Malaysia

Organizing Chairs

Mohd Saiful Azimi Mahmud	Universiti Teknologi Malaysia, Malaysia
Noorhazirah Sunar	Universiti Teknologi Malaysia, Malaysia
Salinda Buyamin	Universiti Teknologi Malaysia, Malaysia

Organizing Co-Chairs

Mohd Shahrizal Rusli	Universiti Teknologi Malaysia, Malaysia
Noor Hanis Izzuddin Mat Lazim	Universiti Sains Islam Malaysia, Malaysia
Sophan Wahyudi Nawawi	Universiti Teknologi Malaysia, Malaysia

Publicity Chairs

Hazriq Izzuan Jaafar	Universiti Teknikal Malaysia Melaka, Malaysia
Herman Wahid	Universiti Teknologi Malaysia, Malaysia

Sponsorship Chairs

Abdul Rashid Husain	Universiti Teknologi Malaysia, Malaysia
Anita Ahmad	Universiti Teknologi Malaysia, Malaysia

Program Committee

Aldo Jonathan Muñoz Vázquez	Texas A&M University, USA
Andi Andriansyah	Universitas Mercu Buana, Indonesia
Awanis Romli	Universiti Malaysia Pahang Al-Sultan Abdullah, Malaysia
Bidyadhar Subudhi	National Institute of Technology Rourkela, India
Byeong-Yun Chang	Ajou University, South Korea
Cristina Stoica	Université Paris-Saclay, France
Gary Tan	National University of Singapore, Singapore
He Chen	Hebei University, China
Jie Huang	Beijing Institute of Technology, China
Le Anh Tuan	Vietnam Maritime University, Vietnam
Liang Li	Ritsumeikan University, Japan
Lin Zhang	Beihang University, China
Mohammad Osman Toki	London South Bank University, UK

Maurizio Palesi	University of Catania, Italy
Mehmet Önder Efe	Hacettepe University, Turkey
Mohammad Hasan Shaheed	Queen Mary University of London, UK
Mohd Khair Hassan	Universiti Putra Malaysia, Malaysia
Muhammad Nasiruddin Mahyuddin	Universiti Sains Malaysia, Malaysia
Mujahed Mohammad Al Dhaifallah	King Fahd University of Petroleum and Minerals, Saudi Arabia
Mustapha Muhammad	Bayero University, Nigeria
Naomie Salim	Universiti Teknologi Malaysia, Malaysia
Ning Sun	Nankai University, China
Pawel Kwiaton	Częstochowa University of Technology, Poland
Ramon Vilanova Arbos	University of Barcelona, Spain
Satoshi Tanaka	Ritsumeikan University, Japan
Shaharin Anwar bin Sulaiman	Universiti Teknologi Petronas, Malaysia
Siti Fauziah Toha	Universiti Islam Antarabangsa Malaysia, Malaysia
Subhas Mukhopadhya	Macquarie University, Australia
Sumeet S. Aphale	University of Aberdeen, UK
Tan Ai Hui	Multimedia Universiti, Malaysia
Umar Shahbaz Khan	National University of Sciences and Technology, Pakistan

Contents

Optimal Connected-Area Distribution for Multi-agent Coverage Path Planning

Rashad Abul Khayr[1], Muhammad Zakiyullah Romdlony[2](✉), Md Abdus Samad Kamal[1], and Kou Yamada[1]

[1] School of Science and Technology, Gunma University, Maebashi, Japan
[2] School of Electrical Engineering, Telkom University, Bandung, Indonesia
zakiyullah@telkomuniversity.ac.id

Abstract. Autonomous unmanned systems have numerous applications in a complex environment. One such application is coverage path planning to explore and cover a known or unknown area. The efficiency of such an application, especially for multi-agents, can be greatly increased by dividing the active area into an exclusive connected area of operation beforehand. Finding an equal division for the coverage path planning can be crucial to the performance of such operation. Therefore, we proposed an optimized region distribution method using Particle Swarm Optimization (PSO) that was modified with repulsion field and inspired by A* to find optimal base placement, and Spanning Tree Coverage (STC) with connectivity-conscious bidding to ensure equal distribution. The proposed method managed to achieve even distribution in an area with or without obstacle outperforming Rapidly Exploring Random Trees (RRT) and Centroid Veronoi Tesselation (CVT), and in an area with large amount of obstacle, managed to outpace both methods in term of iteration or average time.

Keywords: Area partioning · PSO · Repulsive force · STC · Multi-agent system

1 Introduction

Autonomous unmanned systems have numerous applications in a complex environment [1–3]. One such application is Coverage Path Planning (CPP) to explore and cover a known or unknown area [4,5]. CPP has become a valuable solution for many industries that require autonomous coverage from cleaning [6], agriculture [7,8], fisheries [9], or search and rescue [10]. The main mission of CPP is to make agent visit a coressponding area. This area can be a simple square area or a complex polynomial area [11,12]. Depending on the size of the area, the CPP method can consumed a lot of resources. One way to eliviate this problem and increase the efficiency of CPP, more agent can be added to divide the workload. The efficiency of CPP can be further increased by adding another agent to help with area covering or called Multi-Agent Coverage Path Planning (mCPP) [13].

N. A. Mohd Subha et al. (Eds.): SMS 2026, CCIS 2956, pp. 1–13, 2026.
https://doi.org/10.1007/978-981-92-0306-2_1

The implementation of mCPP can be rather difficult as the agents were prone to revisiting Some inspiration can be taken from [12] to solve this problem by dividing the active area into multiple exclusive regions. This solution was implemented in [13,14] for mCPP by reducing it into a single robot CPP problem by dividing the area into an equal-sized exclusive region. The method was proven to be effective even in a complex environment with obstacles [15]. [15] also shows another problem that can occur when dividing the active area which is disconnected area that need to be considered. While the mCPP method, as mentioned before, can be improved by dividing the active area into multiple regions. The distribution of said region should be able to be optimized by moving the starting point of each region. This movement is similar in concept to Centroid Voronoi Tesselation (CVT) [16] to decrease the iteration needed to find a balanced distribution for each region. Therefore, we proposed an optimized region distribution method using Particle Swarm Optimization that was modified with repulsion field and inspired by A* to find optimal base placement, and Spanning Tree Coverage with connectivity-conscious bidding to ensure equal distribution.

The rest of this paper is organized as follows. Section 2 provides a description of the method for finding the optimal position and dividing the region alongside the overall algorithm of the proposed method. Section 3 describes the simulation results for multiple methods in environments with and without obstacles. The conclusions from the comparative evaluation are summarized alongside the future work, which was written in Sect. 4.

2 Method and Algorithm

2.1 Problem Formulation

As mention briefly in Introduction, the main objective of the proposed method is to create an even connected area distribution. To ensure that an optimal solution can be achieved, there should not exist an overlap between set of grid in region i (L_I) and set of grid in region j (L_j) for every i and j in set of region (N). There also should not exist an overlap for set L_i and set of obstacle D for every i in set of region (N). The combination of every regions should completely cover the entirerity of grid in an area except obstacle (L). Each grid the region hold should approximate others region. Lastly, there should not exist a disconnected grid or group of grid in a region. The equivalence conditions required to establish an optimal regions solution similar to [15] are summarized as follows:

- $L_i \bigcap L_j = \emptyset, \quad \forall i, j \in N, \quad i \neq j;$
- $L_i \bigcap D = \emptyset, \quad \forall i \in N,$
- $L_1 \bigcup L_2 \bigcup ... \bigcup L_n \supseteq L,$
- $L_1 \approx L_2 \approx ... \approx L_n,$
- $L_i \quad \textit{is connected} \quad \forall i \in N$

2.2 Optimal Base Grid Placement

Before the area can be distributed to each region, the base grid placement need to be moved into an optimal position. By preemptively moving the base grid, the amount of iteration needed to find an equal distribution for each region can be minimized. In order to find the optimal position, a Particle Swarm Optimization (PSO) [17] that can be defined in Eq. (1) was employed.

$$v_{(t+1)} = wv_{(t)} + c_1 r_1 (P_1 - x_{i(t)}) + c_2 r_2 (P_2 - x_{i(t)}) \tag{1}$$

with v as the base region velocity, w as the inertia factor, and x_i as the base region i position. P_1 and P_2 are the positions, c_1 and c_2 are the fixed gain, r_1 and r_2 are the random gain for personal and global best respectively. The range random gain r_1 and r_2 itself are between 0 and 1.

The new velocity of particle was influenced by the particle own best position and the overall system best position. This method promote cooperation between particle to find the optimal position that every particle will converge into. To find the optimal base grid however, each particle nead to diverge to the farthest position possible rather than converge into one. Therefore a change in PSO algorithm need to be done for this implementation. First the global best was changed with repulsive force from other particles to find the farthest position possible. Equation (2) can be used to calculate the force.

$$f_r = -c_2 r_2 \sum_{n=1}^{j} e^{-(\frac{x_n - x_i}{a})^2} \tag{2}$$

with j as the total region and a a positive definite real number to control repulsive force effective distance. Since the optimal base placement only concern with distance to others, the personal base from previous iteration no longer hold similar value as the base grid move away from each others. Inspired by A* method [18], each base grid search their adjacent grid for the furthest possible position to other particle for personal best instead. Equation (3) shows the total Euclidean distance from all other particles with f_s as total distance and x_s as the position. Then the best position was searched using Eq. (4) with f_s^* as the farthest total distance and x_s^* as the best position.

$$f_s = \sum_{n=1}^{j} \|x_n - x_s\| \tag{3}$$

$$f_s^* = \max f_s \tag{4}$$

By substituting Eqs. (2) and (3) into (1), a new optimal position can be defined in Eq. (5).

$$v_{(t+1)} = wv_{(t)} + c_1 r_1 (x_{s(t)}^* - x_{i(t)}) - f_{r(t)} \tag{5}$$

In summary, the process to find the optimal base grid position can be achieved by Algorithm 1.

Algorithm 1: Calculate optimal base grid placement for one region

While **Base Grid Position Not Optimal**
  ForEach $k \in n$
    $f_r = f_r - c_2 r_2 e^{-(\frac{x_k - x_i}{a})^2}$
  EndFor
  ForEach $s \in$ **Adjecent Grid**
    $f_s = \sum_{n=1}^{j} \|x_n - x_s\|$
    If $f_s > f_s^*$
      $f_s^* = f_s$
    EndIf
  EndFor
  $v_{(t+1)} = wv_{(t)} + c_1 r_1 (x_{s(t)}^* - x_i) + c_2 r_2 f_r$
  $x_i = x_i + v_{(t+1)}$
EndWhile

2.3 Balanced Region Distribution

The distribution of the grid was started using a Spanning Tree Coverage (STC) similar to [19] from the base grid in previous subsection. Once every active grid was claimed, each region identify every grids that border other regions and assigned them a point. Assuming that grid i is the element of region l, Eq. (6) shows the point assignment for each grid.

$$B = \left(\frac{1}{C_i} + \sum_{n=1}^{k} \frac{1}{C_n}\right) \frac{S - A_l}{F_l} \tag{6}$$

with B as the grid point, C_i and C_n as the connectivity of grid i and n, k is the total adjacent grid i in region l, S is the total available grid, A_l is the total grid covered by region l, and F_l is the total grid in region l that borders other regions. The inclusion of adjacent connectivity is to deflect any grid acquisition to the outermost part of the region, as they naturally have less connection and therefore fewer points to defend it. The connectivity also prevents grid acquisition to completely divide a region. After assigning a point to the grid, they can take their adjacent grid as long as the following condition is fulfilled.

$$B_d < bB_a \tag{7}$$

with B_d and B_a as the points for defending and attacking the grid, respectively. b itself is a definite positive to adjust the aggressiveness of the distribution. The algorithm to equalize the area distribution can be seen in Algorithm 2.

3 Simulation Result

To see the effectiveness of the proposed method, two kinds of simulations were done in an environment with and without obstacles. The proposed method was tasked to distribute an equal amount of area into five regions in a square map

Algorithm 2: Calculate area distribution for one region

While **Distribution Not Equal**
  Define Grid Bordering Other Region
  ForEach **Grid Bordering Other Region**
    $B_a = (\frac{1}{C_i} + \sum_{n=1}^{k} \frac{1}{C_n}) \frac{S - A_l}{F_l}$
    ForEach **Adjecent Grid** $\notin$ **Same Region**
      If $B_d < bB_a$
        Take Adjecent Grid
      EndIf
    EndFor
  EndFor
EndWhile

that was divided into a 15 by 15 grid map, with the base grid for each region placed at the center of the map at the start. To validate the proposed method, the result was compared with optimized Rapidly exploring Random Trees (RRT) [15] and CVT [16] methods using similar parameters, with the exception of the optimized RRT method that required a predefined base grid.

3.1 Region Distribution

Based on the task and environment as previously mentioned, a simulation was done without obstacle. Each methods was tasked to evenly distribute the area into five regions and terminate when even distribution or steady state was achieved with steady state defined as a point where there are no more grid being exchange. The simulation was done five times with iteration limited at 20. Using GNU Octave with Gen 11th Intel i5 and RTX 3050 Mobile, the simulation produced a visual representation of the grid distribution for each method with each region were differently colored and the darker color was used to represent base grid for given regions that can be seen in Fig. 1, with the distribution for each iteration that can be seen in Fig. 2.

The proposed method managed to evenly divide the area into five regions with 45 grids each. This division was met after five iterations with average time of 9.37 s with 0.89 s was dedicated to finding the optimal base grid. The other methods on another hand, was unable to evenly divide the area and managed to achieved unoptimal steady state after five iterations or an average time of 7.55 s for RRT and four iterations or an average time of 6.11 s for CVT. While the speed was not impressive, the fact that the proposed method was able to divide the area evenly while other methods does not need to be noted. The unequal distribution of the optimized RRT is due to the two regions that wanted to trade grids to achieve equilibrium but were blocked by another region. As for the CVT, since all of the base grid managed to reach the centroid of the region, this method no longer moved the base grid despite the unbalanced distribution.

Shown in Fig. 1, each methods managed to maintain connectivity of each grid in their respective regions. another thing of note for the proposed method is that the base grid did not find the most optimal position for the green and yellow region. As the optimal placement of the base grid only concern with distance to others, a simple calculation can show that the optimal placement for the two base grid were at the corner of the map. This problem is due to repulsion of other regions, especially from the purple region that denied the corner position for both of them. The local search was supposed to solve this problem; however, the search area was too small to find a better solution, resulting in suboptimal placement. Despite the problem, the proposed method still shows its effectiveness as the only one to achieve equal distribution without disconnected regions.

3.2 Region Distribution with Obstacle

In order to demonstrate the robustness of our proposed method, another simulation was conducted using similar parameters as the previous one with some obstacles as an addition to the simulation. Three obstacles were added with placement that can be seen in Fig. 3.

Comparing once more between the three methods with added obstacles, we managed to get a visual representation of the grid distribution for each method

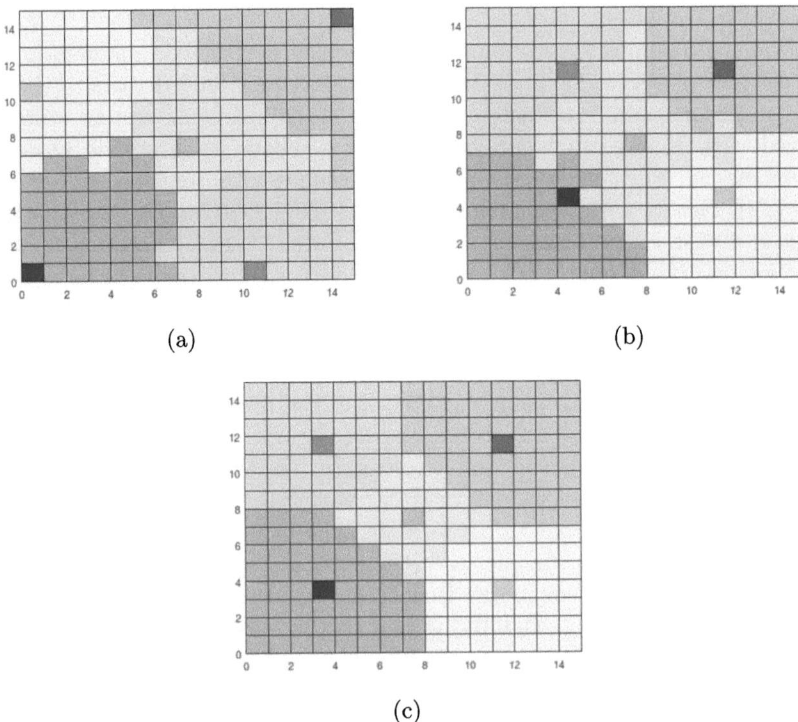

Fig. 1. Area distribution visual representation after 20 iteration in an obstacle free environment for (a) proposed method, (b) optimized RRT, and (c) CVT.

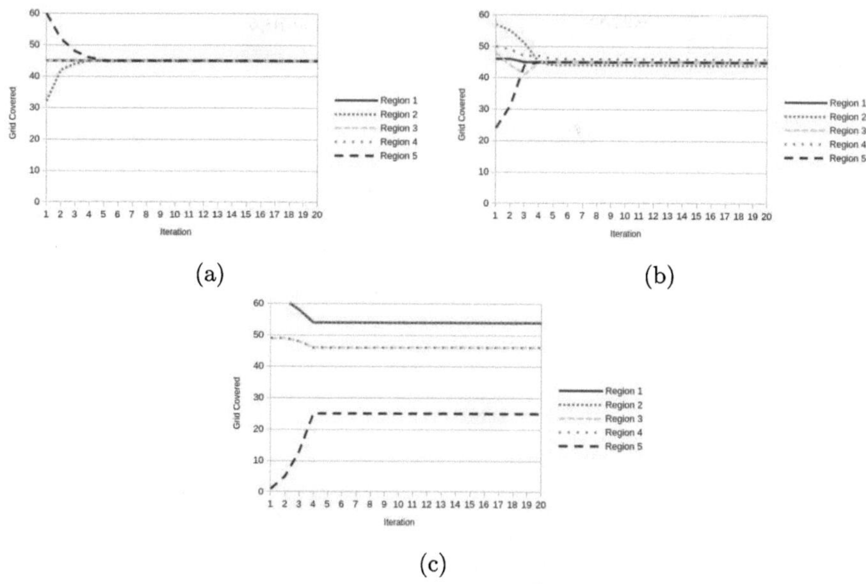

Fig. 2. Area distribution for each iteration in an obstacle free environment for (a) proposed method, (b) optimized RRT, and (c) CVT.

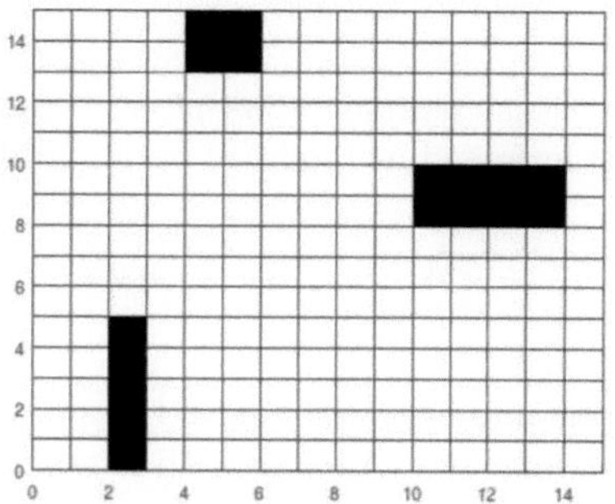

Fig. 3. Obstacle placement in the grid map.

with each region were differently colored for clarity, which can be seen in Fig. 4, and the distribution for each iteration can be seen in Fig. 5.

This simulation shows that none of the methods can achieved an even distribution. This problem arise because of the obstacles reducing the available grid from 225 to 208 which was not divisible by five. Because of this problem, both proposed method and RRT were stuck trying to exchange grid for even distribution while the CVT method managed to achieved steady state. To compared each methods, a standard deviation between each region was used to find the methods required time. The lowest standard deviation was used to determined

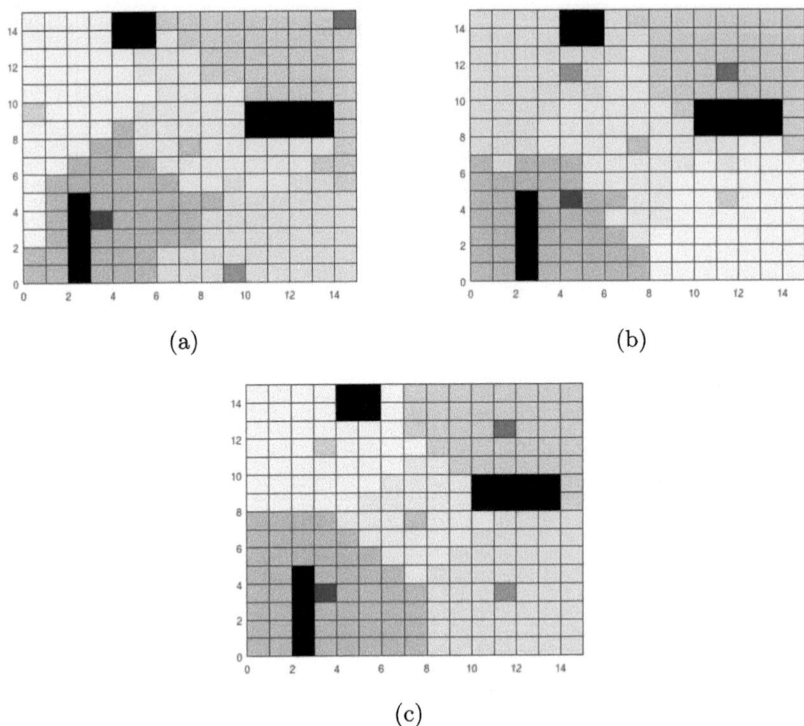

Fig. 4. Area distribution visual representation after 20 iteration in an environment with some obstacle for (a) proposed method, (b) optimized RRT, and (c) CVT.

time required for the proposed methods and RRT. The standard deviation for all methods can be seen in Fig. 6.

The proposed methods managed to achieved standard deviation of 0.55 after four iterations for an average time of 8.24 s with 1.64 s was used to find the optimal base grid. The RRT managed a 0.7 standard deviation after seven iterations or an average time of 10.3 s. The CVT achieved the worst standard deviation of 9.32 while being the fastest with four iteration for an average time of 7.61 s. The proposed methods managed to outperform the others while outpacing the RRT. To shows that the uneven distribution was because of the amount of grid, two more simulations with only the proposed method were done with available grid of 220 and 205, both divisible by five. The results of both simulations can be seen in Fig. 7.

Both simulation shows that the proposed method able to evenly divide the area. In 220 available grid, the proposed method evenly divide the area into five regions with 41 grids each after four iteration for an average time of 9.4 s with 1.11 s was used for finding the optimal base grid placement. In 205 available grid, the proposed method divide the area evenly with 41 grids for each regions

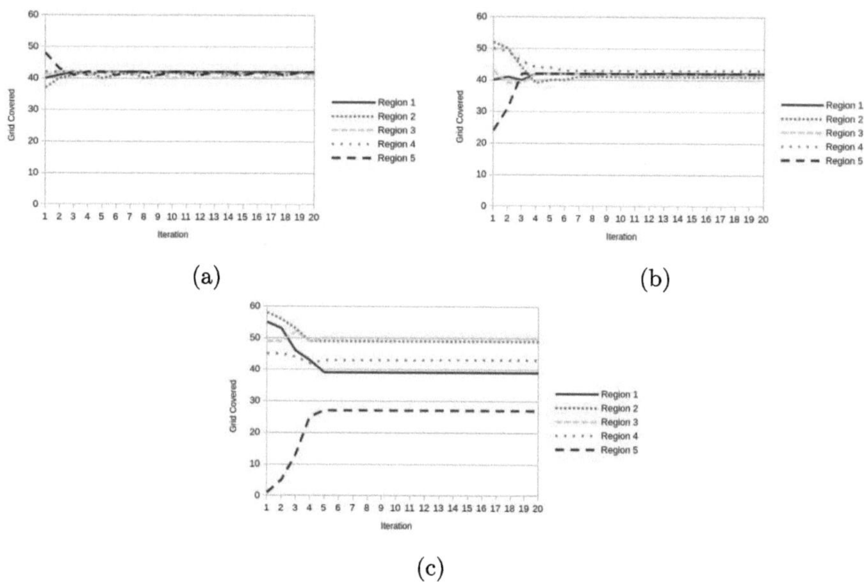

Fig. 5. Area distribution for each iteration in an environment with some obstacle for (a) proposed method, (b) optimized RRT, and (c) CVT.

after three iterations for an average time of 7.85 s with 1.75 s dedicated to base grid placement. The proposed method perform considerably better with more obstacle added as shown in time required as the available grids was shrinking. If the other methods was compared again in 205 available grids, the proposed methods considerably outperform and outpace both of them. The RRT only managed to achieved steady state after 19 iterations for an average time of 33.79 s and CVT after six iterations for an average time of 9.95 s. The results for both RRT and CVT can be seen in Fig. 8.

The problem with local minima due to repulsion force can be seen in this simulation as well as the base grid for yellow and green. Similar to previous simulation, the optimal placement of the base grid only concern with distance to others, a simple calculation can show that the optimal placement for the two base grid were at the corner of the map. Despite the less than optimal grid base placement, the proposed method still managed to outperform the other methods in term of grid distribution and for area with large amount of obstacle outpacing the other methods while maintaining connectivity.

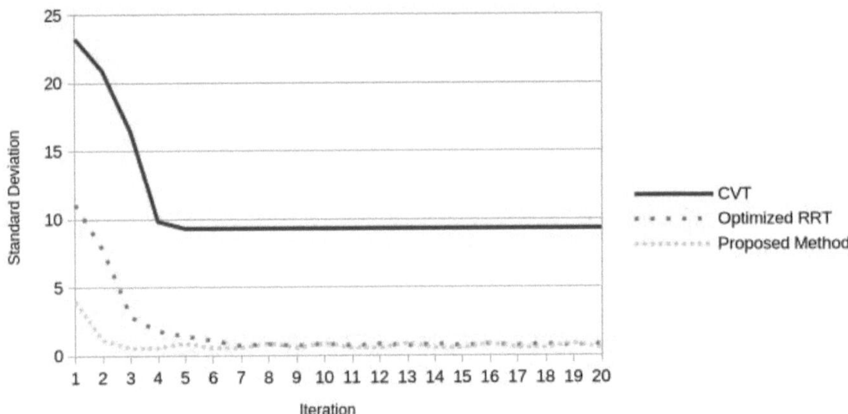

Fig. 6. Standard deviation for each iteration in an environment with some obstacle for proposed method, optimized RRT, CVT.

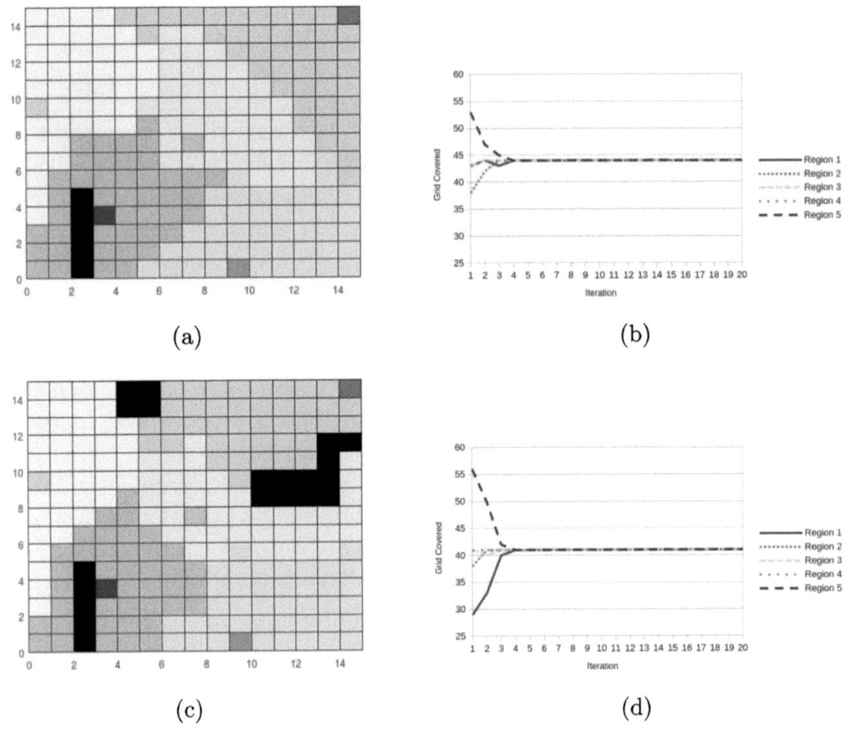

Fig. 7. The proposed method area distribution results (a) visual representation and (b) grid amount graph for 220 grid available, and (c) visual representation and (d) grid amount graph for 205 available.

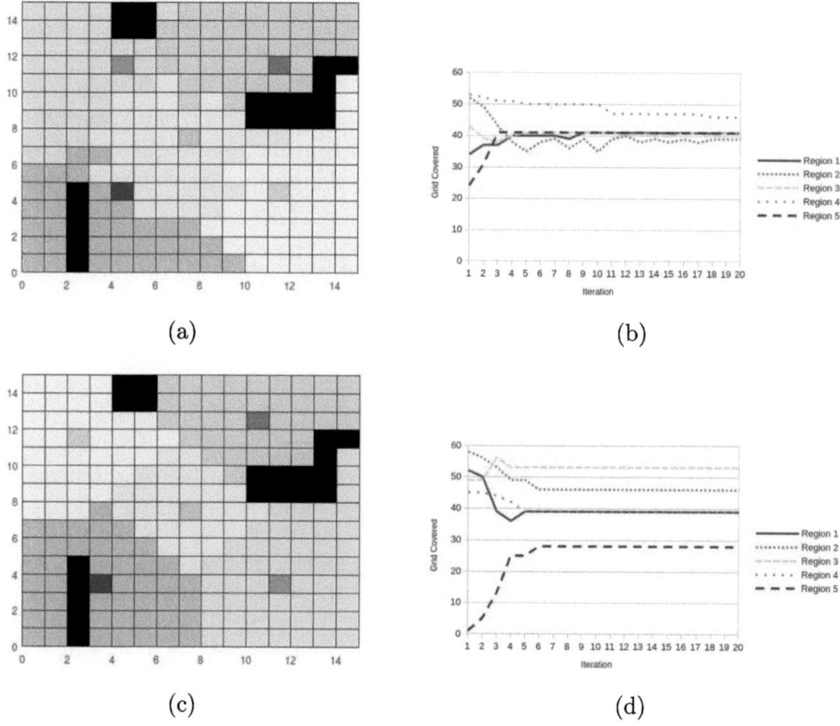

Fig. 8. Area distribution results (a) visual representation and (b) grid amount graph for CVT, and (c) visual representation and (d) grid amount graph for Optimized RRT with 205 grid available.

4 Conclusion

The efficiency of coverage path planning, especially for multi-agents, can be greatly increased by dividing the active area into an exclusive connected area of operation beforehand. By finding an equal distribution of area, the workload of each agent were in turn equally balaced to the lowest resourced required possible. Therefore finding an equal division for the coverage path planning can be crucial to the performance of said operation. We proposed an optimized region distribution method using Particle Swarm Optimization that was modified with repulsion field and inspired A* to find optimal base placement, and Spanning Tree Coverage with connectivity-conscious bidding to ensure equal distribution. The proposed method managed to achieved even distribution in an area with or without obstacle outperforming Optimized Rapidly Exploring Random Trees and Centroid Veronoi Tesselation. Even in area with grid amount undivisible the the specified amount of region, the proposed methods still outperform the other methods with the lowest standard deviation. The proposed methods also perform much better with increasing amount of obstacle while others perform much worse. This method, however, still required some improvement. When finding an optimal place for the base grid, the proposed method was trapped in local minima due to repulsion from other base grids, and the search area was not enough to find a more optimal place. A larger search area can solve this problem; however, a adaptive search area that changes depending on the situation might be more beneficial for this method. Beside this problem, an implementation with path planning can be done to see the effect of the proposed method in a coverage path planning.

Acknowledgments. Authors thank the Gunma University for providing facility to help this research.

References

1. Jin, Y., Qian, Z., Yang, W.: UAV cluster-based video surveillance system optimization in heterogeneous communication of smart cities. IEEE Access **8**, 55654–55664 (2020). https://ieeexplore.ieee.org/document/9040423/
2. Pasha, J., et al.: The drone scheduling problem: a systematic state-of-the-art review. IEEE Trans. Intell. Transp. Syst. **23**(9), 14224–14247 (2022). https://ieeexplore.ieee.org/document/9733971/
3. Jayalakshmi, K.P., Nair, V.G., Sathish, D.: A comprehensive survey on coverage path planning for mobile robots in dynamic environments. IEEE Access **13**, 60158–60185 (2025). https://ieeexplore.ieee.org/document/10946186/
4. Chen, J., Ling, F., Zhang, Y., You, T., Liu, Y., Du, X.: Coverage path planning of heterogeneous unmanned aerial vehicles based on ant colony system. Swarm Evol. Comput. **69**, 101005 (2022). https://linkinghub.elsevier.com/retrieve/pii/S221065022100167X
5. Tan, C.S., Mohd-Mokhtar, R., Arshad, M.R.: A comprehensive review of coverage path planning in robotics using classical and heuristic algorithms. IEEE Access **9**, 119310–119342 (2021). https://ieeexplore.ieee.org/document/9523743/
6. Pham, H.V., Lam, T.N.: A new method using knowledge reasoning techniques for improving robot performance in coverage path planning. Int. J. Comput. Appl. Technol. **60**(1), 57 (2019). http://www.inderscience.com/link.php?id=99503
7. Wang, Y., et al.: Coverage path planning for kiwifruit picking robots based on deep reinforcement learning. Comput. Electron. Agric. **205**, 107593 (2023). https://linkinghub.elsevier.com/retrieve/pii/S0168169922009012
8. Marwah, N., Singh, V.K., Kashyap, G.S., Wazir, S.: An analysis of the robustness of UAV agriculture field coverage using multi-agent reinforcement learning. Int. J. Inf. Technol. **15**(4), 2317–2327 (2023). https://doi.org/10.1007/s41870-023-01264-0
9. Huang, H., et al.: Robotic organism targets regional coverage capture path planning for marine aquafarm based on value iteration network. Ocean Eng. **281**, 114981 (2023). https://linkinghub.elsevier.com/retrieve/pii/S0029801823013653
10. Cai, C., Chen, J., Yan, Q., Liu, F.: A multi-robot coverage path planning method for maritime search and rescue using multiple AUVs. Remote Sens. **15**(1), 93 (2022). https://www.mdpi.com/2072-4292/15/1/93

11. Muñoz, J., López, B., Quevedo, F., Monje, C.A., Garrido, S., Moreno, L.E.: Multi UAV coverage path planning in urban environments. Sensors **21**(21), 7365 (2021). https://www.mdpi.com/1424-8220/21/21/7365
12. Majeed, A., Hwang, S.O.: A multi-objective coverage path planning algorithm for UAVs to cover spatially distributed regions in urban environments. Aerospace **8**(11), 343 (2021). https://www.mdpi.com/2226-4310/8/11/343
13. Apostolidis, S.D., Kapoutsis, P.C., Kapoutsis, A.C., Kosmatopoulos, E.B.: Cooperative multi-UAV coverage mission planning platform for remote sensing applications. Auton. Robot. **46**(2), 373–400 (2022). https://doi.org/10.1007/s10514-021-10028-3
14. Kapoutsis, A.C., Chatzichristofis, S.A., Kosmatopoulos, E.B.: DARP: divide areas algorithm for optimal multi-robot coverage path planning. J. Intell. Rob. Syst. **86**(3–4), 663–680 (2017). https://doi.org/10.1007/s10846-016-0461-x
15. Qin, Y., Fu, L., He, D., Liu, Z.: Improved optimization strategy based on region division for collaborative multi-agent coverage path planning. Sensors **23**(7), 3596 (2023). https://www.mdpi.com/1424-8220/23/7/3596
16. Lee, K., Lee, K.: Adaptive centroidal voronoi tessellation with agent dropout and reinsertion for multi-agent non-convex area coverage. IEEE Access **12**, 5503–5516 (2024). https://ieeexplore.ieee.org/document/10382493/
17. Shami, T.M., El-Saleh, A.A., Alswaitti, M., Al-Tashi, Q., Summakieh, M.A., Mirjalili, S.: Particle swarm optimization: a comprehensive survey. IEEE Access **10**, 10031–10061 (2022). https://ieeexplore.ieee.org/document/9680690/
18. Lin, Z., Wu, K., Shen, R., Yu, X., Huang, S.: An efficient and accurate a-star algorithm for autonomous vehicle path planning. IEEE Trans. Veh. Technol. **73**(6), 9003–9008 (2024). https://ieeexplore.ieee.org/document/10376315/
19. Xu, R., Yao, S., Gouhe, W., Zhao, Y., Huang, S.: A spanning tree algorithm with adaptive pruning with low redundancy coverage rate. Comput. Ind. Eng. **186**, 109745 (2023). https://linkinghub.elsevier.com/retrieve/pii/S0360835223007696

Novel Cascade Model for Designing Tunable Filters with Cascaded Low-Order Sections

Tian-Bo Deng(✉)

Dept. Information Science, Toho University, Miyama 2-2-1, Funabashi, Japan
deng@is.sci.toho-u.ac.jp

Abstract. The paper reveals a novel model (structure) for realizing a tunable digital filter (TDF) having tunable frequency bandwidths. This model is comprised of cascaded low-order sections that contain 2nd-order denominators, while their numerators can be higher orders. Since this cascade model consists of only low-order sections, the low-order sections can be utilized as fundamental building elements for implementing the entire cascade model. The cascade structure also features low implementation noise induced from coefficient quantization. Since the low-order sections have 2nd-order denominators, it is easy to guarantee the stability by exploiting an explicit stability condition. Based on the explicit stability condition, the paper also reveals a stability-ensuring scheme, which uses a uniquely defined function to represent the denominator parameters of the low-order sections. Representing the denominator parameters as the functions enables the stability condition to be definitely satisfied, and thus keeps the whole cascade model stable during bandwidth tuning. This paper includes the simulation results of designing a highpass bandwidth-tunable filter for exemplifying the achieved high accuracy as well as the ensured stability. The simulations verify that exploiting the novel cascade model can obtain a highly accurate TDF with tunable highpass responses. The design results also demonstrate that the highpass TDF remains stable during frequency-bandwidth tuning.

Keywords: Tunable digital filter (TDF) · Tunable bandwidth · Cascade model · Low-order section · Stability · Stability-ensuring scheme · Design simulation

1 Introduction

A tunable digital filter (TDF) is a general signal processing system that features quick tuning of its frequency response. The tunable frequency response can be amplitude response [1–12] as well as delay response [13–22]. The tunability offers high flexibility to tune various responses on the fly whenever such a tuning is needed.

N. A. Mohd Subha et al. (Eds.): SMS 2026, CCIS 2956, pp. 14–26, 2026.
https://doi.org/10.1007/978-981-92-0306-2_2

To design a TDF, the designer needs to find an optimal mathematical model called transfer function so that the model's frequency-domain response gets closest to a given ideal one. This model can be non-recursive or recursive, while the latter may suffer instability despite its low complexity. That is, a recursive model may become unstable if its parameters take inappropriate values violating the stability condition [23]. Therefore, it is necessary to take a proactive measure in designing a recursive tunable filter for preventing instability. This ensures the TDF stability during frequency-response tuning.

The paper proposes a novel cascade model for designing a TDF. The model is composed of low-order cascaded sections, leading to a cascade structure. Since the model employs low-order sections, it incurs low implementation noise when implemented using hardware. This is because the low-order section's parameters cause low quantization noise as compared to those in high-order direct-form models [7–12]. On the other hand, since the low-order sections are recursive, it is necessary to ensure the stability. Fortunately, the novel model (structure) adopted in this paper facilitates the development of a stability-ensuring scheme so as to avoid instability. This is because the model cascades low-order sections with 2nd-order denominators, and the explicit stability condition for the 2nd-order denominators can be skillfully employed. Based on the explicit condition, this paper develops a stability-ensuring scheme to achieve the goal. By exploiting both the novel model and the stability-ensuring scheme, the paper reveals a least-squares technique for designing a stable cascade-form TDF.

To verify the efficacy of this novel model, the paper details the simulation results of designing a highpass TDF. The highpass design exemplifies that a highly accurate highpass TDF can be obtained by using the model-based least-squares design technique. The simulations further demonstrate that the highpass TDF maintains stability during frequency-bandwidth tuning.

2 Tunable Cascade Model

This section introduces a tunable cascade model for implementing a cascade-form TDF. The model cascades multiple low-order sections with 2nd-order denominators. As noted in the preceding section, this cascade model has two advantages. One is the easy stability guarantee, while the other features low parameter-quantization noise.

Let $G(\omega, p)$ be the ideal bandwidth-tunable gain response. Here, p tunes the bandwidths of $G(\omega, p)$, taking values within

$$p \in [p_{\min}, p_{\max}].$$

Furthermore, ω stands for normalized angular frequency, taking values in $\omega \in [0, \pi]$.

The design aims to use TDF's gain response to fit $G(\omega, p)$ through minimizing the sum of squared errors. This is done by introducing the novel cascade model

$$T(z,p) = \prod_{m=1}^{M} T_m(z,p) \tag{1}$$

where all sections

$$T_m(z,p) = \frac{\sum_{i=0}^{I_m} b_{m,i}(p) z^{-i}}{1 + a_{m,1}(p)z^{-1} + a_{m,2}(p)z^{-2}} \tag{2}$$

are connected in series to construct $T(z,p)$, and

$$b(m,0)(p) = \begin{cases} \text{undetermined unknown}, & m = 1 \\ 1 & m \neq 1. \end{cases} \tag{3}$$

In (2), the order of $T_m(z,p)$ is I_m, which is low. Thus, $T(z,p)$ represents a tunable model that cascades all the low-order sections $T_m(z,p)$. Since the coefficients $b_{m,i}(p)$, $a_{m,1}(p)$, and $a_{m,2}(p)$ are parameterized using p, all those sections have changeable responses. Figure 1 shows the structure of $T(z,p)$, which cascades all $T_m(z,p)$.

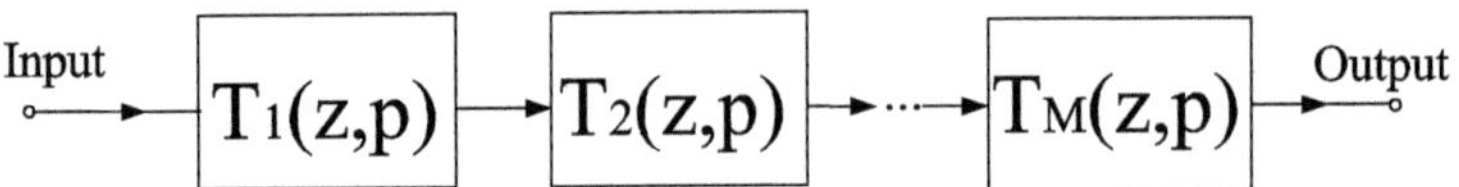

Fig. 1. Tunable cascade model.

The approximation of the ideal gain $G(\omega, p)$ needs to find the optimal functions

$$\{b_{1,0}(p), b_{m,i}(p), a_{m,1}(p), a_{m,2}(p)\} \tag{4}$$

for

$$m = 1, 2, \cdots, M \tag{5}$$
$$1 \leq i \leq I_m. \tag{6}$$

It should be noted that only $a_{m,1}(p)$ and $a_{m,2}(p)$ impact the stability of $T_m(z,p)$. Hence, a proactive measure must be taken to ensure $T_m(z,p)$ stable. This requires

$$\begin{cases} |a_{m,1}(p)| < 1 + a_{m,2}(p) \\ |a_{m,2}(p)| < 1 \end{cases} \tag{7}$$

for any value of $p \in [p_{\min}, p_{\max}]$. After introducing a stability-ensuring scheme, the next section details an effective strategy to get the functions in (4).

3 Designing TDF and Guaranteeing Stability

This section introduces a 2-phase procedure for designing the TDF $T(z,p)$. That is, our objective is to find the functions in (4). To maintain $T(z,p)$'s stability, a stability-ensuring scheme using a unique function $U(x)$ is also revealed.

3.1 Phase-I: Constant-Filter Design and Stability

The approximation of $G(\omega, p)$ starts from sampling $p \in [p_{\min}, p_{\max}]$. This leads to the samples p_k, $k = 1, 2, \cdots, K$. Each p_k defines a design specification $G(\omega, p_k)$. Then it is approximated via using the gain response of the filter

$$T^{(k)}(z) = \prod_{m=1}^{M} T_m(z) \tag{8}$$

with

$$T_m(z) = \frac{\sum_{i=0}^{I_m} b_{m,i} z^{-i}}{1 + a_{m,1} z^{-1} + a_{m,2} z^{-2}} \tag{9}$$

but

$$b(m, 0) = \begin{cases} \text{unknown (undetermined)}, & m = 1 \\ 1 & m \neq 1. \end{cases} \tag{10}$$

Here, $T^{(k)}(z)$ cascades low-order sections $T_m(z)$, while each $T_m(z)$ has fixed coefficients. Thus, $T^{(k)}(z)$ is a cascade-form constant filter. Figure 2 illustrates the overall configuration of $T^{(k)}(z)$ with cascaded constant sections $T_m(z)$.

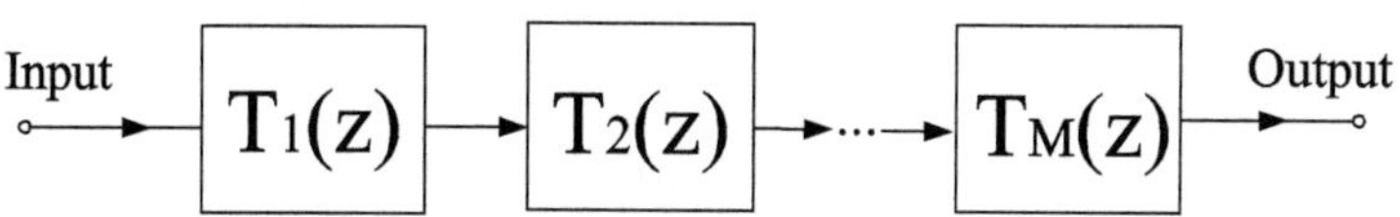

Fig. 2. Fixed cascade model.

The stability guarantee of $T^{(k)}(z)$ needs to constrain

$$\begin{cases} |a_{m,1}| < 1 + a_{m,2} \\ |a_{m,2}| < 1. \end{cases} \tag{11}$$

This is accomplished by transforming $a_{m,1}$ and $a_{m,2}$ to the functions of $x_{m,1}$ as well as $x_{m,2}$ as

$$\begin{cases} a_{m,1} = U(x_{m,1})(1 + a_{m,2}) \\ a_{m,2} = U(x_{m,2}). \end{cases} \tag{12}$$

The above transformations utilize the function

$$U(x) = \begin{cases} \dfrac{2x}{1 + x^2}, & x \neq \pm 1 \\ (1 - \lambda)x, & \text{otherwise} \end{cases} \tag{13}$$

where λ denotes machine epsilon

$$\lambda = 2.220446049250313 \times 10^{-16}.$$

As justified in [9], although this machine epsilon is very small, computer simulations recognize $(1 - \lambda)$ as a value below one. Therefore, the design based on the transformations in (12) leads to a stable filter. Figure 3 depicts $U(x)$, which features $-1 < U(x) < 1$.

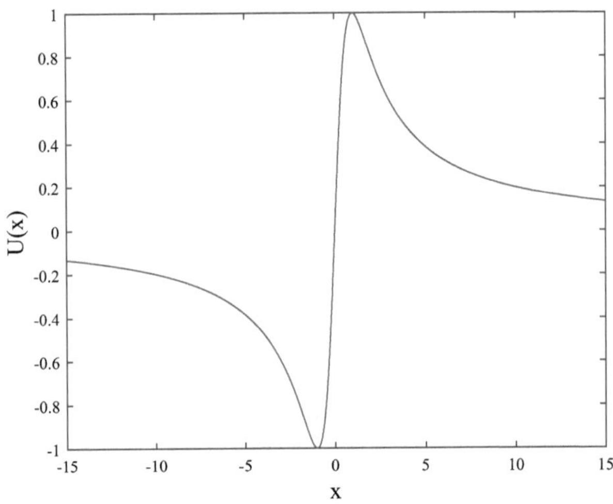

Fig. 3. $U(x)$ for ensuring stability.

Replacing $a_{m,1}, a_{m,2}$ in (9) with the transformations in (12) yields

$$T_m(z) = \frac{\sum_{i=0}^{I_m} b_{m,i} z^{-i}}{1 + U(x_{m,1})[1 + U(x_{m,2})]z^{-1} + U(x_{m,2})z^{-2}}. \tag{14}$$

Consequently, the approximation of $G(\omega, p_k)$ requires optimizing

$$\{b_{1,0}, b_{m,i}, x_{m,1}, x_{m,2}\} \tag{15}$$

for

$$1 \leq m \leq M, \quad 1 \leq i \leq I_m. \tag{16}$$

Here, the optimization minimizes the sum of squared errors

$$E_2 = \sum_{l=1}^{L} e^2(\omega_l) \tag{17}$$

where

$$e(\omega_l) = G(\omega_l, p_k) - T^{(k)}(\omega_l)$$

denotes the approximation error at frequency ω_l, and $T^{(k)}(\omega_l)$ denotes the gain response of $T^{(k)}(z)$.

The transformations in (12) absolutely meet the stability condition, regardless of the values of $x_{m,2}, x_{m,1}$. All the specifications $G(\omega, p_k)$ are approximated one after another, leading to the constant filters $T^{(k)}(z)$, $k = 1, 2, \cdots, K$.

As can be outlined below, designing each $T^{(k)}(z)$ results in a set of optimal values of the parameters in (15),

$$\begin{aligned} T^{(1)}(z) &\longleftrightarrow b_{1,0}, b_{m,i}, x_{m,2}, x_{m,1} \\ T^{(2)}(z) &\longleftrightarrow b_{1,0}, b_{m,i}, x_{m,2}, x_{m,1} \\ T^{(3)}(z) &\longleftrightarrow b_{1,0}, b_{m,i}, x_{m,2}, x_{m,1} \\ &\vdots \\ T^{(K)}(z) &\longleftrightarrow b_{1,0}, b_{m,i}, x_{m,2}, x_{m,1}. \end{aligned} \tag{18}$$

3.2 Phase-II: Least-Squares Curve Fitting

After obtaining the above constant filters $T^{(k)}(z)$, the next procedure is to utilize distinctive polynomials to fit the parameter values in (15). This is curve fitting, which can be accomplished by minimizing the sum of squared errors. The fitting yields the polynomials

$$\left\{b_{1,0}(p), b_{m,i}(p), x_{m,2}(p), x_{m,1}(p)\right\}. \tag{19}$$

One more step is to produce

$$\begin{cases} a_{m,1}(p) = U(x_{m,1}(p))[1 + a_{m,2}(p)] \\ a_{m,2}(p) = U(x_{m,2}(p)). \end{cases} \tag{20}$$

The above transformations from $x_{m,2}(p)$ and $x_{m,1}(p)$ to $a_{m,2}(p)$ and $a_{m,1}(p)$ definitely meet the stability condition in (7). Thus, Phase-I together with Phase-II gets a stable cascade-form TDF $T(z, p)$.

4 Highpass TDF Design Simulations

The highpass TDF design specification is

$$G(\omega, p) = \begin{cases} 0, & \omega \in [0, 0.6\pi + p] \\ 1, & \omega \in [0.7\pi + p, \pi] \end{cases} \tag{21}$$

with p varying within

$$-0.1\pi \le p \le 0.1\pi.$$

The continuous specification $G(\omega, p)$ is first sampled to a set of discrete ones $G(\omega, p_k)$, $k = 1, 2, \cdots, 21$. Then each $G(\omega, p_k)$ is approximated by setting

$$\begin{cases} M = 2 \\ I_1 = I_2 = 3 \\ L = 1001 \end{cases} \tag{22}$$

where L is the number of frequencies ω_l. Figure 4 plots all the discrete specifications $G(\omega, p_k)$.

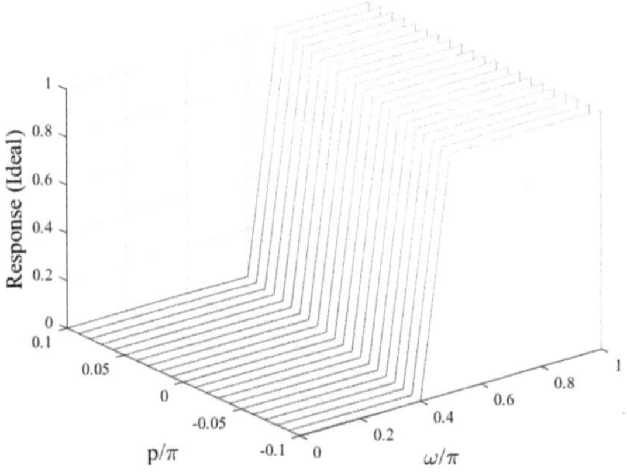

Fig. 4. Highpass design specifications for $T^{(k)}(z)$.

The designs of all highpass filters $T^{(k)}(z)$ begin with designing $T^{(1)}(z)$ first, adopting the initial values

$$\begin{bmatrix} b_{1,0} \\ b_{1,1} \\ b_{1,2} \\ b_{1,3} \\ b_{2,1} \\ b_{2,2} \\ b_{2,3} \\ x_{1,2} \\ x_{2,2} \\ x_{1,1} \\ x_{2,1} \end{bmatrix} = \begin{bmatrix} 1.282280747092566 \\ -0.582630964347252 \\ 0.222614480633598 \\ 0.779451472862399 \\ 0.384707157585715 \\ 0.696366922883312 \\ -0.112715559173794 \\ -0.038823700928052 \\ 0.088088629112916 \\ -0.789655973216349 \\ 1.422960618799131 \end{bmatrix} \tag{23}$$

After $T^{(1)}(z)$ is obtained, its coefficient values are exploited as initial values to start designing $T^{(2)}(z)$. The same procedure continues until finishing the design

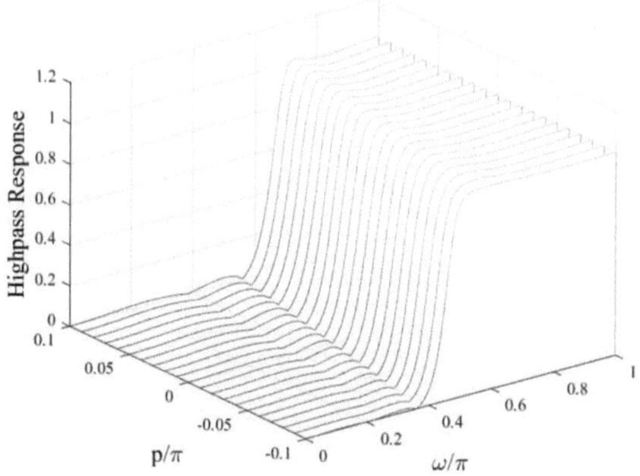

Fig. 5. Highpass responses of $T^{(k)}(z)$.

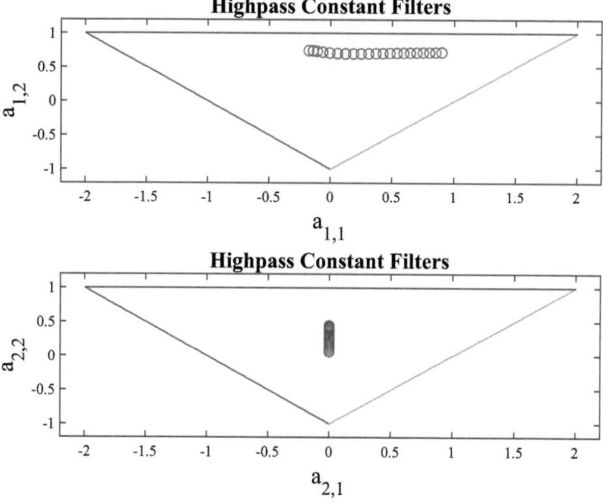

Fig. 6. Checking the stability of $T^{(k)}(z)$.

of $T^{(K)}(z)$. Their highpass responses are shown in Fig. 5. The average normalized RMS (root-mean-squared) error is 2.332479%. Figure 6 checks the stability of $T^{(k)}(z)$, where the points $(a_{1,1}, a_{1,2})$ and $(a_{2,1}, a_{2,2})$ remain inside the triangles, indicating the ensured stability.

The next step is to get the polynomials in (19). It is achieved by using 3rd-order polynomials to fit the parameter values in (18) separately. Figure 7 shows the polynomials

$$\left[b_{1,0}(p)\ b_{1,1}(p)\ b_{1,2}(p)\ b_{1,3}(p)\ b_{2,1}(p)\ b_{2,2}(p)\ b_{2,3}(p)\right] \tag{24}$$

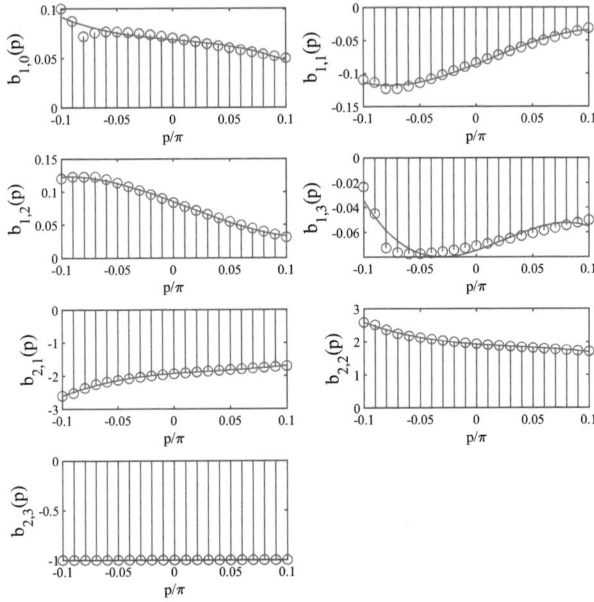

Fig. 7. Polynomials for fitting $b_{1,0}, b_{m,i}$.

while Fig. 8 shows the polynomials

$$\left[x_{1,2}(p)\ a_{2,2}(p)\ x_{1,1}(p)\ x_{2,1}(p)\right]. \tag{25}$$

Using those polynomials gets the highpass TDF with tunable highpass responses shown in Fig. 9. In Fig. 9, doubly dense samples p_k are used for checking the details of the responses. For comparison, the ideal highpass tunable responses are plotted in Fig. 10, alongside the errors in Fig. 11. The average normalized RMS error is 2.576963%, which is comparable to that of highpass constant filters $T^{(k)}(z)$. This indicates that the fitting polynomials can produce highly accurate TDF. Moreover, Fig. 12 checks $T(z,p)$'s stability, where both $(a_{1,1}(p), a_{1,2}(p))$ and $(a_{2,1}(p), a_{2,2}(p))$ remain moving inside the triangles. This concludes that the designed highpass TDF maintains its stability.

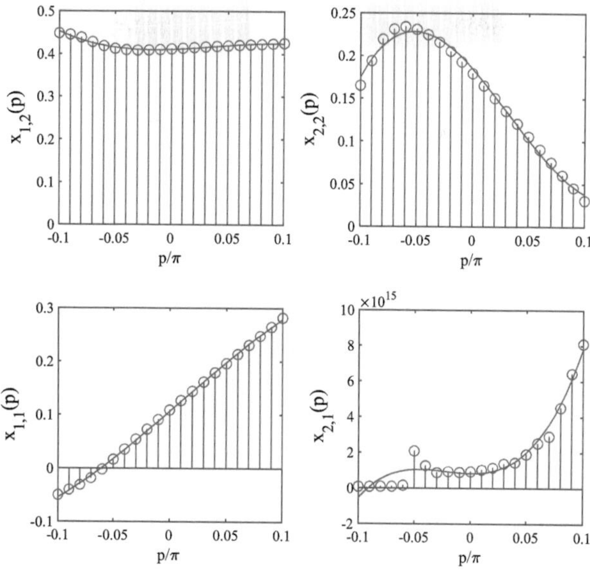

Fig. 8. Polynomials for fitting $x_{m,2}, x_{m,1}$.

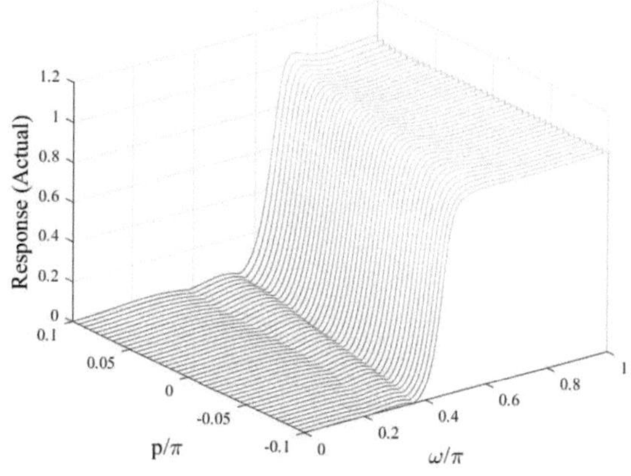

Fig. 9. Highpass tunable responses of $T(z, p)$.

5 Conclusions

This paper has proposed a novel cascade model for simulating the design of cascade-form TDFs. As compared to the direct-form structures [7–12], this novel model cascades low-order sections, and thus features low quantization noise in hardware implementation. Furthermore, since the low-order sections have 2nd-order denominators, the existing explicit relation between the denominator

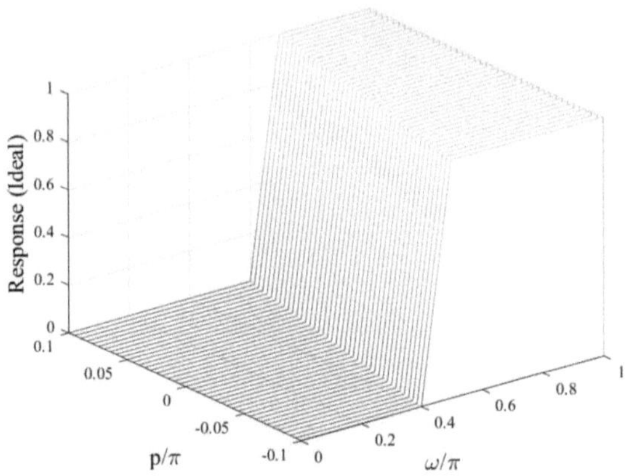

Fig. 10. Ideal highpass tunable responses for $T(z,p)$.

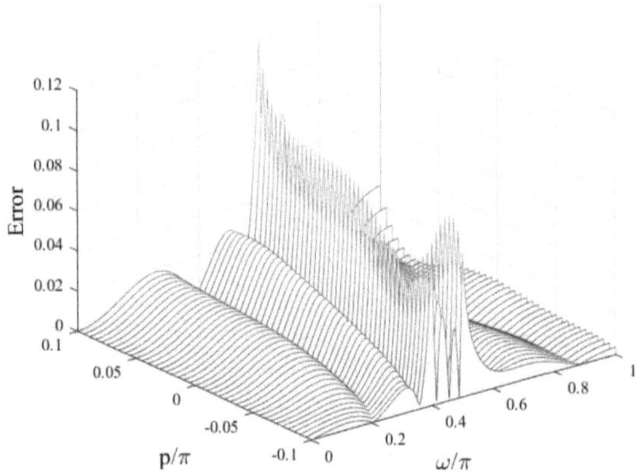

Fig. 11. Highpass tunable response errors.

parameters and stability can be exploited in developing the stability-ensuring scheme.

This paper has revealed a stability-ensuring scheme that exploits a unique function $U(x)$ to transform the denominator parameters to the functions of a set of unconstrained parameters. This enables the unknown parameters to be optimized without concerning the stability issue. Based on the novel model, a least-squares technique has been developed for designing the cascade-form TDF. The design technique comprises two phases, and both phases skillfully incorporate the stability-ensuring scheme in the design process, resulting in a stable TDF.

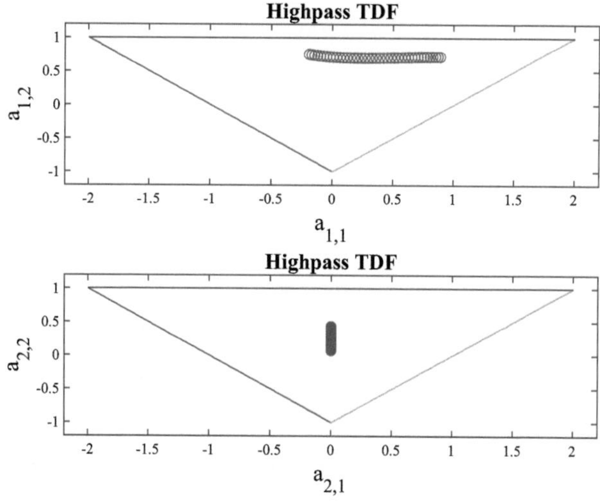

Fig. 12. Checking the stability of $T(z, p)$.

The design simulations using the novel cascade model have verified that a highly accurate highpass TDF can be obtained, and the stability is also ensured. Therefore, the design technique can produce a high-performance cascade-form TDF with definitely ensured stability. As compared with the designs using direct-form models [7–12], the novel cascade model suffers low coefficient-quantization noise in hardware implementation because it cascades only low-order sections.

References

1. Zarour, R., Fahmy, M.M.: A design technique for variable digital filters. IEEE Trans. Circ. Syst. **36**(11), 1473–1478 (1989)
2. Deng, T.-B.: Design of recursive 1-D variable filters with guaranteed stability. IEEE Trans. Circ. Syst. II, Analog Digit. Signal Process. **44**(9), 689–695 (1997)
3. Deng, T.-B.: Design of linear phase variable 2-D digital filters using real-complex decomposition. IEEE Trans. Circ. Syst. II, Analog Digit. Signal Process. **45**(3), 330–339 (1998)
4. Deng, T.-B.: Design of variable 2-D linear phase recursive digital filters with guaranteed stability. IEEE Trans. Circ. Syst. I, Fund. Theor. Appl. **45**(8), 859–863 (1998)
5. Deng, T.-B.: Variable 2-D FIR digital filter design and parallel implementation. IEEE Trans. Circ. Syst. II, Analog Digit. Signal Process. **46**(5), 631–635 (1999)
6. Sutthikarn, P., Chivapreecha, S., Trirat, A., Jongsataporn, T.: A simple tunable biquadratic digital bandpass filter design for spectrum sensing in cognitive radio. Proc. IEEE ECTI-CON **2020**, 71–75 (2020)
7. Deng, T.-B.: Design of recursive variable-digital-filters with theoretically-guaranteed stability. Int. J. Electron. **103**(12), 2013–2028 (2016)
8. Deng, T.-B.: The Lp-norm-minimization design of stable variable-bandwidth digital filters. J. Circ. Syst. Comput. **27**(7), 1850102 1-18 (2018)

9. Deng, T.-B.: L1-norm-minimized variable-center-frequency bandpass filters with widened-range tunability. Proc. IEEE ICFTSS **2024**, 12–16 (2024)
10. Deng, T.-B.: Typical benchmark specifications for designing stable variable filters using novel unity-bounded functions. Viet. J. Comput. Sci. **11**(1), 53–74 (2024)
11. Deng, T.-B.: Stability-guaranteed odd-order variable-bandwidth filters using stabilized odd-order transfer function. J. Inf. Telecommun. **8**(3), 384–398 (2024)
12. Deng, T.-B.: Generalized strategy for stabilizing minimax-type variable bandpass filter. Int. J. Electron. **111**(12), 2279–2302 (2024)
13. Laakso, T.I., Valimaki, V., Karjalainen, M., Laine, U.K.: Splitting the unit delay: tools for fractional delay filter design. IEEE Signal Process. Mag. **13**(1), 30–60 (1996)
14. Deng, T.-B.: Discretization-free design of variable fractional-delay FIR digital filters. IEEE Trans. Circ. Syst. II, Analog Digit. Signal Process. **48**(6), 637–644 (2001)
15. Deng, T.-B.: Closed-form design and efficient implementation of variable digital filters with simultaneously tunable magnitude and fractional-delay. IEEE Trans. Signal Process. **52**(6), 1668–1681 (2004)
16. Deng, T.-B.: Symmetry-based low-complexity variable fractional-delay FIR filters. Proc. IEEE ISCIT **2004**, 194–199 (2004)
17. Deng, T.-B., Lian, Y.: Weighted-least-squares design of variable fractional-delay FIR filters using coefficient symmetry. IEEE Trans. Signal Process. **54**(8), 3023–3038 (2006)
18. Huang, Y.-D., Pei, S.-C., Shyu, J.-J.: WLS design of variable fractional-delay FIR filters using coefficient relationship. IEEE Trans. Circ. Syst. II, Exp. Briefs **56**(3), 220–224 (2009)
19. Deng, T.-B., Qin, W.: Coefficient relation-based minimax design and low-complexity structure of variable fractional-delay digital filters. Signal Process. **93**(4), 923–932 (2013)
20. Deng, T.-B., Qin, W.: Improved bi-equiripple variable fractional-delay filters. Signal Process. **94**(1), 300–307 (2014)
21. Deng, T.-B., Soontornwong, P.: Delay-error-constrained minimax design of all-pass variable-fractional-delay digital filters. Signal Process. **120**(3), 438–447 (2016)
22. Soontornwong, P., Deng, T.-B., Chivapreecha, S.: Low-complexity and high-modularity structure for implementing transient-free Pascal-delay filter. IEEE Trans. Signal Process. **65**(23), 6233–6243 (2017)
23. Jackson, L.B.: Digital Filters and Signal Processing. Kluwer Academic Publishers, Amsterdam (1992)

LLM-Based Simultaneous Waypoint Generation for Multi-robot Planning in 2D Corridor Environments

Marouane El Massine Lfellah, Pei Ling Leow(✉), and Nurulaqilla Khamis

Department of Control and Mechatronics Engineering, Faculty of Electrical Engineering, Universiti Teknologi Malaysia, Johor Bahru, Johor, Malaysia
leowpl@utm.my

Abstract. Multi-robot systems (MRS) require efficient path planning to navigate complex environments while avoiding collisions. While traditional heuristic algorithms and modern learning-based methods have made significant strides, they often face challenges with computational scalability, adaptability, or semantic reasoning in dynamic environments. This paper presents a novel framework that leverages a Large Language Model (LLM) for waypoint generation in a 2D corridor environment with two mobile robots. The environment consists of axis-aligned corridors, junctions, and obstacles, all encoded into a structured textual prompt. The LLM generates coordinate-based waypoints for both robots simultaneously, and a closed-loop feedback mechanism validates these waypoints against spacing, collision, and junction constraints. Invalid waypoints trigger iterative refinement until all constraints are satisfied. Experimental results demonstrate that the proposed approach achieves high waypoint generation success rates while maintaining path feasibility in complex scenarios.

Keywords: Multi-robot systems. Large Language Model. Waypoint Generation. Closed-loop Feedback

1 Introduction

The evolution of Multi-Robot Systems (MRS) from pre-programmed routines to autonomous collective planning is a critical development in industrial automation. While MRS deployments in dynamic environments such as logistics centers and healthcare facilities offer significant efficiency gains, simultaneous path planning in constrained spaces remains an unresolved challenge. A primary limitation is the lack of high-level semantic reasoning; existing robots often adhere strictly to local navigational rules but fail to interpret complex spatial contexts, leading to logical deadlocks at shared resources like corridor junctions [1, 2].

Today, ensuring that two or more robots can safely navigate past each other in a narrow hallway requires complex joint planning that traditional mathematical models struggle to provide. Most current solutions rely on complex engineering to define every possible "what-if" scenario, resulting in systems that are brittle and slow to adapt [3].

N. A. Mohd Subha et al. (Eds.): SMS 2026, CCIS 2956, pp. 27–43, 2026.
https://doi.org/10.1007/978-981-92-0306-2_3

This leaves the field polarized between two extremes: Traditional algorithms that are safe but inflexible, and advanced Artificial Intelligence (AI) and Deep Learning (DL) models that are flexible but data-hungry and often opaque [4–6]. There is a clear need for a navigation planner that bridges this gap which offers both the intuitive reasoning of natural language and the rigorous precision of geometric planning, allowing robots to interpret environmental constraints through semantic understanding rather than purely numerical or statistical representations.

The core challenge addressed in this paper is the simultaneous planning of two robots in a shared 2D corridor environment populated with static obstacles where the aim is to bridge the gap between high-level reasoning and low-level control, ensuring collision-free navigation and lawful junction. To achieve this, a Large Language Model (LLM) is proposed as a centralized planner [7]. By encoding the physical boundaries of corridors and static obstacles into structured text, it enables the LLM to be aware of the workspace and generate precise waypoints for both robots simultaneously. In addtion, the planner reasoning is wrapped in a robust closed-loop feedback mechanism by validating the LLM-proposed waypoints against spacing violations, obstacles collision, and junction constraints and provides corrective textual description [8]. This allows the LLM to refine its plan iteratively, learning from its mistakes until all safety constraints are met. By prioritizing semantic logic over exhaustive training, LLMs can serve as effective, adaptable spatial reasoners for real-world robotic motion planning.

This paper introduces a language-driven system for multi-robot coordination, centered on three key contributions. First, it details a structured prompt engineering methodology that explicitly grounds the LLM in the environment, precisely encoding geometric elements such as corridors, junctions, and static obstacles. Second, it implements a closed-loop feedback mechanism, which provides the LLM with granular, constraint-specific semantic diagnostics after each planning attempt, enabling systematic and iterative trajectory refinement. Finally, a comprehensive experimental evaluation demonstrates the practical efficacy of this integrated approach, demonstrating its high success rates in generating collision-free, coordinated waypoint trajectories for multiple agents.

The remainder of this paper is structured as follows. Section 2 reviews existing approaches to LLM-based robot planning, environment representation, and state feedback mechanisms. Section 3 details the proposed methodology, including environment encoding, constraint definitions, the closed-loop feedback mechanism, and prompt design. Section 4 describes the experimental setup, covering the simulation environment, obstacle scenarios, and LLM configuration. Section 5 presents quantitative and qualitative results, and Sect. 6 concludes with directions for future work.

2 Related Work

2.1 Centralized LLM Planning

One of the debates in this field centers on the choice between centralized and decentralized LLM architectures. Chen et al. [7] systematically compared four distinct approaches: a fully decentralized framework (DMAS) where each robot operates with its own LLM agent; a centralized architecture (CMAS) with a single LLM orchestrating all agents; and two hybrid variants; HMAS-1, where the central planner initialized decentralized agents with an initial central plan. In addition to HMAS-2, which augments central planning with local LLM feedback from individual robots. Their evaluation demonstrated that HMAS-2 achieved the highest planning success rates, particularly in complex scenarios with numerous agents. However, this hybrid architecture incurs significant computational overhead, as the central LLM must process long textual feedback from each local agent for iterative re-planning.

The proposed approach offers a more efficient alternative with utilizing a single central LLM for high-level reasoning but replace the resource-intensive local LLM agents with a deterministic feedback loop. This approach minimizes token consumption while ensuring the central plan is validated against rigorous physical constraints.

Transitioning from planning architectures to trajectory generation, waypoint generation methods that translate high-level semantic instructions into executable robot trajectories are examined.

2.2 LLM-Based Waypoint Generation

The application of LLMs to waypoint generation has recently moved from theoretical concepts to working prototypes. Tariq et al. [9] established a strong baseline with a robust framework where natural language commands are translated into dynamic waypoints within corridor maps. However, a key limitation of their approach is its reliance on an incremental approach, which generates and validates waypoints one-by-one. This local optimization fails to consider the global trajectory. Consequently, the incremental path can be inefficient and prone to robot collision and failure to reach the target positions. Similarly, Meng et al. [10] utilized LLMs to propose intermediate target waypoints for A* search, which is effective for single agents but less so for coordinated teams.

This paper bridges this gap by adopting a global perspective: The LLM is prompted to generate the entire waypoint sequence for both agents in a single pass. This allows the system to explicitly reason about long-range patterns such as anticipating a potential deadlock at a shared junction and coordinate movements proactively.

To facilitate this global reasoning capability, a robust prompt structure is required to properly ground the LLM in the spatial environment. The following subsection reviews the structured prompt design and spatial grounding methodologies that enables precise semantic-to-geometric translation.

2.3 Prompt Design and Spatial Grouding

Effective LLM navigation requires a prompt structure that bridges natural language with environment constraints. Pham et al. [12] introduced the CO-STEP framework, which standardizes prompt engineering into structured Context, Objective, Stylization, Tone, Execution, and Polish components. While CO-STEP establishes a robust syntax for complex tasks, it remains a general-purpose framework lacking specific measures for grounding the geometric constraints of a physical environment. To address this spatial gap, Azkune et al. [11] demonstrated that language models can ground spatial relations by feeding them the output of a specific object detector (VinVL), providing the LLM with precise labels and bounding box coordinates of scene objects. However, this method focuses on object-centric relations rather than the navigable space itself.

The proposed apporach utilizes the structural framework of CO-STEP [12] to organize the prompt and populate it with the environments constraints coordinates similar to Azkune's bounding boxes [11] but applied to corridors and junctions to provide the necessary spatial framework for simultaneous multi-agent planning.

For this structured prompt to be operational, the underlying simulation environment and dynamic robot states must first be converted into natural language. The following subsection reviews existing techniques for converting geometric environments and dynamic robot states into textual descriptions suitable for LLM-based planning systems.

2.4 Converting Simulation Environment and State to Natural Language

Bridging the gap between a robot's digital reality and an LLM's text-based understanding requires translating both the static environment and the dynamic state of the robots. Park et al. [13] laid the groundwork for this by representing static environments as hierarchical trees converted into natural language descriptions. In the domain of dynamic control, Xue et al. [14] and Bhat et al. [8] demonstrated the critical role of state feedback, converting system constraints and execution outcomes into textual updates to guide planner reliability.

The proposed framework builds upon these static and dynamic approaches a unified navigation loop. To achieve this, the simulation environment is represented as a dictionary similar to hierarchical trees [13], that is later translated into natural language. In addtion, the content of state feedback reports specific geometric violations (e.g., exact obstacle intersection coordinates, spacing deviations.) and they're complemented with navigational hints (e.g., suggesting lateral movements). This focuses the LLM's reasoning on robust spatial constraints, effectively extending the state-verbalization frameworks of prior work into the continuous domain of coordinate-based planning.

3 Proposed Methodology

3.1 Environment Encoding

To enable precise spatial reasoning, we abstract the 2D simulation environment E into a structured representation, which serves as the ground truth for the LLM:

$$E = \{C, J, O\} \tag{1}$$

where C, J and O represent corridors, junction and obstacles respectively.

Corridors (C): The navigable space is decomposed into a set of axis-aligned rectangular regions. Each corridor $c_i \in C$ is defined by its geometric bounds where:

$$c_i = \left\{x_{i,\min}, x_{i,\max}, y_{i,\min}, y_{i,\max}\right\} \tag{2}$$

Junctions (J): Topological connectivity is governed by fixed spatial connection points called junctions. Each junction $j_k \in J$ is defined by a centroid coordinate, p_k:

$$p_k = (x_k, y_k) \tag{3}$$

and a connectivity list enabling lawful transition between specific corridors:

$$j_k = \{p_k, \{c_i, c_{i+1}, \dots\}\} \tag{4}$$

Obstacles (*O*)
Static constraints are modeled as forbidden zones. Each obstacle $o_m \in O$ is defined similarly to a corridor but represents a strictly non-navigable area, augmented by a safety margin, δ:

$$o_m = \{\text{position},\ \text{bounds},\ \delta\} \tag{5}$$

This structured format discretizes the continuous environment into logical rules, allowing the LLM to process spatial validation checks as textual propositions.

To ensure proper waypoint generation inside this structured environment, critical constraints are defined and discussed in the following subsection.

3.2 Constraint Definitions

The LLM must generate waypoints in a Python dictionary format with keys 'robot_1' and 'robot_2' pairs to ensure consistent parsing and direct validation. Each key maps to a list of coordinates formatted as: $[[x_1, y_1], [x_2, y_2], \ldots, [x_N, y_N]]$.

To validate these waypots, the following constraints are enforced for validation and defined in Table 1.

Table 1. Waypoints constraints defintion and values

Constraint	Definition	Values
spacing	consecutive waypoints must respect a minimum and maximum spacing.	Min: 0.5 m Max: 4.0 m
corridor	waypoints must lie within corridor bounds with a robot margin.	Margin: 0.05 m
obstacle	waypoints must not intersect with obstacle bounds and must lie within robot radius and margin measured from obstacle's center for safe distance.	Radius: 0.3 m Margin: 0.05 m
junction	corridor transitions are allowed only via valid junctions	–
target	last waypoint must match the robot's target coordinates with a tolerance.	Tolerance: 0.1 m

Spacing constraint defines a minimum of 0.5 m and maximum of 4.0 m spacing for consecutive waypoints to maintain a safe distance. In addition, corridor constraint keeps the waypoints inside safe bounds with a safety margin of 0.05 m, while obstacle constraint ensures each waypoint maintains a minimum safe distance including robot radius of 0.3 m and margin of 0.05 m (Minimum safe distance: 0.35 m). The junction constraint enforces lawful transitions between corridors to avoid collision with corridors corner points. Finally, the target constraint requires the last waypoint to match robot's target position within 0.1 m tolerance.

These constraints are embedded within the LLM prompt to ground its reasoning and reduce hallucination problems.

3.3 Closed-Loop Feedback Mechanism

After the LLM generates waypoints, a validation module checks waypoints against each constraint from subsection 3.2 and producesfeedback as illustrated in Table 2.

Table 2. Closed-Loop feedback definitions

Feedback	Definition
spacing_feedback	identifies waypoint pairs violating minimum/maximum spacing requirement
corridor_feedback	identifies waypoints outside of the corridor bounds
obstacle_feedback	identifies waypoints inside obstacles regions and suggest navigational hints
junction_feedback	Identifies unsafe corridor transitions
targets_feedback	computes the distance between the final waypoint and the target position

Spacing feedback calculates the euclidean distance between consecutive waypoints and identifies waypoints pairs violatin the minimum and maximum spacing requirement defined in subsection 3.2. Corridor feedback identifies waypoint location while considering robot radius and margin for safe localization. Obstacle feedback identifies waypoints close to obstacles regions inflated with a safe margin of 0.35 m. Junction feedback checks if one of the waypoints of each robot is a junction point. Finally, targets feedback computes the distance between the final waypoint and target position.

For example, obstacle_feedback is generated when a waypoint is inside an obstacle's safety perimeter. Considering a robot (radius 0.3 m, plus a 0.05 m margin) near 'obstacle_1' at (10.5, 4.0) with square shape and size of 0.2 m. The resulting forbidden x-range is [10.05, 10.95]. A waypoint at (10.6, 4.1) within this range and which is located on the right side of the obstacle triggers the feedback: "robot_1: Waypoint 2 at (10.60, 4.10) is inside obstacle 'obstacle_1' forbidden zone. Move waypoint 2 to the right (increase its x value)."

This feedback is injected into the next LLM prompt, enabling iterative refinement. The loop continues until all constraints are satisfied or a maximum iteration count is reached.

3.4 System and User Prompts Design

The LLM inputs are structured into two distinct components: a System Prompt and a User Prompt. The System Prompt is static and incorporates the spatial encoding from Eq. (1) as illustrated in Fig. 1. In contrast, the User Prompt functions as a dynamic control signal within the planning loop. It is updated at every iteration to reflect the current robot state and any constraint violations, thereby enabling iterative refinement, as depicted in Fig. 1 as well.

The system prompt starts by defining the role of the LLM as a navigation planning agent, and the environment specifications which consists of corridors, junctions and obstacles. Next, the constraints from subsection 3.2 along with the feedback definitions from subsection 3.3 are included, in addition to planning rules for enforcing waypoints correct order (from start to target) and preserve valid waypoints while correcting violations.

The user prompt specifies the task as simultaneously planning for both robots while also using the feedback report consisting of spacing violations, collisions, junction and targets checks, forming a closed-loop feedback scheme.

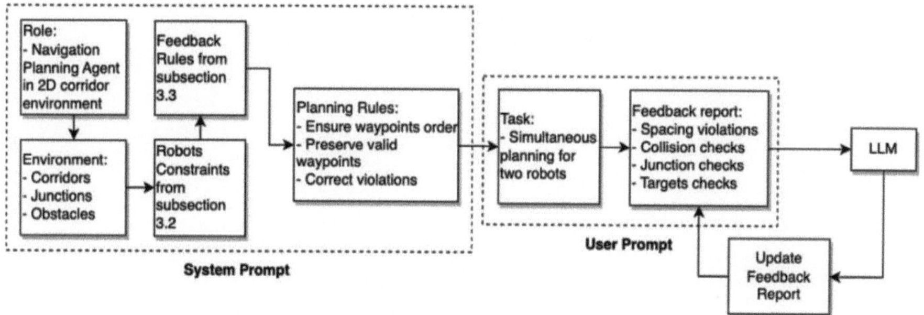

Fig. 1. Prompts block diagram

4 Design and Implementation Setup

4.1 Simulation Environment

The simulation is implemented in Python using Matplotlib for 2D visualization. The environment map is displayed in Fig. 2. It consists of a main corridor and two side corridors (corridor_1 and corridor_2) forming a U-corridors shape. Two junctions connect the main corridor with each side corridor for safe robot planning. In addition, the robots start and target position are initialized.

The obstacles configuration includes three test scenarios as illustrated in Fig. 3. In Fig. 3. (a), there are no obstacles in the environment, while (b) features one obstacle positioned at the middle of each side corridor. (c) however, include another obstacle near the target positions of the robots. The obstacle size in all scenarios is a rectangle of size 0.4 m x 0.4 m.

The symmetrical U-corridor tests if the LLM, due to the environment's identical mirrored halves, will generate symmetrical waypoints all the time, even when the optimal path requires an asymmetrical solution across the three obstacle scenarios.

4.2 LLM Configuration and Waypoints Validation Functions

The planner uses GPT-4.1 mini to ensure a consistent and parsable output accessed via the OpenRouter API. The temperature is set to 0.75 to favor deterministic behaviour over creativity for safety constraints and output formatting rules. The model is prompted with structured system and user messages as described in subsection 3.4.

The waypoint validation functions use the feedback definitions from subsection 3.3 consisting of spacing, corridors, obstacles, junctions and final target checks. LLM-proposed waypoints are evaluated by all validation functions, each outputting a boolean value (True/False) indicating constraint satisfaction along with corresponding feedback. With this setup, a closed-loop feedback mechanism is implemented, which iteratively refines trajectories by providing the LLM with constraint-specific corrections until all constraints are satisfied or a maximum iteration threshold **(30 iterations)** is reached.

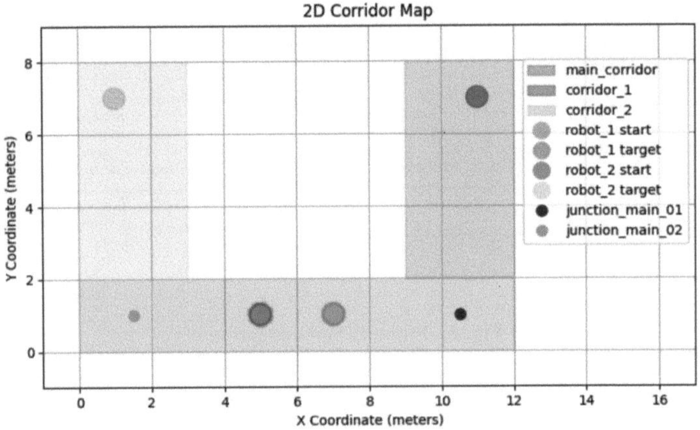

Fig. 2. 2D Corridor Map

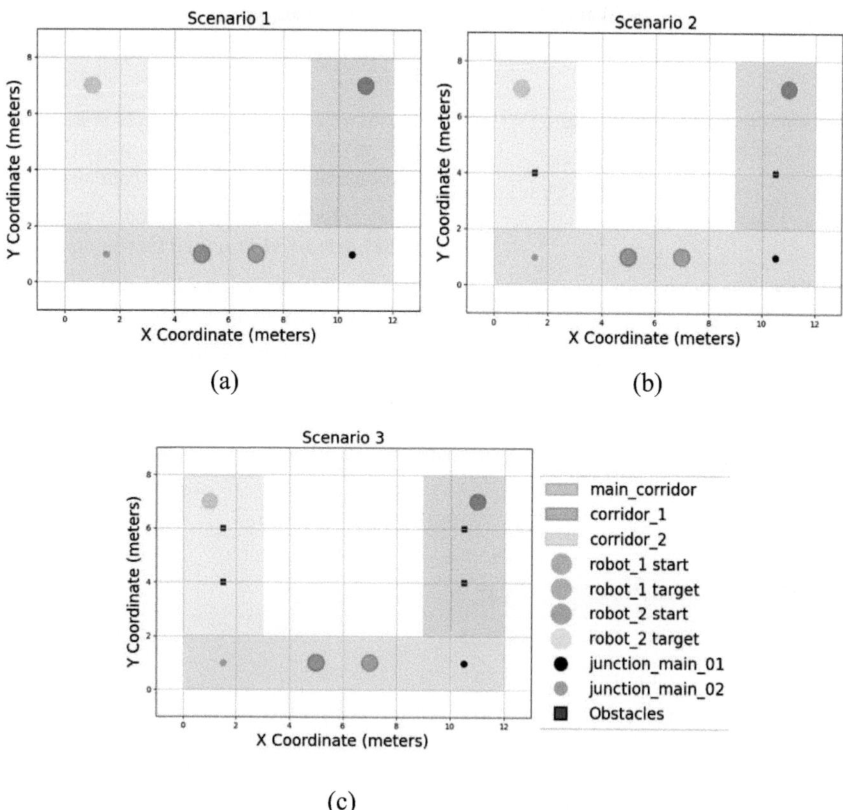

Fig. 3. Obstacles configuration, (a) no obstacles, (b) one obstacle at the middle and (c) obstacle at the middle and near the target positions

5 Results and Discussion

5.1 Evaluation Metrics

To quantify the performance of the proposed framework, five metrics shown in Table 3 were measured across all test scenarios.

The first metric objective is to evaluate the LLM success rate in generating and validating the waypoints. Average planning time is used to observe LLM latency in different obstacles scenarios. While total token consumption measures the LLM cost and the average path length compute the mean distances travelled by the robots.

These metrics give insights into the LLM accuracy, latency and cost in generating and validating waypoints for successful robots' navigation.

5.2 Data Analysis

Five attemps were conducted for each of the theree obstacle scenarios. The results are summarized in Table 4.

Table 3. Framework evalution metrics

Metric	Definition
Waypoint Generation Success Rate (WGSR)	Percentage of successful attempts in generating valid waypoints for both robots within maximum allowed attempts (30 attempts)
Average Planning Time	Mean computation time required to generate and validate LLM-proposed waypoints
Total Token Consumption	Prompt and Completion tokens processed by the LLM across all attempts
Average Path Length	Mean distance traveled by each robot

Table 4. Framework performance across all obstacles scenarios

Metric	Scenario 1	Scenario 2	Scenario 3
WGSR (%)	100	100	100
Avg. Planning Time (s)	3.58	3.65	3.72
Total tokens	1,979	11,870	47,857
Robot 1 avg. path length (m)	9.56	9.82	10.23
Robot 2 avg. path length (m)	9.56	9.80	10.61

The results demonstrate the robustness of the proposed framework, achieving a 100% success rate across all levels of environmental complexity. As expected, token consumption increases significantly (from approximately 2 k to 48 k) as obstacles are added, reflecting the accumulating context required for the iterative feedback loop to resolve constraints. Similarly, path lengths slightly increase in Scenarios 2 and 3 as the robots deviate from straight-line trajectories to navigate around obstacles. Interestingly, the average planning time remained relatively stable at approximately 3.6 s, suggesting that the closed-loop feedback effectively guides the LLM to a solution without requiring exponentially more time, even as the textual context grows.

To assess the repeatability of the proposed framework, 5 independent attempts were conducted for each scenario. In Scenario 1 (Figs. 4, 5 and 6), the obstacle-free environment revealed a consistent baseline behavior: the LLM tends to propose symmetrical paths, which are then streamlined by post-processing. In contrast, Scenarios 2 and 3 (Figs. 7, 8, 9, 10, 11 and 12) tested the system's ability to break symmetry in the presence of obstacles, forcing the planner to generate valid trajectories that navigated around obstacles from either the left or the right. Despite higher variability in Scenario 3 due to increased complexity, the iterative feedback loop consistently converged on collision-free solutions across all attempts.

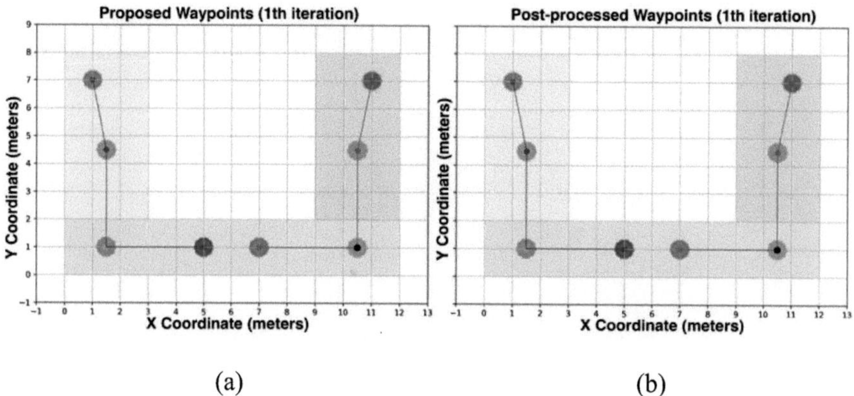

(a) (b)

Fig. 4. Scenario 1: first attempt visiualization, (a) proposed waypoints, (b) processed waypoints

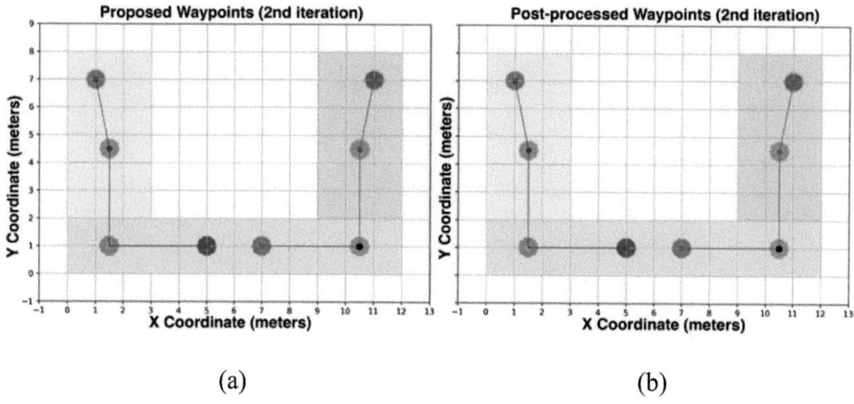

(a) (b)

Fig. 5. Scenario 1: second attempt visiualization, (a) proposed waypoints, (b) processed waypoints

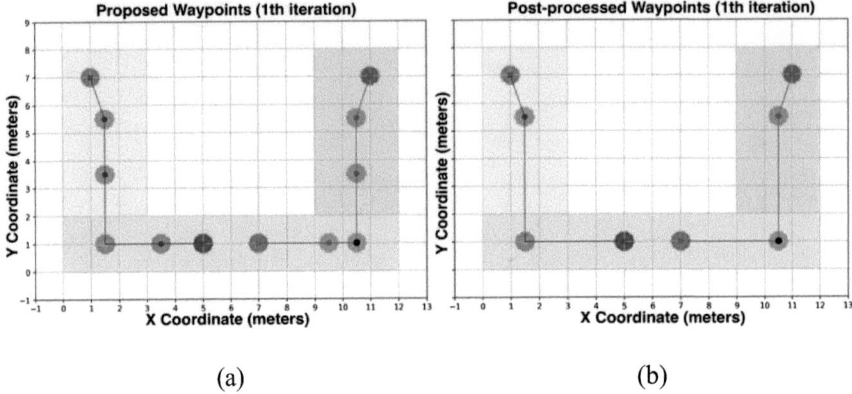

(a) (b)

Fig. 6. Scenario 1: fifth attempt visiualization, (a) proposed waypoints, (b) processed waypoints

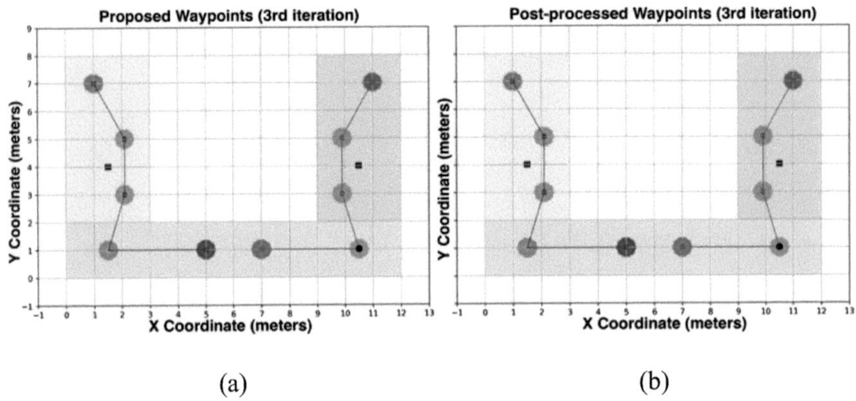

(a) (b)

Fig. 7. Scenario 2: first attempt visiualization, (a) proposed waypoints, (b) processed waypoints

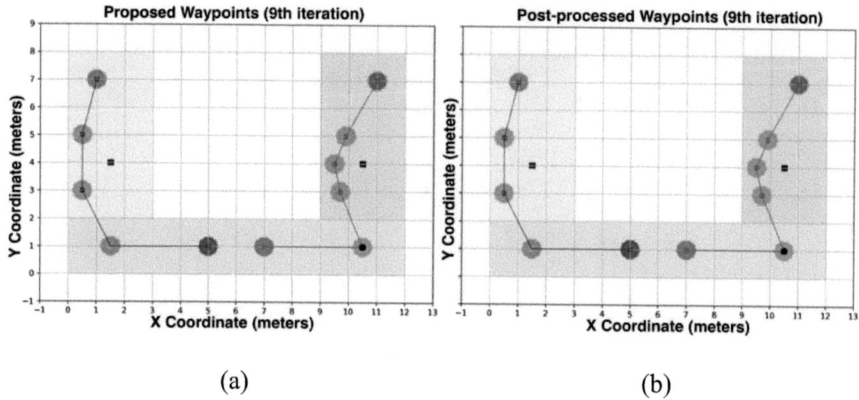

(a) (b)

Fig. 8. Scenario 2: second attempt visiualization, (a) proposed waypoints, (b) processed waypoints

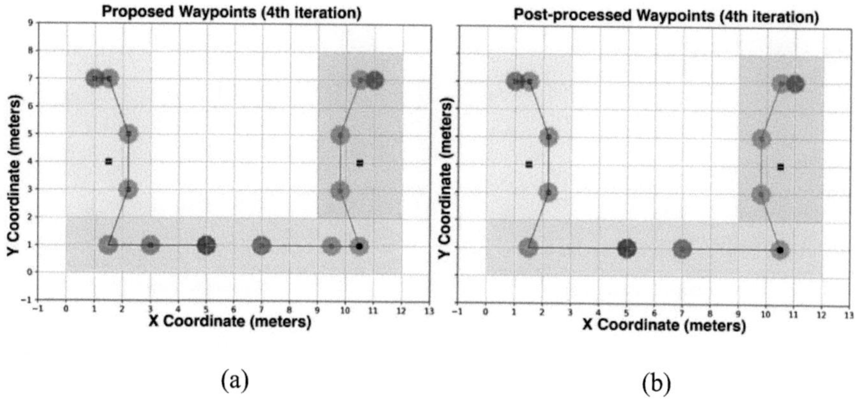

(a) (b)

Fig. 9. Scenario 2: fourth attempt visiualization, (a) proposed waypoints, (b) processed waypoints

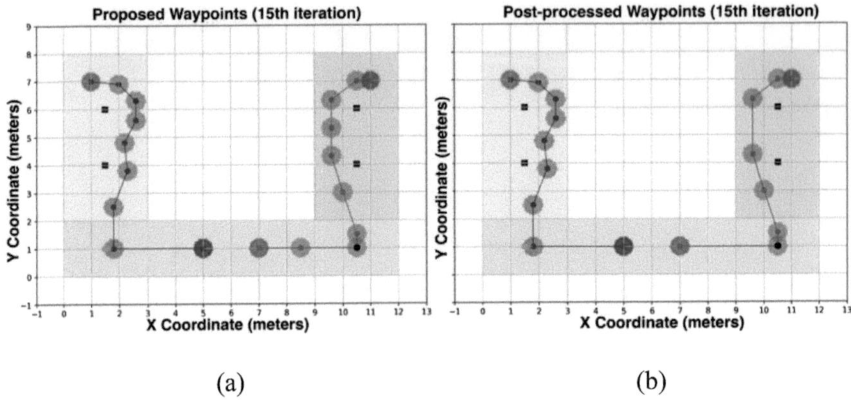

(a) (b)

Fig. 10. Scenario 3: first attempt visiualization, (a) proposed waypoints, (b) processed waypoints

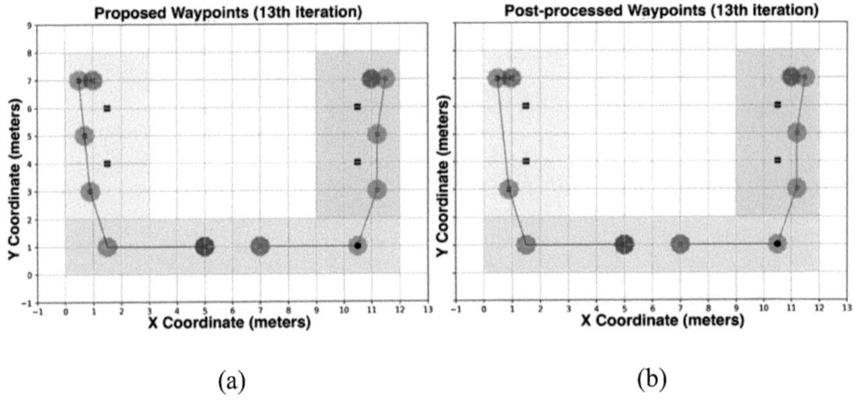

(a) (b)

Fig. 11. Scenario 3: second attempt visiualization, (a) proposed waypoints, (b) processed waypoints

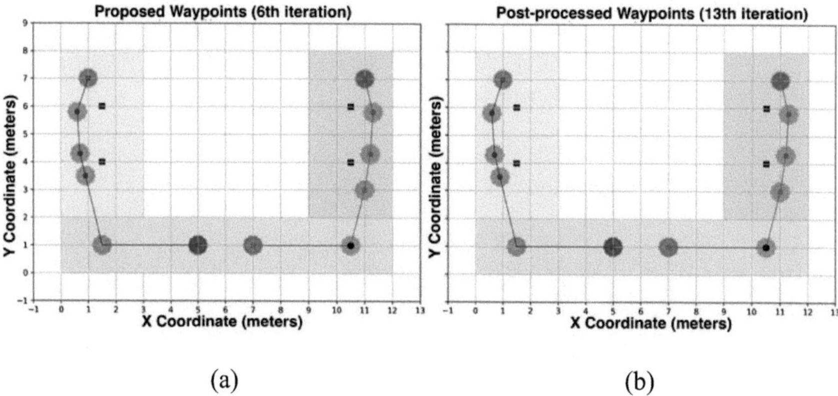

(a) (b)

Fig. 12. Scenario 3: fifth attempt visiualization, (a) proposed waypoints, (b) processed waypoints

6 Conclusion

This paper presented an LLM-based framework for multi-robot waypoint generation in 2D corridor environments with static obstacles. By combining structured prompt engineering with closed-loop feedback, the planner achieves high waypoint generation success rates while respecting spatial constraints. The iterative feedback mechanism proved essential in translating geometric violations into actionable linguistic guidance, enabling the LLM to refine trajectories without requiring domain-specific fine-tuning. Experimental validation across three obstacle scenarios demonstrated consistent repeatability, with all attempts successfully converging on collision-free solutions despite varying environmental complexity. Future work will focus on extending the approach to heterogeneous robot teams [16] and more complex 3D environments.

Acknowledgements. The authors express their gratitude to Universiti Teknologi Malaysia for their generous financial support through the UTM Fundamental Research Grant (Vote: Q.J130000.3823.23H97).

References

1. Sun, J., Huang, J., Pompili, D.: LLM-based multi-agent decision-making: challenges and future directions. IEEE Rob. Autom. Lett. **10**(6), 3562–3569 (2025). https://doi.org/10.1109/LRA.2025.3562371
2. Dong, K.: Large language model applied in multi-agent system—a survey. Appl. Comput. Eng. **109**, 9–16 (2024). https://doi.org/10.54254/2755-2721/109/20241330
3. Jestel, C., et al.: MuRoSim – a fast and efficient multi-robot simulation for learning-based navigation. In: Proceedings of the IEEE International Conference on Robotics and Automation (ICRA), pp. 4311–4317. IEEE Press (2024). https://doi.org/10.1109/ICRA57147.2024.10610375
4. Athira, K.A., Divya Udayan, J., Umashankar, S.: A systematic literature review on multi-robot task allocation. ACM Comput. Surv. **57**(3), 68 (2025). https://doi.org/10.1145/3700591
5. Ma, S., Ruan, J., Du, Y., Bucknall, R., Liu, Y.: An end-to-end deep reinforcement learning based modular task allocation framework for autonomous mobile systems. IEEE Trans. Autom. Sci. Eng. **21**(1), 1–15 (2024). https://doi.org/10.1109/TASE.2024.3367237
6. Wang, H., Tan, A.H., Nejat, G.: NavFormer: a transformer architecture for robot target-driven navigation in unknown and dynamic environments. IEEE Rob. Autom. Lett. **9**(8), 6902–6909 (2024). https://doi.org/10.1109/LRA.2024.3412638
7. Chen, Y., Arkin, J., Zhang, Y., Roy, N., Fan, C.: Scalable multi-robot collaboration with large language models: centralized or decentralized systems? In: Proceedings of the IEEE International Conference on Robotics and Automation (ICRA), pp. 4311–4317. IEEE Press (2024). https://doi.org/10.1109/ICRA57147.2024.10610676
8. Bhat, V., Kaypak, A.U., Krishnamurthy, P., Karri, R., Khorrami, F.: Grounding large language models for robot task planning using closed-loop state feedback. Adv. Robot. Res., e202500072 (2025). https://doi.org/10.1002/adrr.202500072
9. Tariq, M.T., Hussain, Y., Wang, C.: Robust mobile robot path planning via LLM-based dynamic waypoint generation. Expert Syst. Appl. **282**, 127600 (2025). https://doi.org/10.1016/j.eswa.2025.127600
10. Meng, S., et al.: LLM-A*: large language model enhanced incremental heuristic search on path planning. Find. Assoc. Comput. Linguist. EMNLP. **2024**, 1087–1102 (2024). https://doi.org/10.18653/v1/2024.findings-emnlp.60
11. Azkune, G., Salaberria, A., Agirre, E.: Grounding spatial relations in text-only language models. Neural Netw. **170**, 215–226 (2024). https://doi.org/10.1016/j.neunet.2023.11.031
12. Pham, T., Cao, C.D.: CO-STEP: a prompt engineering framework improving LLM'S response. In: Proceedings of the ATiGB, pp. 588–593. Springer (2025). https://doi.org/10.1109/ATIGB66719.2025.11142099
13. Park, J.S., et al.: Generative agents: interactive simulacra of human behavior. In: Proceedings of the ACM Symposium on User Interface Software and Technology (UIST), pp. 1–12. ACM (2023). https://doi.org/10.1145/3586183.3606763
14. Xue, D., et al.: Formation control and path planning of multi-robot systems via large language models. SCIENCE CHINA Inf. Sci. **68**(5), 150205 (2025). https://doi.org/10.1007/s11432-024-4290-4

15. Waga, A., et al.: Waypoint-guided trajectory planning for mobile robots using GPT-4.1 mini and ensemble learning-based action prediction. Array. **29**, 100636 (2026). https://doi.org/10.1016/j.array.2025.100636
16. Do, H., Jang, J., Kim, J.: Heterogeneous multi-robot system mission planning with cooperative replenishment through data-driven rendezvous point selection. Intell. Serv. Robot. **18**, 61–73 (2025). https://doi.org/10.1007/s11370-024-00568-9

Performance Evaluation of Fuzzy Logic and Nonlinear PID Controllers for Antenna Positioning Systems

Abdulganiyu Mukhtar[1(✉)], Muhammad Sani Gaya[1,2], Norhaliza Abdul Wahab[3], and Sulaiman Muhammad Gana[1]

[1] Department of Physics, Bayero University, Kano BUK,, Nigeria
agmukhtar001@gmail.com

[2] Department of Electrical Engineering, Aliko Dangote University of Science and Technology, (ADUSTECH), Wudil, Kano, Nigeria

[3] Department of Control and Mechatronics Engineering, Universiti Teknologi, Johor Bahru, Malaysia

Abstract. Accurate alignment and efficient tracking of the satellite by an Earth station antenna to maintain the desired line of sight are critical for optimal power transfer and reliable signal reception. Incorrect alignment leads to instability and poor reception, which in turn increases operational costs. While conventional control strategies often fail to deliver the desired performance, many advanced control techniques are computationally expensive. This research presents a comparative performance analysis of a two-input Fuzzy Logic Controller (FLC) and a variable gain Nonlinear Proportional Integral Derivative (NPID) control strategy applied to an antenna azimuth positioning system. Simulations in MATLAB/Simulink evaluated both controllers under various scenarios, including fixed and variable reference signals, as well as external disturbances. The results show that the NPID controller provides a faster initial response of 0.1807s but with a significantly high overshoot of 33.22%. Conversely, the FLC demonstrates superior overall performance, characterized by minimal overshoot (2.4%), settling time (1.785 s), enhanced disturbance attenuation, and more stable tracking. The findings confirm that the FLC is a more effective and robust control solution for ensuring precise antenna alignment and reliable operation amidst system nonlinearities and external disturbances.

Keywords: Antenna · Control · Earth-Station · Satellite

1 Introduction

In recent years, geostationary satellite systems have undoubtebly become crucial in various fields of human endeavor, including telecommunications and broadcasting, navigation, global positioning systems, metrology, as well as disaster and environmental monitoring [1]. These satellite systems are structured to primarily function as microwave repeaters, receiving electromagnetic wave signals from Earth station antennas (uplink), amplifying the signals, and transmitting them back to other Earth station antennas (downlink) without any modifications [2].

N. A. Mohd Subha et al. (Eds.): SMS 2026, CCIS 2956, pp. 44–57, 2026.
https://doi.org/10.1007/978-981-92-0306-2_4

Large Earth station antennas widely used in commercial communications employ a very narrow beamwidth to maximize the power transmitted to the satellite and ensure efficient signal reception. Since these systems rely on a clear line of sight (LOS) for maximum power transmission and reception, a position control mechanism is critical for maintaining precise alignment of the antenna's Look Angles (i.e., azimuth and elevation angles) with the target satellite, ensuring stable and effective communication [3]. As a result, there is a growing need to design and develop advanced control strategies that can continuously adjust the antenna's position in response to satellite movements in real time [4].

Several control strategies have been developed, including enhanced classical control, proportional-integral-derivative (PID), Tuned Compensator (TC), and Neural Network (BPNN)-based Proportional Integral Derivative (PID) [5, 6]. The enhanced classical control strategies performed better than the classical PID. However, despite these enhancements, the controller remains primarily ineffective in handling the system's nonlinear behaviour. Model Predictive Control (MPC) was proposed by [4, 7, 8]. They presented a detailed comparison of MPC with various control strategies, which includes PID, Fractional Order PID (FOPID), Sliding Mode (SM), employing the Camel Optimizations Algorithm (COA) to optimize the controller's parameters for controlling the azimuth position of earth station antennas. The results show that the MPC outperformed the other three control strategies, achieving a faster settling time and very minimal overshoot. However, MPC is more computationally intensive, therefore costly. Moreover, MPC tends to have a slower rise time when compared to the PID. Intelligent control approaches have evolved to offer better alternatives for handling the nonlinearities, uncertainties, and imprecision of a system, as they do not require an accurate mathematical model of the system unlike advance control techniques such as MPC. Furthermore, they are flexible and require fewer computations [9].

Fuzzy logic, as one of the key pillars of intelligent control techniques, has received considerable attention owing to its wide range of applications. The fuzzy logic technique imitates the human decision-making process and is quite easy to comprehend and implement. The frequent failure of conventional approach to handle the complex, nonlinear, and uncertain nature of real-world control problems makes fuzzy logic a preferred tool [10]. Conventional control methods, such as PID, can also be enhanced to effectively handle nonlinearities and uncertainties. There are various approaches to overcoming the limitations of PID, but improvement via the nonlinearity method offers several advantages, including simplicity, high robustness, reduced reliance on an accurate mathematical model, and ease of implementation [11].

This research focuses on comparing the effectiveness of Fuzzy Logic Controller (FLC) and Non-linear Proportional-Integral-Derivative (NPID) in controlling the earth satellite antenna position on ground station using various control measures, including rise time, settling time, peak time, and maximum overshoot.

2 Methodology

2.1 System Description and Mathematical Modelling

The antenna azimuth positioning system is a closed-loop, automated antenna control system that converts the input position angle $\theta_i(s)$ in rad, into a corresponding position output response $\theta_o(s)$. The system's two potentiometers act as transducers to control the azimuth position, as shown in Fig. 1.

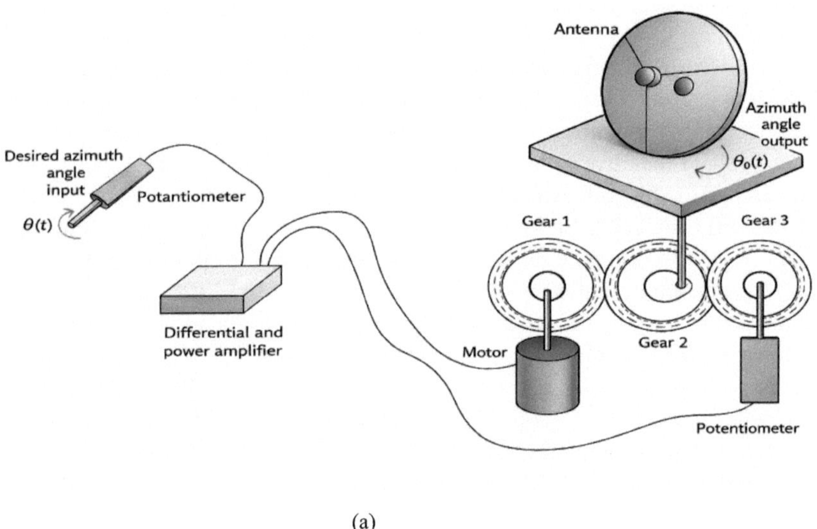

(a)

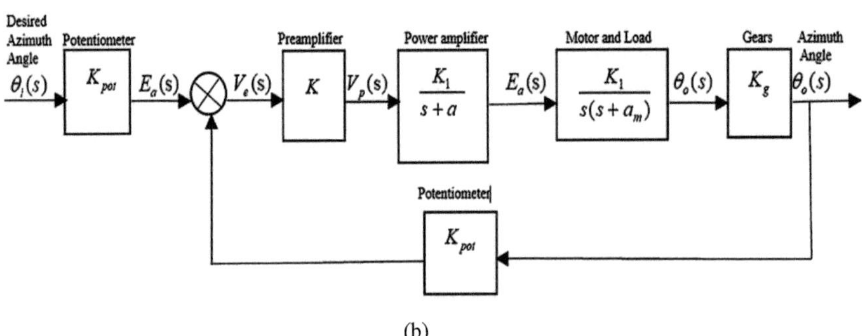

(b)

Fig. 1. (a) A detailed layout of the antenna azimuth positioning control system (b) Functional block diagram of the antenna azimuth positioning control system [12].

The antenna azimuth positioning control consists of five sub-blocks: the Input and Output potentiometers, the Preamplifier, the Power amplifier, the Motor and Load, and the Gears, which can be modelled based on the block diagram shown in Fig. 1(b). The first potentiometer is manually adjusted to set the desired antenna angle position, converting it into an input voltage. In contrast, the second potentiometer measures the actual antenna position and converts it into a voltage signal for the system's feedback path. A differential preamplifier amplifies the difference between the input and output voltages, representing the system error. This amplified error signal is then fed into a power amplifier, which adjusts the voltage level appropriately to drive the motor through a gear system, ensuring accurate antenna positioning [13]. The mathematical model of the system can be obtained by computing the Motor and Load, as well as the Gear sub-blocks transfer functions respectively.

The motor is connected to the load, which in this case is the antenna, it receives the output of the power amplifier, $E_a(s)$, as its input and generates a mechanical output $\theta_m(s)$. Considering the input voltage $E_a(s)$, in the schematic design in Fig. 2 and applying Kirchhoff's Voltage Law, the input voltage of the motor will be given by [15];

$$E_a(s) = I_a(s)R_a + sI_a(s)L_a + V_b(s) \tag{1}$$

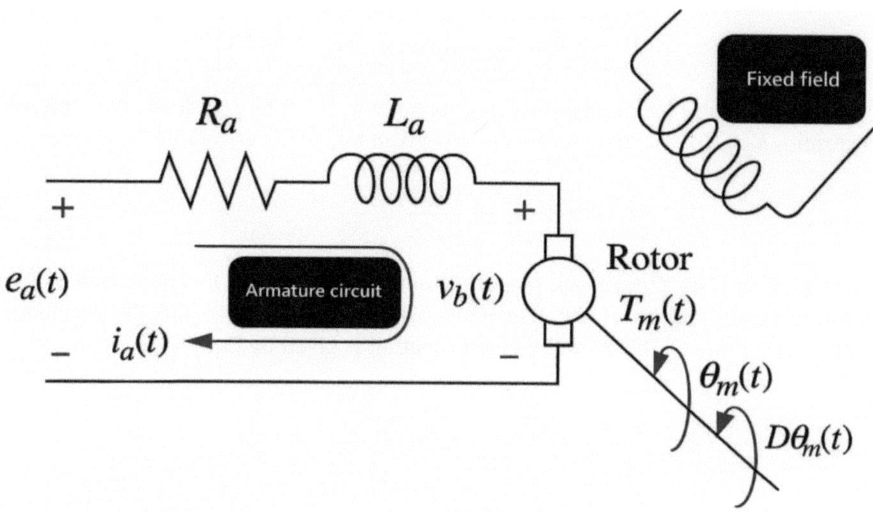

Fig. 2. Schematic design and block diagram of motor and load [14]

The motor torque constant $T_m(s)$ is proportional to the input current of the armature motor with a constant of proportionality K_t. The equation gives;

$$I_a(s) = \frac{T_m(s)}{K_t} \tag{2}$$

By taking the Laplace transform of the back electromotive force(back emf) equation that was derived from the proportionality relation of voltage and speed when the current passes through the armature and rotates in a magnetic field, the back emf V_b (s) is given by Eq. (3) where K_b is the constant of proportionality and $s\theta_m(s)$ is the angular velocity;

$$V_b(s) = K_b s\theta_m(s) \tag{3}$$

Considering the equivalent inertia, , and damping, , of the motor, which are linked with the torque constant, is given by the equations;

$$J_m = J_a + J_l\left(\frac{N_1}{N_2}\right)^2 \tag{4}$$

$$D_m = D_a + D_l\left(\frac{N_1}{N_2}\right)^2 \tag{5}$$

$$T_m(s) = \left(J_m s^2 + D_m s\right)\theta_m(s) \tag{6}$$

J_a and J_l are the inertia constants of the motor and load, D_m and D_l are the damping constants of the motor and load. After substituting Eq. (2), Eq. (3), and Eq. (6) into Eq. (1), ignoring the effect of the inductance L_a, the motor's transfer function is shown in Eq. (7).

$$G(s) = \frac{\theta_m(s)}{E_a(s)} = \frac{\frac{K_t}{R_a J_m}}{s\left[s + \frac{1}{J_m}\left(D_m + \frac{K_t K_b}{R_b}\right)\right]} \tag{7}$$

The gears sub-block consists of gear one teeth, N_1, gear two teeth, N_2, and has a gear ratio, K_g [16]. The transfer function is given by;

$$K_g = \frac{\theta_o(s)}{\theta_m(s)} = \frac{N_1}{N_2} \tag{8}$$

K_g is the gear ratio, $\theta_o(s)$ is the output azimuth angle, $\theta_m(s)$ is the angle of the motor. Substituting the values of the schematic parameter values [4] into the sub-blocks' mathematical model, the overall transfer function is given by Eq. (9)

$$G_T = \frac{\theta_o(s)}{\theta_i(s)} = \frac{20.83}{s^3 + 101.71s^2 + 171s} \tag{9}$$

2.2 Fuzzy Logic Controller and Nonlinear PID Design

Fuzzy Logic Controller. Fuzzy logic is a problem-solving control system that is intrinsically capable of handling nonlinearities by emulating human reasoning when dealing with vague, ambiguous, imprecise, noisy, or missing input information. The core concept involves processing crisp sensor inputs into membership values within a specified range of 0 and 1, which define the degree to which an input corresponds to a membership function [17, 18].

The FLC is designed based on the specific nature and characteristics of the given plant and reference signal. The controller consist of five fundamental stages: fuzzification, knowledge base, inference, rule base, and defuzzification, which ensures the closed-loop fuzzy control system functions effectively for the intended application [20, 21]. The membership functions used to define the degree of membership for the corresponding input crips values are in Fig. 3.

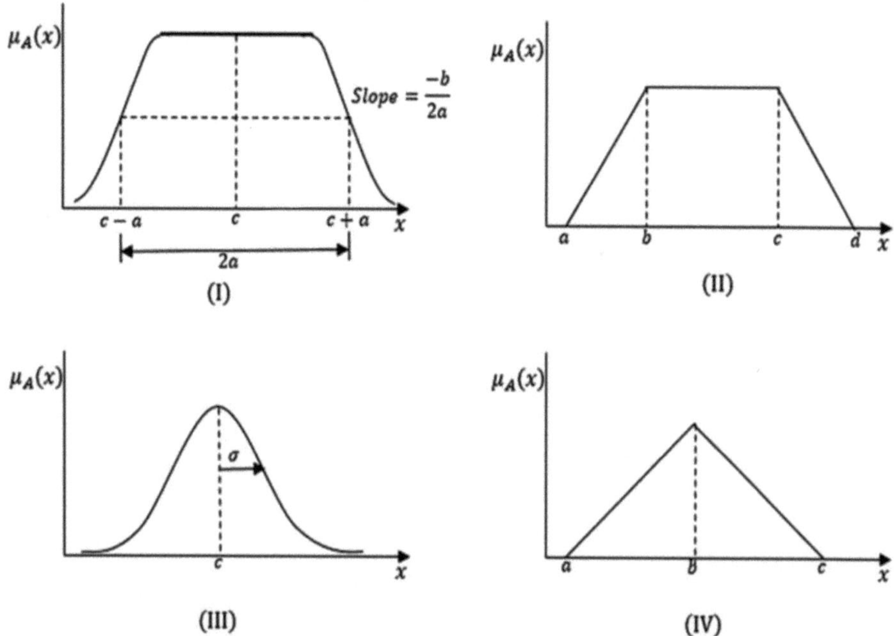

Fig. 3. Membership Functions: (I) Bell function, (II) Trapezoidal function, (III) Gaussian function, (IV) Triangular function, [19].

Fuzzy Logic Controller Design. The fuzzy logic controller uses a two-input control strategy that accepts the actual system error *(e)* and the change in error *(Δe)*. This enabled it to adapt dynamically to different conditions, anticipate future error trends, and make better decisions. As a result, the controller achieved a faster response, improved stability, enhanced tracking accuracy, reduced overshoot, and better handling of nonlinearities. The input membership functions for e and Δe range from -10 to 10 and $-$ 3 to 3, respectively, while the control output ranges from -25 to 25, as shown in Fig. 4.

The error and change in error each consist of five membership functions: three triangular and two trapezoidal (one at each end). These membership functions are used to construct a set of 25 linguistic rules, as shown in Table 1. The linguistic terms used are NB ('Negative Big'), NS ('Negative Small'), ZR ('Zero'), PS ('Positive Small'), and PB ('Positive Big').

2.3 Nonlinear Proportional-Integral-Derivative (NPID)

The simple structure, easy implementation, and favourable cost-to-benefit ratio of PID controllers have made their applications long dominate the control industry [22, 23]. Ranging from the traditional fields of Chemical, Electrical, Aerospace, and Mechanical engineering to applications such as process control, motion control, manufacturing, hydraulics, pneumatics, and power electronics to modern and emerging fields such as self-driving vehicles, robotics, and automated aerial vehicles [24, 25]. However, since

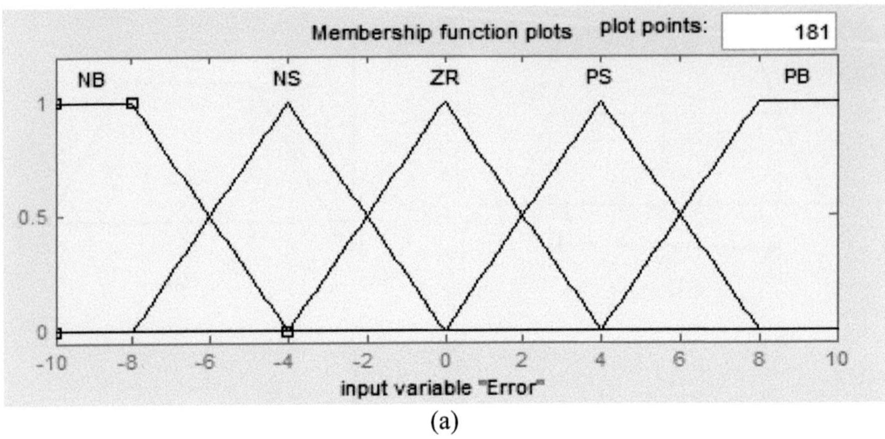

(a)

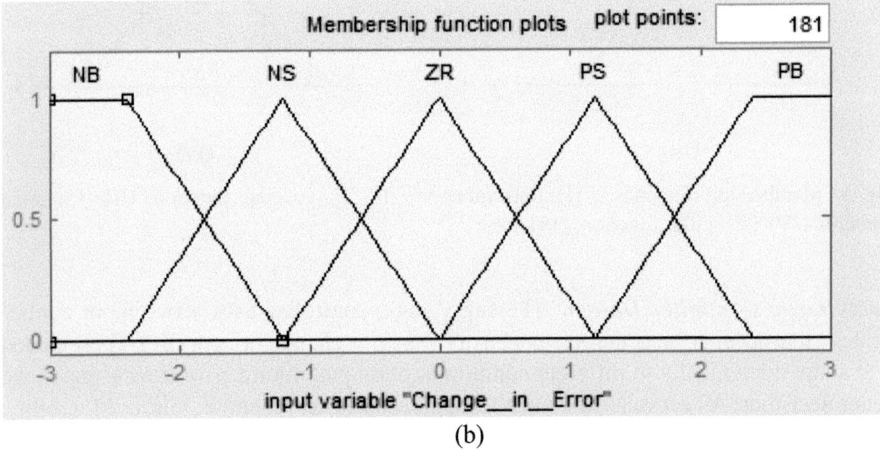

(b)

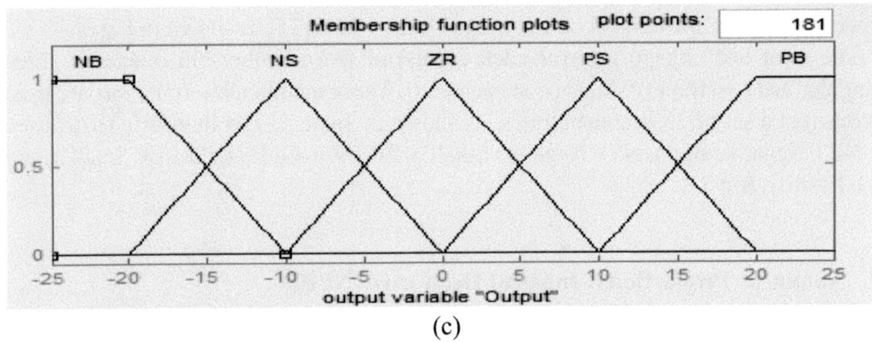

(c)

Fig. 4. (a) Input Error MFs (b) Change in Error MFs (c) Control Output MFs

Earth station antenna systems are inherently nonlinear, a conventional PID controller, being purely linear, cannot effectively control the system. Hence, there is a need for a

Table 1. Fuzzy Linguistic Rules.

E\CE	NB	NS	ZR	PS	PB
NB	NB	NB	NB	NS	ZR
NS	NB	NB	NS	ZR	PS
ZR	NB	NS	ZR	PS	PB
PS	NS	ZR	PS	PS	PB
PS	ZR	PS	PB	PB	PB

Nonlinear PID (PID) controller. The PID and NPID control strategies are represented by [26];

$$u(t) = k_p e + k_i \int e(\tau)d\tau + k_d \frac{de(t)}{dt} \tag{10}$$

$$u(t) = k_p(t)e + k_i(t) \int e(\tau)d\tau + k_d(t) \frac{de(t)}{dt} \tag{11}$$

$$k_{eff} = \frac{e^{K_o * e} + e^{-K_o * e}}{2} \tag{12}$$

$$k_p(\bullet) = k_{eff} * K_p \tag{13}$$

$$k_i(\bullet) = k_{eff} * K_i \tag{14}$$

$$k_d(\bullet) = k_{eff} * K_d \tag{15}$$

The proportional, integral, and derivative gains of the NPID controller are denoted by $K_p(\bullet)$, $K_i(\bullet)$, and $K_d(\bullet)$, respectively. Unlike the constant gains of the PID controller, these gains were design to be time-varying and are dependent on the effective gain (k_{eff}) as shown in Eq. (13), Eq. (14), and Eq. (15). This adaptive gain control strategy enables the controller to dynamically adjust its response based on the instantaneous system error making the controller more robust, accurate, and capable of handling system nonlinearities and uncertainties.

3 Results and Discussions

3.1 Simulation of the Open Loop System

The mathematical model of an earth station antenna was implemented using MATLAB Simulink. When a step response of magnitude 1° was applied, the response was entirely erroneous, with the output increasing continuously over time, as illustrated in Fig. 5. This demonstrated the need for a controller to not only minimize the error but also ensure precise alignment and accurate tracking of the target satellite.

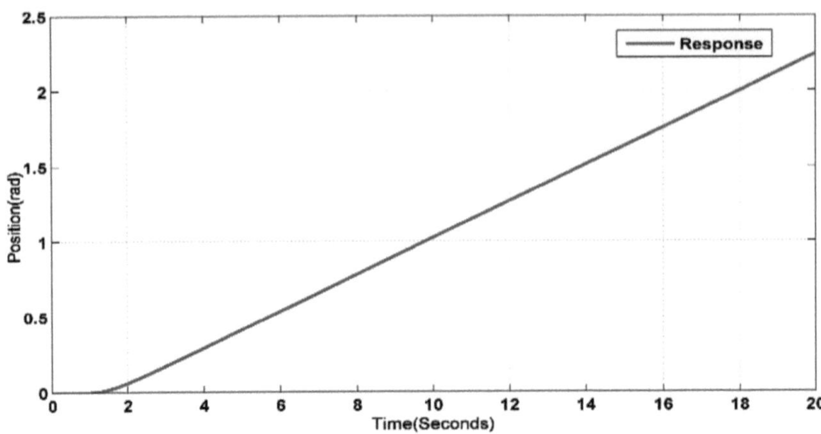

Fig. 5. Open Loop Response.

In this paper, FLC and NPID control strategies are employed to investigate how each of these strategies can be utilized to control the earth station antenna effectively based on rise time (Tr), settling time (Ts), overshoot (%OS), Peak time (Tp), Mean Absolute Error (MAE), and Mean Squared Error (MSE).

3.2 FLC and NPID Simulation Comparison

To effectively evaluate the controllers, their performance was assessed based on their effects on system response, tracking accuracy, and robustness to nonlinearities and external disturbances. The simulations and comparisons carried out are as follows;

Fixed Reference. When a fixed reference step signal of magnitude 1° was applied. The NPID Controller demonstrated a faster rise time of 0.18 s and a better peak time of 1.33 s, but it achieved this at a significantly high 33% overshoot. While the FLC controller provided a more damped response with minimal Overshoot of 2.4% indicating that it quickly stabilizes and has considerably lower settling time. The detailed step response analysis is provided in Fig. 6 and Table 2.

Variable Reference Signals For effective satellite communication, the earth station antenna should maintain high tracking accuracy. To evaluate this, a variable reference signal was used as input in Fig. 7 to assess each controller's performance in tracking the input.

The results show that the FLC achieves a faster rise time of 0.13 s, a settling time of 15.4 s, and a lower 2% overshoot. Although the NPID controller achieves a 0.69 s peak time, it does so at a significant cost of 24% Overshoot, as shown in Table 3. Both the FLC and NPID controllers demonstrated robust tracking capabilities. However, the FLC offers better tracking accuracy, as it responds more quickly and stabilizes with fewer deviations from the reference signal (Fig. 8).

External Disturbance. External disturbances from wind, mechanical vibrations, and satellite orbital deviations affect the optimal transmission and reception of the signal;

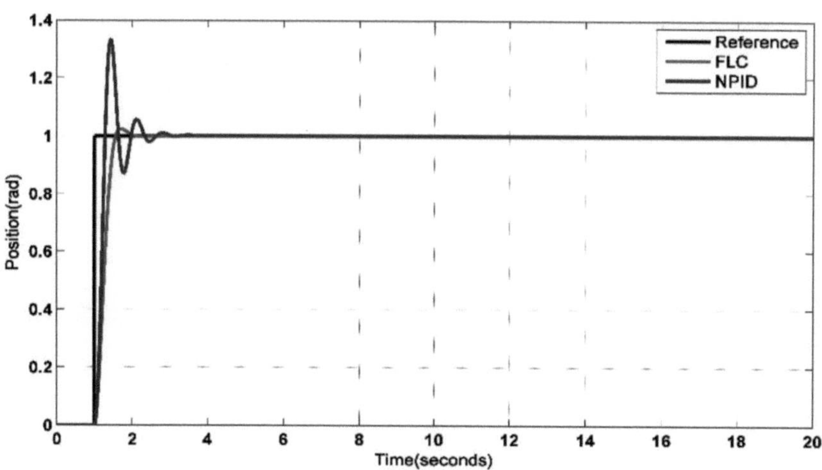

Fig. 6. Step Response for FLC and NPID Controllers.

Table 2. Time Domain Step Response Specifications.

Controller	Tr(s)	Ts(s)	%OS	Tp(s)	MAE	MSE
FLC	0.3493	1.7851	2.4011	1.7061	0.127146	0.114859
NPID	0.1807	2.4894	33.2200	1.3335	0.124561	0.107105

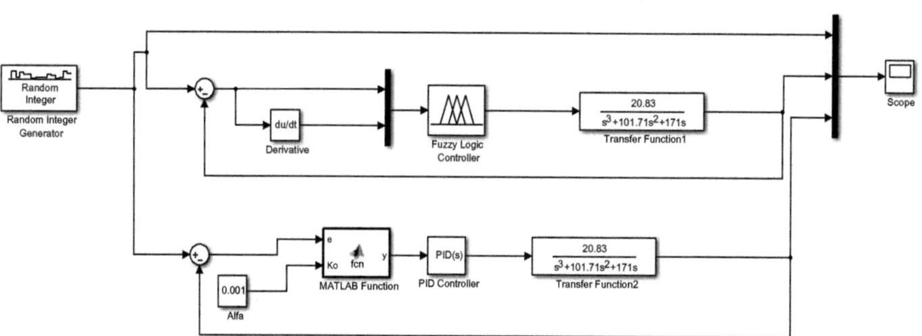

Fig. 7. Tracking Simulation Block Diagram.

Table 3. Time Domain Tracking Response.

Controller	Tr(s)	Ts(s)	%OS	Tp(s)	MAE	MSE
FLC	0.1339	15.4197	2.105633	0.8002	0.66651	1.527904
NPID	0.1766	16.1661	24.68416	0.6979	0.719814	1.711391

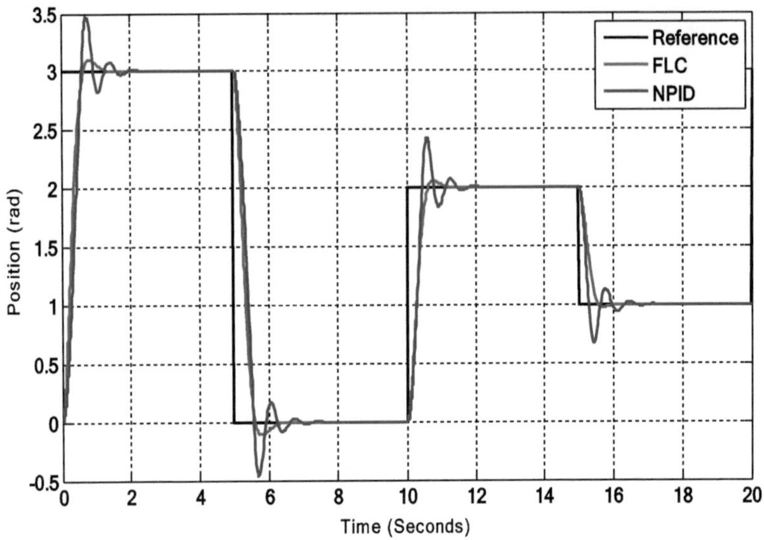

Fig. 8. Tracking Response for FLC and NPID Controllers.

hence, the system's robustness in handling these disturbances is critical. To evaluate this, we injected a pulse signal into the system at 5-s intervals as shown in Fig. 9.

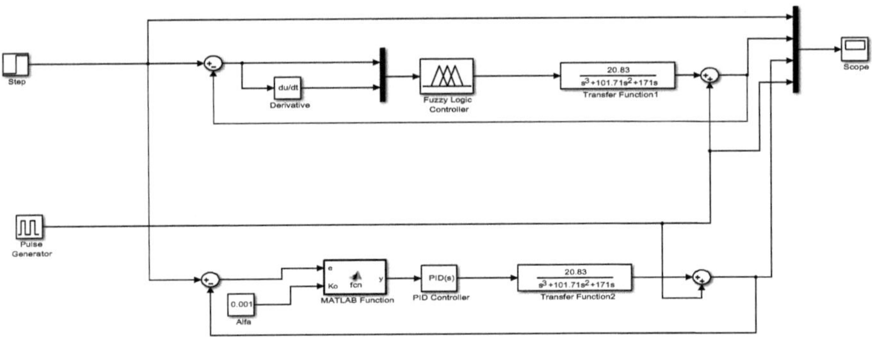

Fig. 9. Block Diagram with External Disturbance.

The 3.054e-5% overshoot of the FLC shows its robustness in attenuating the injected external disturbances compared to the NPID controller in Fig. 10. However, due to the additional disturbance input, the controllers tend to simultaneously reject the disturbance, maintain stability, prevent overshoot, and force the system to follow the actual input signal, thereby reducing the aggressiveness of the FLC and NPID and temporarily leading to a slightly higher rise time. The test results in Table 4 show that both the FLC and NPID effectively respond, handle, and attenuate the injected disturbances.

Table 4. Time Domain Response with External Disturbance.

Controller	Tr(s)	Ts(s)	%OS	Tp(s)	MAE	MSE
FLC	3.7093	19.9983	3.0594e-05	5	0.049098	0.038148
NPID	3.7958	19.9982	0.0338	5	0.050791	0.035294

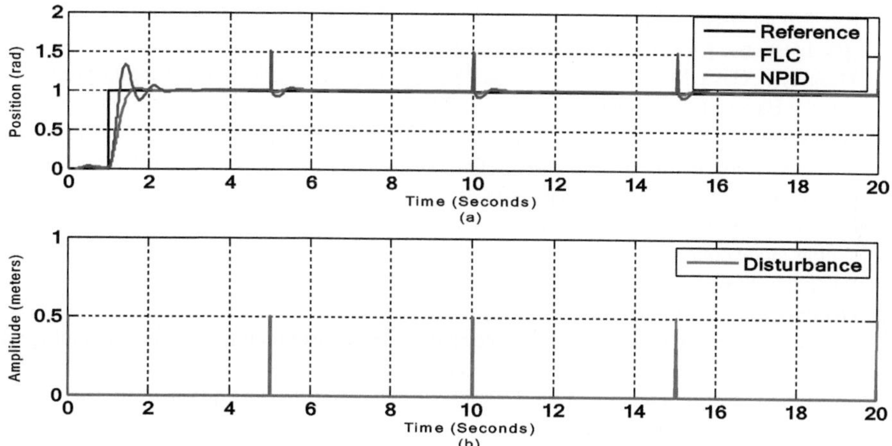

Fig. 10. (a) FLC and NPID with External Disturbance Response. (b) Injected External Disturbance Signal.

4 Conclusions

The research clearly demonstrates how both the FLC and NPID controllers can effectively improve the Earth station antenna system's response and tracking accuracy, and how they provide robustness in handling system nonlinearities and in attenuating external disturbances, which address the underlying challenge of maintaining effective communication between the Earth station antenna and the geostationary satellite by precisely aligning and tracking the antenna's azimuth angle. However, the FLC demonstrates slightly better disturbance attenuation, minimal overshoot, and a faster rise time with fewer oscillations than the NPID controller. This research provides valuable insights into the performance of FLC and NPID controllers and demonstrates their application in the control of large Earth station antennas widely used in commercial communications.

Acknowledgement. The authors acknowledge the support provided by Department of Physics, Bayero University Kano, Nigeria, in facilitating this research.

References

1. Tao, M., Li, S., Jia, Y., Zhang, G.: Design of antenna control system P-fuzzy-PI controller. In: Proceedings of the 2015 International Conference on Applied Science and Engineering Innovation, pp. 1565–1569 (2015)

2. Minoli, D.: Introduction to Satellite Communications, in Satellite Systems Engineering in an IPv6 Environment. Auerbach Publications, pp. 1–34 (2009)
3. Roddy, D., Sinha, R., Verma, A., Choudhary, S.: General analysis of cube satellite technology: an overview. Int. J. Comput. Appl. **178**(21), 19–22 (2019)
4. Al-Mayyahi, A., Aldair, A., Zainab, A.K., Wang, W.: Control system design for azimuth position of earth station antennas. J. Eng. Appl. Sci. **71**(1), 199 (2024)
5. Eze, P.C., Ugoh, C.A., Inaibo, D.S.: Positioning control of DC servomotor-based antenna using PID tuned compensator. J. Eng. Sci. **8**(1), E9–E16 (2021)
6. Eze, P., Obichere, J., Mbonu, E., Onojo, O.: Positioning control of satellite antenna for high-speed response performance. IPTEK J. Eng. **10**(2), 117 (2024)
7. Musa, F.H., Gaya, M.S., Bisu, A.A.: Antenna azimuth position control using model predictive control. Bayero J. Phys. Math. Sci, 1–11 (2021)
8. Ahlawat, H.D., Prasad, M.P.R.: Chauhan, antenna azimuthal position control using model predictive control. In: 2019 IEEE International Conference on Electrical, Computer and Communication Technologies, pp. 1–6 (2019)
9. Fandakli, S.A., Okumus, H.I.: Antenna azimuth position control with PID, fuzzy logic and sliding mode controllers. In: Proceedings 2016 International Symposium on Innovations in Intelligent Systems and Applications INISTA (2016)
10. Okumus, H.I., Sahin, E., Akyazi, O.: Antenna azimuth position control with classical PID and fuzzy logic controllers. In: 2012 International Symposium on Innovations in Intelligent Systems and Applications, pp. 1–5. IEEE (2012)
11. Nise, N.S., Katsuhiko, O.: Control System Engineering Sixth Edition. California State Polytechnic University, Pomona. (2012)
12. Chishti, A.R., Bukhari, S.F., Khaliq, H.S., Khan, M.H., Bukhari, S.Z.H.: Radio Telescope ANTENNA Azimuth Position Control System Design and Analysis in MATLAB/SIMULINK Using PID & LQR Controller. No, LXIV (2014)
13. Mahmood, A., Al-Bayati, K.Y.A., Szabolcsi, R.: Optimizing Antenna Azimuth Position Control using Fuzzy PD, Fuzzy PD-I, and Fuzzy PD-plus-I Controllers. Nanotechnol. Perceptions. **20**(3), 18–32 (2024)
14. García-Martínez, J.R., Cruz-Miguel, E.E., Rodríguez-Reséndiz, J., Ramírez-González, L.D., Rojas-Hernández, M.A.: PID-Like Fuzzy Controller Design for Anti-Slip System in Quarter-Car Robot. Advances in Fuzzy Logic Systems, IntechOpen (2023)
15. Nise, N.S.: Control System Engineering, 48(8). Productivity Press, New York (2021)
16. Patnaik, K.S.A., He, Y.: A Comparison of Hydraulic and Electric Trailer Steering Actuators for Articulated Heavy Vehicles, pp. 1–5, (2019)
17. Dadions, E.P., Biliran, J.J.C., Gercia, R.G., Johnson, D., Valenci, A.R.B.: Humanoid Robot: Design and Fuzzy Logic Control Technique for Its Intelligent Behaviors. Fuzzy Log. - Control. Concepts Theor. Appl., 3–22 (2012)
18. Karthikeyan, N., Kaliappan, E., Dinakaran, K.P., Dharmaprakash, R.: Independent swarm grid vehicle charging by a logical DC-DC converter with hybrid boundary conduction mode. **32**(3), 1318–1326 (2023)
19. Sosimi, AA., Gbenebor, O.P., Oyerinde, O., Bakare, O.O., Adeosun, S.O., Olaleye, S.A., Analysing Wear Behaviour of Al–CaCO3 Composites Using ANN and Sugeno-Type Fuzzy Inference Systems, (2020)
20. Dadios, E.P.: Fuzzy logic - controls, concepts, theories and applications 287(1). INTECH. (2012)
21. Hamrouni, C.: A new fuzzy controlled antenna network proposal for small satellite applications. Comput. Mater. Contin. **72**(3), 4233–4248 (2022)
22. Ross, T.J.: Fuzzy Logic with Engineering Third Edition, University of new Mexico. WILEY, USA (2010)

23. Díaz-Rodríguez, I.D., Han, S., Bhattacharyya, S.P.: Analytical Design of PID Controllers. Springer International Publishing, Cham (2019)
24. Borase, R.P., Maghade, D.K., Sondkar, S.Y., Pawar, S.N.: A review of PID control, tuning methods and applications. Int. J. Dyn. Control. **9**(2), 818–827 (2021)
25. Gaya, M.S., Bature, A., Yusuf, L.A., Madugu, I.S., Abubakar, U., Abubakar, S.A.: Comparison of control strategies applied to nonlinear quarterly car passive suspension system. Int. Rev. Autom. Control. **8**(3), 203 (2015)
26. Rasheed, L.T., Yousif, N.Q., Al-Wais, S.: Performance of the Optimal Nonlinear PID Controller for Position Control of Antenna Azimuth Position System. Math. Model. Eng. Probl. **10**(1), 366–375 (2023)

Data-Driven Agent-Based Evacuation Simulation in a Campus Digital Twin

Ryo Kishimoto(✉), Shady Salama, Valentinus Roby Hananto, and Victor Kryssanov

College of Information Science and Engineering, Ritsumeikan University, Osaka, Japan
ryokmoto@gmail.com

Abstract. This study introduces a data-driven agent-based evacuation framework implemented in a building digital-twin environment to analyze the performance of static and dynamic evacuation guidance strategies under varying congestion conditions. The first three floors of Building H at the Osaka Ibaraki Campus (OIC) of Ritsumeikan University are used as a real-world testbed and are reconstructed, using Blender and Unity. Human movements are simulated via NavMesh-based navigation, incorporating empirically grounded heterogeneity in reaction times and walking speeds, together with a lightweight mechanism that adaptively reduced speed, based on local density. Seven guidance strategies are evaluated under measured off-peak and peak occupancy conditions. Results show that strategy effectiveness depends strongly on the congestion level. In off-peak scenarios, the Dynamic-Individual strategy achieved the shortest evacuation time, demonstrating the benefit of individual-level rerouting when viable alternative paths exist. Under peak conditions, however, dynamic strategies deteriorated markedly, particularly Dynamic-Individual, and evacuation times increased by approximately 72%, as saturated exits and staircases constrained rerouting opportunities. In contrast, pre-balanced static strategies, particularly Static-Phased, exhibited greater robustness under heavy congestion. While Static-Naive enables rapid early evacuation, Static-Phased reduced extreme tail delays and shortened the maximum evacuation completion time by approximately 10% compared to baseline, Static-Naive. These findings clarify when the dynamic guidance is effective and when the static guidance is preferable, and they underscore the broader value of digital-twin-based simulation as a generalizable tool for evacuation analysis and building safety planning.

Keywords: Digital Twin · Agent-based evacuation simulation · Smart Building · Static and Dynamic Routing

1 Introduction

1.1 Research Background

Safe evacuation planning for large-scale facilities has become increasingly critical due to the frequency of natural and human-caused disasters. Contemporary

N. A. Mohd Subha et al. (Eds.): SMS 2026, CCIS 2956, pp. 58–72, 2026.
https://doi.org/10.1007/978-981-92-0306-2_5

research on crowd dynamics indicates that evacuation behavior is complex and depends on factors, such as congestion, social interactions, and information presentation, rather than on simple shortest-path choices [4].

Crowd management is particularly challenging in complex buildings, such as university campuses, where the occupancy and spatial distribution of people fluctuate strongly with class periods and schedules. When static evacuation plans are developed mainly from the maximum capacity and typical usage patterns, ignoring spatio-temporal fluctuations in human flow, the resulting evacuation routes may deviate from actual conditions. This can cause large discrepancies between predicted and observed locations of congested areas and bottlenecks. To address this issue, frameworks have been proposed that use Internet of Things (IoT) sensors and digital twins to obtain real-time information on occupancy and hazardous areas, and to dynamically update evacuation routes and guidance policies [10].

Against this background, evacuation simulations using Building Information Modeling (BIM) and game engines have been actively studied. BIM-based fire and evacuation analysis systems have been developed for various real-world buildings, including large-scale infrastructures [3]. A framework for updating evacuation plans, based on real-time pedestrian flow, has also been proposed by combining IoT sensors and cellular automata [8]. In addition, systematic reviews that summarize human behavior and route optimization in large-scale evacuation planning, as well as studies integrating BIM/3D models with agent-based simulations have accumulated significant knowledge in the related fields [1]. However, most of these studies assumed a single representative occupancy state or a single evacuation strategy, while relatively few focused on temporal fluctuations in the number of people or on performance differences across multiple strategies.

This study uses Building H (OIC), Ritsumeikan University, an education and research facility with a complex indoor layout that is used simultaneously by large numbers of students and faculty members. The building accommodates diverse environments, including classrooms, self-study areas, and laboratories. The objective of this study is to conduct evacuation simulations of the Building H within a digital-twin framework designed for integration with the campus IoT infrastructure. As an initial step towards real-time evacuation support, this paper investigates multiple evacuation strategies, using occupancy datasets obtained from the smart-building platform.

1.2 Problem Statement and Research Gap

In conventional evacuation planning, building-wide procedures are often designed mainly based on maximum occupant capacity and for a few typical usage patterns [1]. For example, evacuation rules may simply state that people in a given room should use the nearest staircase and the exit on the first floor. However, in a university environment, the number of people and their distribution on each floor change significantly over time according to class schedules and various events. As a result, evacuation plans based on fixed, time-invariant assumptions about occupancy can concentrate people at specific staircases or exits at certain times, creating bottlenecks and prolonging overall evacuation time.

Many past evacuation simulation studies with BIM and 3D models have focused on detailed representation and visualization of building geometry, while the evacuation strategy itself often relied on simple rules, such as take the shortest route to the nearest exit [3]. Reviews on large-scale evacuation planning highlight the importance of human behavior modeling and route optimization, but the existing work does still not sufficiently explore variations in evacuation strategies or explicitly handle individual differences among agents [1]. Real-time evacuation guidance systems with IoT sensors and dynamic signage have been proposed [8], targeting offices and public facilities but not educational buildings, where usage patterns change markedly over the day.

Although several studies have examined variability in agent parameters such as reaction time and movement speed [4], relatively limited work has systematically evaluated multiple evacuation strategies under common conditions while explicitly accounting for individual differences. In particular, few studies have compared static allocation, phased guidance, and dynamic reallocation at both room and individual levels within a unified framework, based on a 3D model of a real campus building. Addressing the gaps, this study aims to compare and evaluate seven static and dynamic evacuation strategies, using agents with heterogeneous reaction times and movement speeds in a 3D simulation environment of the Building H, under multiple dynamic occupancy scenarios.

1.3 Research Objectives and Projected Contributions

The primary objective of this study is to quantitatively examine how different evacuation strategies and individual-level behavioral heterogeneity influence evacuation performance in a 3D simulated environment representing a university campus building. More specifically, this study aims to:

1. Construct a 3D evacuation simulation system for the Building H by linking floors, classrooms, staircases, and exits on floors 1–3, using occupancy scenarios derived from the campus smart-building platform.
2. Develop an agent-based model incorporating individual differences in reaction times and walking speeds, and systematically compare various evacuation strategies under identical building and demand conditions, including static allocation, room-based exit assignment, and individual-level dynamic rerouting.

In line with these objectives, the projected contributions of this study can be summarized as follows:

1. The development of a data-driven 3D digital twin of an educational building, leveraging measured occupancy data to represent peak and off-peak demand conditions in a university campus environment.

2. Models of heterogeneous agents with individual parameters governing evacuation start time and walking speed, enabling the impact of behavioral variability on evacuation performance to be quantitatively assessed under realistic usage patterns.
3. A unified evaluation framework for comparing different evacuation strategies under common conditions.

The remainder of this paper is organized as follows. Section 2 reviews related work on crowd and evacuation modeling, BIM/3D model-based simulations, and digital-twin approaches for smart buildings. Section 3 presents the simulation framework, including the digital twin of the Building H, the agent-based model, the congestion-aware navigation mechanism, and the evacuation strategies. Section 4 outlines the implementation details, the occupancy scenarios derived from smart-building logs, and the experimental settings. Section 5 presents simulation results and compares the performance of the strategies under off-peak and peak conditions. Section 6 discusses the findings, offers practical implications, and formulates limitations of the study. Section 7 concludes the paper and gives directions for future work.

2 Related Work

2.1 Crowd and Evacuation Modeling

Crowd dynamics and evacuation behavior have been studied in various fields, including traffic engineering, safety engineering, and physics. The social-force model reproduces high-density phenomena, such as arching near exits, bottleneck congestion, and the "faster-is-slower" effect, and it also shows how self-organization can turn into dangerous crowding [5]. Empirical reviews further suggested that exit and route choices rarely follow a simple shortest path rule; instead, they depend on the perceived crowding, the movement of others, and how information is presented to evacuees [4]. From a simulation perspective, there is a need to represent heterogeneity, at minimum, in the reaction time and walking speed when comparing different evacuation strategies. A review of 134 agent-based evacuation studies summarized common model structures and verification practices, and noted that relatively few studies compared multiple evacuation strategies under truly identical conditions [13]. For studies aiming to compare guidance policies, systematic like-for-like comparisons with explicit individual differences remain particularly scarce.

BIM-based evacuation simulations have been used to evaluate evacuation performance in real-world buildings, including university libraries and large-scale infrastructures [3]. BIM models have also been integrated with agent-based approaches for university buildings and dormitories, showing that start-delay and speed distributions affect bottlenecks and completion time [7]. At the same time, game engines, such as Unity have been widely adopted for evacuation modeling and visualization, and many implementations rely on the built-in navigation component (`NavMeshAgent`) for pathfinding and basic local avoidance.

While the latter approach is practical and generally robust, it offers limited flexibility for explicitly representing high-density effects. In our implementation, we keep `NavMeshAgent` for global navigation and add a lightweight congestion-aware speed-control layer that modulates desired speed using local density, aligning with density-based crowd simulation approaches [15]. The proposed mechanism enables simple density-induced deceleration and stop–go behavior without substantially increasing computational costs.

2.2 Congestion, Guidance, and Digital Twins

Bottleneck experiments have demonstrated nonlinear relationships between density and throughput, highlighting strong effects of exit width and crowd composition on evacuation performance [6,14]. These findings underscore the fact that evacuation outcomes depend critically on how occupants are distributed across exits, and that routing strategies also account for congestion patterns rather than distance alone. At the building scale, real-time or near-real-time guidance systems that combine sensor networks with building models have been proposed to update routes in real time and mitigate overloading of specific exits [8]. Also, digital twin-based frameworks for emergency evacuation and fire safety management have been reported, integrating building status, occupant information, and route-finding algorithms [10]. In contrast to these online systems, the present study develops an offline digital-twin environment for an actual campus building, using measured occupancy logs to define representative off-peak and peak scenarios. The environment is utilized as a benchmark-style testbed to compare various static and dynamic evacuation strategies under controlled and repeatable conditions, while explicitly accounting for heterogeneous agent parameters.

3 Simulation Model

3.1 Digital Twin of Building H

In this study, a 3D evacuation simulation environment was constructed for Building H by importing Unity 3D models of each floor and staircase. The models were created in Blender based on blueprints and field surveys. The resulting digital-twin model of floors 1–3 is shown in Fig. 1: walkable corridor areas are shown in blue, staircases in green, and exits in red. Although the target building has nine floors, this study focuses on floors 1–3, which host most classes and self-study activities and are expected to be the most congested during evacuation. In contrast, upper floors mainly contain laboratory areas with relatively low usage, and occupancy sensors are not installed in many rooms, limiting time-specific occupancy logs. In total, 34 lecture rooms and self-study spaces were modeled as spawn zones, and 13 exits were considered. For path planning, Unity's NavMesh was used to generate walkable paths, with floors and staircases defined as separate navigation areas to apply a staircase-specific speed reduction model.

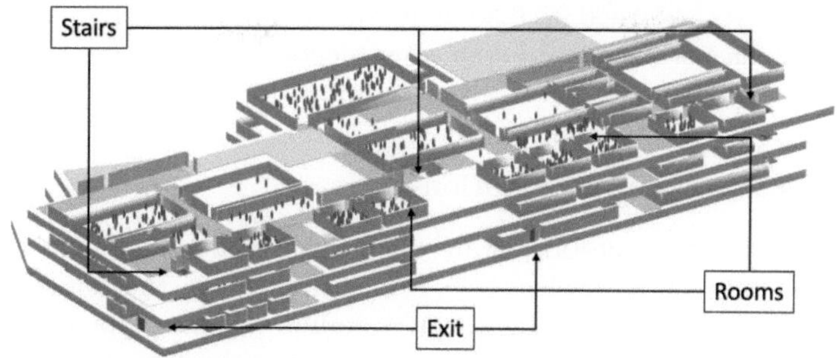

Fig. 1. A digital-twin model of the Building H (floors 1–3) in Unity.

3.2 Agent Model and Individual Parameters

Evacuation agents have two types of individual parameters: reaction time, which is the delay between the issuance of an evacuation order and the start of movement, and movement speed, which is the walking speed within passable areas. Both parameters are sampled from probability distributions, with distinct values assigned to each agent to account for individual variability.

Reaction Time: Li et al. analyzed pre-evacuation behavior in complex traffic environments and reported that approximately 20% of people can initiate evacuation immediately, while the rest begin moving after a period of observation and preparation [9]. Based on these findings, reaction time (RT) was modeled as a random variable following a log-normal distribution. In particular, samples were drawn from a log-normal distribution with log-space parameters $(\mu, \sigma) = (-0.2, 0.6)$ and clamped to the range $RT \in [0.1, 8.0]$ s for numerical stability.

Movement Speed: A meta-analysis of outdoor walking speed showed that the average normal speed of healthy adults is about 1.31 m/s, while low-mobility pedestrians average around 0.82 m/s [11]. A library evacuation study reported an average stair-descending speed of 0.97 m/s with a standard deviation of 0.36 m/s [12].

In this study, agents are therefore assigned to three different walking-speed categories, each modeled, using a normal distribution:

- Student male: $v \sim \mathcal{N}(1.31, 0.25^2)$ [m/s]
- Student female: $v \sim \mathcal{N}(1.16, 0.25^2)$ [m/s] (slightly lower, based on observed female walking speeds at pedestrian crossings [2])
- Low mobility: $v \sim \mathcal{N}(0.82, 0.20^2)$ [m/s]

Based on university enrollment statistics (May 1, 2025), the student gender ratio is set to 74% male and 26% female, and 1 to 3 low-mobility agents are added per lecture room to represent slower users, such as faculty members. On stairs, all agents use $v_{\text{stairs}} \sim \mathcal{N}(0.97, 0.36^2)$ [m/s], so that the model captures the difference between horizontal and stair movements, using the simple three-class scheme.

3.3 Congestion-Aware Navigation

We use a lightweight congestion-aware navigation model built on Unity's `NavMeshAgent`. A congestion level $c \in [0,1]$ is estimated on the horizontal (XZ) plane by counting nearby agents within a radius r and computing the local density $N/(\pi r^2)$. With the mean speed $\bar{v}$ of those neighbors, we define a raw congestion sample as

$$\text{cong} = \frac{N}{\pi r^2} / \max(\bar{v}, 0.1), \tag{1}$$

which is a heuristic congestion indicator (not a physical quantity) used only after normalization to $c = \min\{1, \text{cong}/s\}$, where s is a fixed normalization scale.

The target walking speed is then updated by

$$v_{\text{target}} = (1 - \alpha c)\, v_{\text{desired}}, \tag{2}$$

where v_{desired} is the free-flow speed and $\alpha \in [0,1]$ controls deceleration; when $c \geq c_{\text{stop}}$, the speed is reduced to near zero.

In the implementation, $r = 2.0$ m for congestion estimation (1.75 m for the avoidance probe), the normalization scale is set to $s = 10$, and the congestion grid is refreshed every 0.25 s.

3.4 Smart-Building Data and Population Scenarios

In the Building H digital twin, zone-wise occupancy logs are stored as CSV files that record timestamps and the number of people in each zone at 1-minute intervals. For the simulation, the occupancy log corresponding to each spawn zone is queried at a user-specified time, and agents are instantiated in that zone accordingly. The total number of agents is determined as the sum across all zones. In all simulation scenarios, the demographic composition (male/female student ratio, percentage of low-mobility users) and the distributions of reaction time and walking speed follow the agent model of Sect. 3.2. The occupancy patterns and total demand can be adjusted to allow comparison of off-peak and peak conditions for the same building and agent models. Figure 2 visualizes the initial occupancy distributions for the two scenarios.

3.5 Evacuation Strategies

Seven evacuation strategies were compared under a common digital-twin environment and the individual-differences models for Building H. They are grouped into static strategies, which fix exit assignments at the beginning of the run (Static-Naive, Static-Phased, Static-Room, Greedy V/C, Manual Plan), and dynamic strategies, which allow rerouting during evacuation (Dynamic-Room,

(a) Off-peak (09:00, 393 agents).

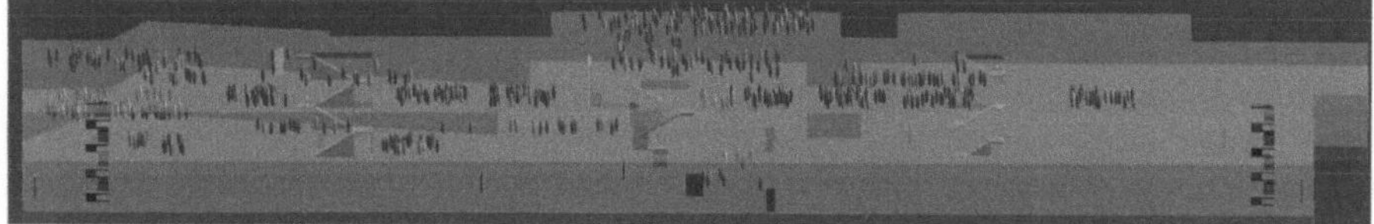

(b) Peak (14:00, 634 agents).

Fig. 2. Initial occupancy distributions (Levels 1–3) for the off-peak and peak scenarios. Walls are hidden and floors are semi-transparent; exits (red) and staircases (green) are shown. (Color figure online)

Dynamic-Individual). In this study, "guidance" refers to the assignment or update of exit choices rather than to a specific communication technology. Static strategies thus correspond to guidance communicated once at the room or building level, whereas dynamic strategies assume a channel capable of updating guidance during evacuation. Room-level and individual-level strategies differ only in the granularity of decision-making: room-level guidance could be delivered via in-room announcements, whereas individual-level updates could be delivered through personal smartphones. These distinctions are introduced to evaluate routing policies under common conditions, not to model a particular deployment.

Together, these seven strategies represent typical policies considered in campus evacuation planning: a nearest-exit baseline (Static-Naive), capacity-aware static allocations with and without phasing (Static-Phased, Greedy V/C), plans based on room-level assignment rules used in practice (Static-Room, Manual Plan), and dynamic rerouting at the room and individual levels (Dynamic-Room, Dynamic-Individual). All strategies share a common implementation for start timing and exit scoring based on free-flow travel time, estimated congestion, and a simple hysteresis term; they differ only in when exit assignments are updated and at what granularity.

Static-Naive: For each agent, the exit with the shortest free-flow travel time is selected once at the start of the simulation. This strategy serves as a simple static baseline.

Static-Phased: This strategy introduces room-wise phasing to avoid an instantaneous surge of evacuees. For each room, a base delay is computed from a target building-wide completion time, free-flow travel time to the best exit, room population, and exit capacity, and is added to the alarm time. The agent's sampled reaction time is adjusted to determine the individual start time. Exit assignments are fixed before the evacuation begins.

Static-Room: For each room, a single exit is selected based on the room's population and travel costs to candidate exits. All agents in a room share the same exit throughout the run. This approximates a scenario in which classroom-level instructions are given so that everyone in the same room uses the same exit.

Greedy V/C: This strategy allocates exits to balance expected loads across exits. An offline greedy algorithm uses the population of each room and travel costs to candidate exits to reduce imbalance in the ratio between expected volume and exit capacity (V/C). This can be interpreted as a planner using sensor-based occupancy estimates and exit capacities to design a near-optimal static evacuation plan.

Manual Plan: Agents follow the evacuation routes defined in the existing manual plan for Building H. Available exits are restricted according to the stairs and exits specified for each floor. This strategy represents the current static policy based on signage and is used as a reference.

Dynamic-Room: A room-level dynamic strategy that periodically re-evaluates the exit assigned to each room by computing an exit score for the current and alternative exits based on estimated travel time and congestion along the route and near the exit. If switching is expected to significantly reduce this score and a cool-down condition is satisfied, all agents in the room are reassigned to the new exit.

Dynamic-Individual: A fully dynamic strategy in which each agent periodically re-evaluates its exit options. At each decision step, the agent compares the score of its current exit with those of alternatives, and switches only if the expected improvement exceeds a predefined threshold. A minimum time between replans, a maximum number of replans per agent, and a compliance rate limit excessive rerouting and prevent unrealistic oscillations.

4 Experimental Setup

4.1 Software and Hardware Environment

All simulations were conducted in Unity 6.0 LTS run on Ubuntu 22.04. Floor and staircase meshes were modeled in Blender and imported into Unity as FBX files. All experiments were run on a workstation with Intel Core i9-10940X CPU, 128 GB RAM, and NVIDIA RTX 3080 GPU. This setup supported simulations of roughly 600 agents at near real-time simulation speed; no hardware-specific optimization was applied.

4.2 Simulation Scenarios and Evaluation Metrics

All experiments used the two occupancy scenarios defined in Sect. 3.4: an off-peak morning at 09:00 with 393 agents and a daytime peak at 14:00 with 634 agents. For each scenario, only the evacuation strategy model was changed; the building model and agent model were kept unchanged. Simulation parameters were set as follows: the time step $\Delta t = 0.1$ s; reaction time followed a log-normal distribution with parameters ($\mu = -0.2, \sigma = 0.6$), clamped to $[0.1, 8.0]$ s; walking speed followed the three-category normal distributions defined in Sect. 3.2. For the dynamic strategies, exits were re-evaluated every 2.0 s, with a stickiness duration of 7.0 s and a maximum of three replans per agent in the Dynamic-Individual strategy.

Each strategy was evaluated using three metrics: (i) total evacuation time (time until the last agent reaches an exit), (ii) T_{80} and T_{90} (time until 80% and 90% of agents have evacuated), and (iii) average individual evacuation duration. Because the behavioral and hazard models are simplified, absolute evacuation times should be treated as approximate; therefore, we focus on relative performance comparisons between the strategies under identical conditions.

5 Results

5.1 Overall Evacuation Performance

Table 1 presents the performance metrics for the seven evacuation strategies under the off-peak and peak occupancy scenarios.

Table 1. Evacuation performance by strategy under the off-peak and peak occupancy conditions.

Scenario	Strategy	Total agents	T_{80} [s]	T_{90} [s]	Overall [s]	Avg. dur. [s]
	Static-Naive		149.44	167.42	233.18	107.03
	Static-Phased		144.60	177.52	263.97	105.09
	Static-Room		155.85	177.07	270.18	108.49
Off-peak	Greedy V/C	393	143.21	168.69	276.27	106.28
	Manual Plan		183.08	217.33	399.45	130.94
	Dynamic-Room		176.97	204.67	305.79	119.46
	Dynamic-Individual		148.61	178.72	**220.43**	105.19
	Static-Naive		176.61	199.15	329.52	130.76
	Static-Phased		202.00	230.85	**291.59**	132.52
	Static-Room		194.88	231.72	379.97	134.93
Peak	Greedy V/C	634	180.82	200.64	298.43	133.02
	Manual Plan		220.02	254.99	533.47	159.33
	Dynamic-Room		233.93	267.75	361.66	155.97
	Dynamic-Individual		232.22	260.88	378.70	155.14

Note: Best strategies (lowest Overall time) are shown in bold.

In the off-peak scenario, Static-Naive yielded an overall evacuation time of 233.18 s. The other static approaches (Static-Phased, Static-Room, and Greedy V/C) ranged between 264 s and 276 s. Dynamic-Individual achieved the shortest overall time (220.43 s), about 6% faster than Static-Naive, with an average individual duration of 105.19 s comparable to the other static strategies. Its T_{80} and T_{90} (148.61 s and 178.72 s) were close to those of Static-Naive and Greedy V/C. Manual Plan showed the poorest performance (399.45 s).

Under the peak scenario, performance differences became more pronounced. Static-Naive required 329.52 s, whereas Static-Phased and Greedy V/C reduced this to 291.59 s and 298.43 s, respectively (roughly 8–11%). Static-Room was significantly slower (379.97 s). Contrastingly, Dynamic-Room and Dynamic-Individual produced longer evacuation times (around 360–380 s), 10–15% slower than Static-Naive. Manual Plan again resulted in the worst performance (533.47 s). In terms of T_{80}, Static-Naive (176.61 s) and Greedy V/C (180.82 s) were faster in the early-to-middle stages than the dynamic strategies (all above 230 s).

To illustrate the evacuation progress over time under peak conditions, Fig. 3 shows the evacuation progress (CDF) for Static-Naive, Static-Phased, Dynamic-Room, and Dynamic-Individual. The two static strategies remain more efficient through most of the process, while the dynamic strategies lag after the early phase (after about 30% evacuated). Although Static-Phased achieved the best overall completion time, Static-Naive evacuated a slightly larger proportion in the mid-to-late stages; however, long delays in a small number of distant rooms extended the tail.

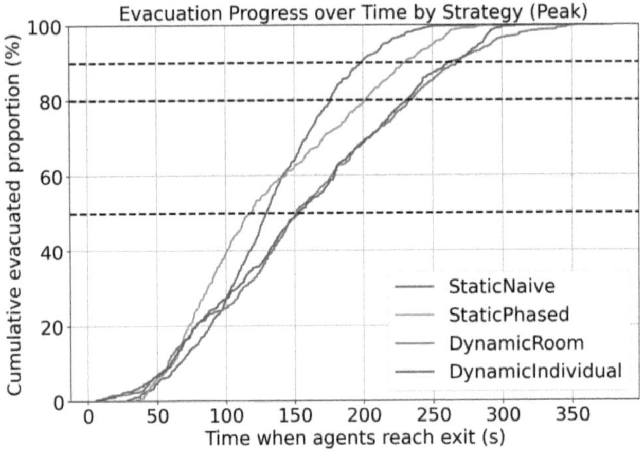

Fig. 3. Evacuation progress (CDF) by strategy under peak occupancy. Dashed lines mark 50%, 80%, and 90% completion.

5.2 Effect of the Occupancy Level

To evaluate strategy robustness with respect to crowd density, we compared outcomes between off-peak and peak occupancy levels. For Static-Naive, the overall time increased by about 41% (233.18 s → 329.52 s). Static-Phased and Greedy V/C were more resilient, with increases limited to roughly 10% and 8%. Dynamic-Individual showed a stark contrast: while competitive under off-peak, its overall time increased by about 72% (220.43 s → 378.70 s) under peak, with T_{80} increasing by more than 50%. Other strategies also increased: Static-Room by about 41%, Manual Plan by about 34%, and Dynamic-Room by approximately 20%.

6 Discussion

6.1 Interpretation of Strategy Differences

In this study, we analyzed seven evacuation strategies and compared their performance under two occupancy regimes derived from occupancy logs: off-peak and peak. As presented in Sect. 5, the best-performing strategy depends strongly on the occupancy regime. Under off-peak conditions, Dynamic-Individual achieved a shorter overall completion time than the static strategies, while its T_{80} and T_{90} remained comparable to those of the best static baseline (Static-Naive). With 393 agents, corridors and staircases were not fully saturated, allowing individual reassignment to avoid emerging queues without large detours or route oscillations. Consequently, some agents followed shorter effective routes than under static pre-assignment, resulting in a modest improvement in the overall evacuation performance. Under peak conditions, however, the dynamic strategies (Dynamic-Individual and Dynamic-Room) underperformed compared with the best static strategy, Static-Phased. The overall evacuation time of Dynamic-Individual increased by 72% relative to the off-peak case. With 634 agents, staircases and exits quickly became saturated, leaving little flexibility for route alternatives. Dynamic-Individual reassigns agents using a local exit score that reflects travel time and congestion. In this regime, however, there are few options that are both clearly less congested and sufficiently close. As a result, it can provide limited congestion relief while adding time in congested areas. Repeated local adjustments then accumulate into longer completion times than with pre-balanced static guidance. By contrast, capacity-aware static strategies (Static-Phased and Greedy V/C) maintained a high performance under peak conditions, reducing overall evacuation time by about 10% compared to Static-Naive. Static-Phased set a start delay for each room based on the shortest travel time to an exit, room occupancy, and exit processing capacity: larger or farther rooms started earlier, while smaller rooms closer to exits start later. Although it delayed some agents, it reduced long waits at staircases and exits and shortened the final completion time. Greedy V/C similarly aimed to balance demand across exits through static allocation that accounts for congestion and capacity. The evacuation progress curves under peak conditions (Fig. 3) highlight a difference in the evacuation dynamics. Static-Naive moved many agents quickly during the early-to-middle phase, whereas Static-Phased reduced extreme tail delays among the last evacuees. In Static-Naive, a small number of agents from distant, high-occupancy rooms tended to remain on the main route for long periods, extending the tail; phased start timing mitigated this left-behind effect and shortened the maximum evacuation completion time. Other strategies (Manual Plan, Static-Room, and Dynamic-Room) tended to perform worse than Static-Phased and Greedy V/C. Manual Plan prescribes specific stair and exit usages, reducing route dispersion and often concentrating flow into one hallway or staircase. Static-Room and Dynamic-Room assign exits at the room level, which provides relatively coarse control and can similarly concentrate demands at particular bottlenecks. Overall, the results suggest that individual-level rerouting can provide modest benefits under medium occupancy, but becomes inefficient once major routes are saturated. In such conditions, frequent rerouting may not yield meaningful congestion reduction but can introduce additional time costs, leading to longer completion times. In contrast, Static-Phased and Greedy V/C balance exit usage and start timing in advance, resulting in stable performance under high occupancy. While the better strategies may delay evacuation for some agents compared to Static-Naive, they reduce extreme tail delays and lead to a more predictable building-wide evacuation outcome.

6.2 Limitations

The presented framework has limitations that should be taken into account when interpreting the results. The simulation scope was limited to floors 1–3 of the Building H due to the lack of occupancy sensors on the higher floors. Including occupants from the higher floors would likely increase congestion on shared staircases; therefore, the finding that phasing-based static strategies perform well under heavy congestion should be interpreted as conservative. The results cannot be directly generalized to vertical evacuation from higher floors or to buildings with different layouts. The agent model is simplified, and does not represent complex social or physical interactions, such as panic-induced pushing, helping behavior, or counterflow movements. Dynamic hazards, including fire and smoke spread or blocked exits, are also not modeled; consequently, absolute evacuation times should be treated as references. In addition, each scenario was simulated, using a single pseudo-random seed for two representative weekday periods. Although multi-seed analysis is generally preferred, single-seed results are reported as preliminary results. Nevertheless, the observed degradation of dynamic strategies under peak conditions (approximately 72%) is large enough for the main qualitative conclusions unlikely to change. Future work will address these limitations by incorporating multi-seed simulations, and extending the spatial scope and hazard modeling.

7 Conclusions and Future Work

This paper presented a data-driven agent-based evacuation simulation and strategy evaluation framework implemented in a digital-twin environment for a smart campus building. Using a Unity-based 3D model of Building H and occupancy datasets obtained from the campus smart-building platform, seven static and dynamic evacuation strategies were systematically evaluated under a common heterogeneous agent model with congestion-aware navigation.

The results obtained showed that evacuation strategy performance depends strongly on the occupancy regime. Under off-peak conditions, Dynamic-Individual achieved a shorter total evacuation time than Static-Naive, indicating that individual-level rerouting can provide modest benefits when congestion is moderate and alternative routes are available. Under peak conditions, however, both dynamic strategies resulted in longer evacuation times than the best-performing static strategies. When major staircases and exits became saturated, locally congestion-aware rerouting rarely identified routes that were both sufficiently short and less congested, and repeated reassignments introduced additional time costs. In contrast, capacity-aware static strategies that balance exit usage and evacuation start timing in advance, such as Static-Phased and Greedy V/C, resulted in a more stable performance. Manual Plan and Static-Room, which rely on fixed room- or plan-based assignments, tended to concentrate flows at specific bottlenecks and consistently underperformed. Overall, the findings indicate that appropriate evacuation guidance policies should be selected, based on occupancy conditions in campus buildings.

As an initial step towards a real-time evacuation support system, this work demonstrated the value of a digital-twin-based framework for validating evacuation strategies using real occupancy data. In particular, future frameworks should investigate hybrid-switching policies that operate in a dynamic mode under moderate congestion, but revert to stabilized static assignments once an occupancy- or congestion-indicator threshold is exceeded. Such a fail-safe mechanism that freezes route reassignment could reduce oscillations and preserve robustness in high-density conditions, while retaining adaptability in normal operations. Future work will extend the simulation scope to include upper floors and additional campus buildings, enabling analyses of vertical evacuation and inter-building interactions under more diverse spatial configurations. Multi-seed simulations will be conducted to quantify stochastic variability and improve the statistical soundness of performance comparisons. Finally, the rule-based strategies investigated in this study can be complemented by optimization and learning-based approaches, such as genetic programming, to discover adaptive meta-strategies that jointly optimize exit assignment, phasing, and rerouting policies in response to evolving occupancy patterns, contributing to future smart-campus safety and resilience systems.

References

1. Aldahlawi, R.Y., Akbari, V., Lawson, G.: A systematic review of methodologies for human behavior modelling and routing optimization in large-scale evacuation planning. Int. J. Disaster Risk Reduction **110**, 104638 (2024). https://doi.org/10.1016/j.ijdrr.2024.104638
2. Chang, X., Huang, H.J., Chen, A.: Empirical study of pedestrian speeds at signalized crosswalks. J. Transp. Eng. **137**(12), 788–798 (2011). https://doi.org/10.1061/(ASCE)TE.1943-5436.0000270
3. Ding, Z., et al.: A building information modeling-based fire emergency evacuation simulation system for large infrastructures. Reliab. Eng. Syst. Safety **244**, 109917 (2024). https://doi.org/10.1016/j.ress.2023.109917
4. Haghani, M., Sarvi, M.: Crowd behaviour and motion: empirical methods. Transp. Res. Part B: Methodol. **107**, 253–294 (2018). https://doi.org/10.1016/j.trb.2017.11.007
5. Helbing, D., Farkas, I., Vicsek, T.: Simulating dynamical features of escape panic. Nature **407**, 487–490 (2000). https://doi.org/10.1038/35035023
6. Hoogendoorn, S.P., Daamen, W.: Pedestrian behavior at bottlenecks. Transp. Sci. **39**(2), 147–159 (2005). https://doi.org/10.1287/trsc.1040.0102
7. Huang, Y., Guo, Z., Chu, H., Sengupta, R.: Evacuation simulation implemented by abm-bim of unity in students' dormitory based on delay time. ISPRS Int. J. Geo Inf. **12**(4), 160 (2023). https://doi.org/10.3390/ijgi12040160
8. Ji, Y., Wang, W., Zheng, M., Chen, S.: Real time building evacuation modeling with an improved cellular automata method and corresponding IoT system implementation. Buildings **12**(6), 718 (2022). https://doi.org/10.3390/buildings12060718
9. Li, Z., Qiu, S., Wang, X., Zhao, L.: Modeling and simulation of crowd pre-evacuation decision-making in complex traffic environments. Int. J. Environ. Res. Public Health **19**(24), 16664 (2022). https://doi.org/10.3390/ijerph192416664
10. Lin, J.R., Chen, K.Y., Song, S.Y., Cai, Y.H., Pan, P., Deng, Y.C.: Digital twin of buildings and occupants for emergency evacuation: Framework, technologies, applications and trends. Adv. Eng. Inform. **66**, 103419 (2025). https://doi.org/10.1016/j.aei.2025.103419
11. Murtagh, E.M., Mair, J.L., Aguiar, E., Tudor-Locke, C., Murphy, M.H.: Outdoor walking speeds of apparently healthy adults: a systematic review and meta-analysis. Sports Med. **51**(1), 125–141 (2021). https://doi.org/10.1007/s40279-020-01351-3
12. Rafi, M.M., Ahmed, S., Lovreglio, R., Dias, C.: Investigating a university library building evacuation in Pakistan during a semi-announced fire drill. Fire Mater. **47**(5), 693–704 (2023). https://doi.org/10.1002/fam.3111
13. Senanayake, G.P.D.P., Kieu, M.L., Zou, Y., Dirks, K.N.: Agent-based simulation for pedestrian evacuation: a systematic literature review. Int. J. Disaster Risk Reduction **111**, 104705 (2024). https://doi.org/10.1016/j.ijdrr.2024.104705
14. Seyfried, A., Steffen, B., Klingsch, W., Boltes, M.: New insights into pedestrian flow through bottlenecks. Transp. Sci. **43**(3), 395–406 (2009). https://doi.org/10.1287/trsc.1090.0263
15. van Toll, W.G., Cook, A.F.I., Geraerts, R.: Real-time density-based crowd simulation. Comput. Animation Virtual Worlds **23**(1), 59–69 (2012). https://doi.org/10.1002/cav.1424

A Unified ML Framework for Cross-Pillar Energy Optimization in 5G Self-organizing Networks

Sharmin Sharmin[1], Ismail Ahmedy[1,2(✉)], Bryan Raj Peter Jabaraj[1], Rafidah Md Noor[1,2], and Muhammad Umair Munir[1]

[1] Faculty of Computer Science and Information Technology, Universiti Malaya, 50603 Kuala Lumpur, Malaysia
ismailahmedy@um.edu.my

[2] Centre for Mobile Cloud Computing Research (C4MCCR), Faculty of Computer Science and Information Technology, Universiti Malaya, 50603 Kuala Lumpur, Malaysia

Abstract. The exponential growth of mobile data traffic and the proliferation of connected devices have positioned energy efficiency as a critical challenge in fifth-generation (5G) networks. While significant research efforts have addressed individual energy-saving pillars—Advanced Sleep Modes (ASM), Dynamic Spectrum Management (DSM), Network Densification (ND), and Energy Harvesting (EH)—these approaches operate in isolation, leading to suboptimal energy utilization and potential conflicts between optimization objectives. This paper introduces a comprehensive framework that combines four energy-efficiency pillars into a single integrated machine learning-based Self-Organizing Network (SON) model. The proposed framework employs Multi-Agent Reinforcement Learning (MARL) in a hierarchical manner to cooperate at different scales, which allows flexible decisions with respect to whether to enter sleep mode, how to share spectrum resources, control cell density and use infrastructure. The framework fills an important gap in the related literature, which is dominated by positioning optimization within the context of a single pillar and fails to consider multi-pillar cooperation. This paper provides an awareness of six crucial pillar interdependencies, a four-tier coordinating architecture and the formal complexity analysis of the mechanisms proposed. In which future work will verify the framework through MATLAB simulation with actual network conditions.

Keywords:: 5G Networks · Self-Organizing Networks · Machine Learning · Energy Efficiency · Cross-Pillar Optimization

1 Introduction

The development of fifth-generation (5G) wireless systems has marked the inception of a new era in connectivity with far-reaching potential applications across enhanced mobile broadband, ultra-reliable low-latency communications and massive machine type communication. Yet, this technological achievement is not without cost to both the

N. A. Mohd Subha et al. (Eds.): SMS 2026, CCIS 2956, pp. 73–85, 2026.
https://doi.org/10.1007/978-981-92-0306-2_6

environment and its usage. The Information and Communication Technology (ICT) sector is currently responsible for an estimated 2–3% of its carbon dioxide emissions worldwide, with mobile networks making a significant contribution to this footprint [1, 2]. Energy consumption increases drastically with the densification of 5G networks to cope with the increasing capacity demand, and an intelligent energy-saving scheme is required. Although this study considers wireless 5G systems in this paper, energy efficiency is also a critical factor when examining other communication technologies such as power line communication for different noise scenarios [3].

Self-Organizing Networks (SON) have been proposed as a promising approach for automation of network management processes and now energy efficiency optimization. Machine Learning (ML) has been extensively leveraged to empower the SON functionality. It allows networks to learn from experience and discover patterns in the data, as well as react to varying conditions. Current studies [4–8] highlighted four main enablers to provide energy efficiency in cellular networks: Advanced Sleep Modes (ASM), which lowers power consumption during low-traffic times; Dynamic Spectrum Management (DMS), aiming the spectrum resource exploitation towards decreasing transmission power, Network Densification (ND) where coverage and capacity are balanced by smart-cell activation approaches and Energy Harvesting (EH) support that couples grid power with green sources.

While considerable progress has been made towards each of these pillars in isolation from the others, a crucial limitation remains: no framework exists that allows one to co-optimize across all four pillars at once. The latest comprehensive studies show that most existing work pays great attention to single-pillar optimization, and less than 5% of works are investigating cross-pillar coordination [4, 9–13]. This uncoordinated approach suffers from several drawbacks. First, there may be a tension between the optimizations in one pillar of energy efficiency and those in another; for example, if sleep mode is used aggressively, it may require higher transmission power within non-sleeping cells than would otherwise be required just to compensate for connectivity and prevent any gains. Second, there is a lost synergistic optimization approach as sleep modes are not optimally combined with energy harvesting to opportunistically consume renewable energy during non-peak hours. Thirdly, the lack of global optimization for networks and makes them unable to minimize energy saving potential from the network level.

This gap is covered in this paper by presenting a universal ML-based framework that unites ASM, DSM, ND and EH under one coherent framework. Our conceptual framework conceptually brings in a hierarchical coordination feature that facilitates the cross-pillar optimization where each pillar operates as independent of one another with their own behavior and timeframe. The contributions of this paper are: (1) studying the coordination issues and interdependence relationships among energy-efficiency pillars, (2) proposing an integrated architectural model with clearly defined interfaces to facilitate cross-pillar communication, (3) defining coordination mechanisms that support holistic energy optimization, and (4) discussing potential benefits along with implementation aspects.

The rest of the paper is organized as follows. Section 2 presents related work on the individual energy-efficiency pillars and current coordination methods. The proposed unified framework architecture is described in Sect. 3. Section 4 covers the cross-pillar coordination mechanisms. Section 5 presents the prospective advantages and challenges. Section 6, discuss prospective validation methods. Section 7 concludes the paper.

2 Related Work

In this section, we survey the existing research in each energy-efficiency pillar and discuss the previous work on multi-pillar cooperation.

2.1 Advanced Sleep Modes (ASM)

Advanced Sleep Mode is a basic mechanism to reduce the energy consumed in idle base stations under low traffic conditions. In contrast to conventional sleep strategies which only turn off the entire base station, the ASM can achieve in-depth component-level control by selectively turning off Power Amplifiers (PAs), Radio Frequency (RF) chains and baseband processing units. Current ASM designs employ traffic prediction algorithms [9, 14–16], such as recurrent neural networks and Long Short-Term Memory (LSTM) networks, to predict demands and drive proactive sleep state configuration. It has been shown that there may be considerable energy savings, when applying ASM" depending on the level of trade-off between energy efficiency and delay constraints [17, 18]. However, wake-up transition times due to sleep mode activation may cause performance degradation in term of QoS (Quality of Service) especially under inaccurate traffic prediction. Furthermore, previous ASM studies have been relatively isolated and ineffective in communication with other acts of optimization: resource space allocation or renewable power production supply [19].

2.2 Dynamic Spectrum Management (DSM)

Dynamic Spectrum Management (DSM) practices were also discussed which focus on optimizing spectrum allocation, scheduling and utility of the spectrum to reduce the consumption of energy without sacrificing quality of service. AMC schemes consider carrier aggregation optimization, bandwidth part configuration in the 5G New Radio or cognitive radio techniques for opportunist spectrum access. Scalable reinforcement learning methods have been proposed in [20, 21] to learn optimal spectrum allocation policies and make trade-offs between fluctuations of channel quality and traffic demand. Enhanced DSM decreases the transmission power by exploiting spectral resources with good propagation properties and collecting transmissions in time. Nevertheless, the current solutions on DSM usually focus only on the spectral efficiency aspect while disregarding network element sleep status and harvested energy availability [22]. Recent progresses in multiple access techniques such as NOMA and hybrid schemes have also extended the scope of spectrum utilization in 5G systems [23]. These emerging access strategies complement DSM in that they allow for greater flexibility of spectrum sharing between users.

2.3 Network Densification (ND)

Network Densification (ND) is the deployment of Heterogeneous Network (HetNet), which contains macro cells, small cells and femtocell as multiple cell tiers. Architecturally, ND optimization is designed to dynamically turn on and off a subset of cells to meet bursty traffic requirements for each time slot with minimum energy consumption. A number of optimization techniques such as convex optimization, game theory and reinforcement learning have been used to address the cell activation/deactivation problem [24, 25]. Nevertheless, though small cells can be more energy efficient due to fewer transmission distances (i.e., higher spectral efficiency gain), the total energy of densified small cells deployments may consume even larger than that of less macro cells under some traffic demands [26]. The available ND optimization techniques besides focusing only on coverage capacity trade-offs rarely consider the temporal alignment with sleep mode schemes or energy harvesting features [27].

2.4 Energy Harvesting (EH)

Energy Harvesting (EH) can help base stations to complement the grid power or even work as an alternative using renewable resources like solar panels, wind turbines and ambient RF signals. The unavailability of renewable energy makes optimization more difficult. So, any algorithm will have to deal with that changing energy condition over time. Prior work has studied battery management, inter-site energy sharing, and energy-aware resource allocation using machine learning to predict renewable generation and determine an optimal battery state-of-charge [5, 28, 29]. EH-based small cells offer great potential in terms of grid energy savings via wind systems during high renewable periods. Nevertheless, the energy harvesting literature provides little consideration about coherency with sleep mode procedures - long sleeping epochs may correspond with low-energy periods - and coordination between spectrum management and transmission schemes adjusted according to the harvested energy levels [30–32].

The discussion above shows a primary deficiency: both the optimization of the pillars can be very sophisticated, but still disconnected (use isolated methods). ASM controllers unaware about the presence of renewable energy [5, 19], DSM algorithms are oblivious to network device's sleep states [22], ND strategies do not cooperate with sleep scheduling [27] and EH systems make resource management decisions without taking into consideration spectral or sleep decisions [29]. Downsides of this kind of fragmentation are inconsistent decisions, as ASM turns off cells even though EH has excess renewable energy [17, 28], loss of opportunities to synergize among sleep scheduling, spectrum allocation and renewable availability [24, 25] and under-performance in its maximum due to the gap between the sum of individual optima and the actual network optimal [4, 26]. These restrictions are the main considerations for a unified framework which we discuss next.

3 Proposed Unified Framework

The proposed model provides a hierarchical structure that respects the operational nature of each pillar, while allowing a unified cross pillar decision making. Unlike the prior methods which calculate ASM, DSM, ND and EH separately in an isolated manner, such a framework could achieve global energy optimization by information exchange and coordination mechanisms as well as resolving conflicting situations.

3.1 Mechanisms Overview and Workflow

Figure 1 presents the logical architecture of the proposed integrated ML-based cross-pillar energy optimization framework, which includes four operational layers. The Pillar Specific Optimization Layer includes the specially designed ML-driven controllers for ASM, DSM, ND and EH whose learning is preserved in each case as domain expertise for local autonomous optimization. The Cross-Pillar Coordination Layer hosts the central engine that serves as an interface between the controllers, provides global network visibility, and coordinates joint decision-making. The Network State Repository is a single location to store information about traffic patterns, energy consumption, renewable generation and configuration. The Policy Management Layer allows operators to specify higher level objectives like maximum acceptable delay, requirements for penetration of renewables or constraints on carbon emissions.

First, operatives of the four pillars set these objectives as part of the optimization process and then during Autonomous Mode, the four pillar controllers constantly monitor local signal states. The Coordination Engine collects and aggregates this information periodically in order to maintain a full network view. When cross-pillar tension is found (such as ASM turning off cells when EH has additional renewable energy), the engine triggers joint optimization based on the multi-agent reinforcement learning or hierarchical decomposition. The Directive Generation module generates the actionable guidelines, with references to constraints or recommendations, or even their exact form.

When receiving a command, controllers change to Coordinated Mode and perform their tasks: the ASM does sleep scheduling, DSM is in charge of spectrum allocation, ND handles cell activation and EH controls energy routing. The Network State Repository is updated with results, and Feedback Processing examines those results to make ML models be better and better all the time. This cycle is repeated at different time frames strategic (hours to days), tactical-level minutes to hours) and operational level (seconds to minutest) obeying the six pillars' interdependencies.

The framework is applicable to deployment with centralized adaptation (all of the components run in one management system) as well as hybrid deployment, where pillar controllers reside in a distributed network element while coordination functions are still central. For cross-pillar coordination, MARL is used in the proposed framework where a character controller from each pillar acts as an individual learning agent. In particular, each agent employs Deep Q-Networks (DQN) to decide how to act from what it observes. The hierarchical scheduling is two levelled:

Local Level: Every pillar agent (ASM, DSM, ND, EH) has its own DQN to learn optimal actions according to the observed local state. The agents receive rewards related to their purpose, for example, conserving energy for ASM, better utilization of the signal source for DSM and optimized coverage for ND or renewable resources usage in EH.

Global Level: The Coordination Engine has a central critic that will be evaluating state-action pairs across those different pillars. This critic finds the potential conflicts between

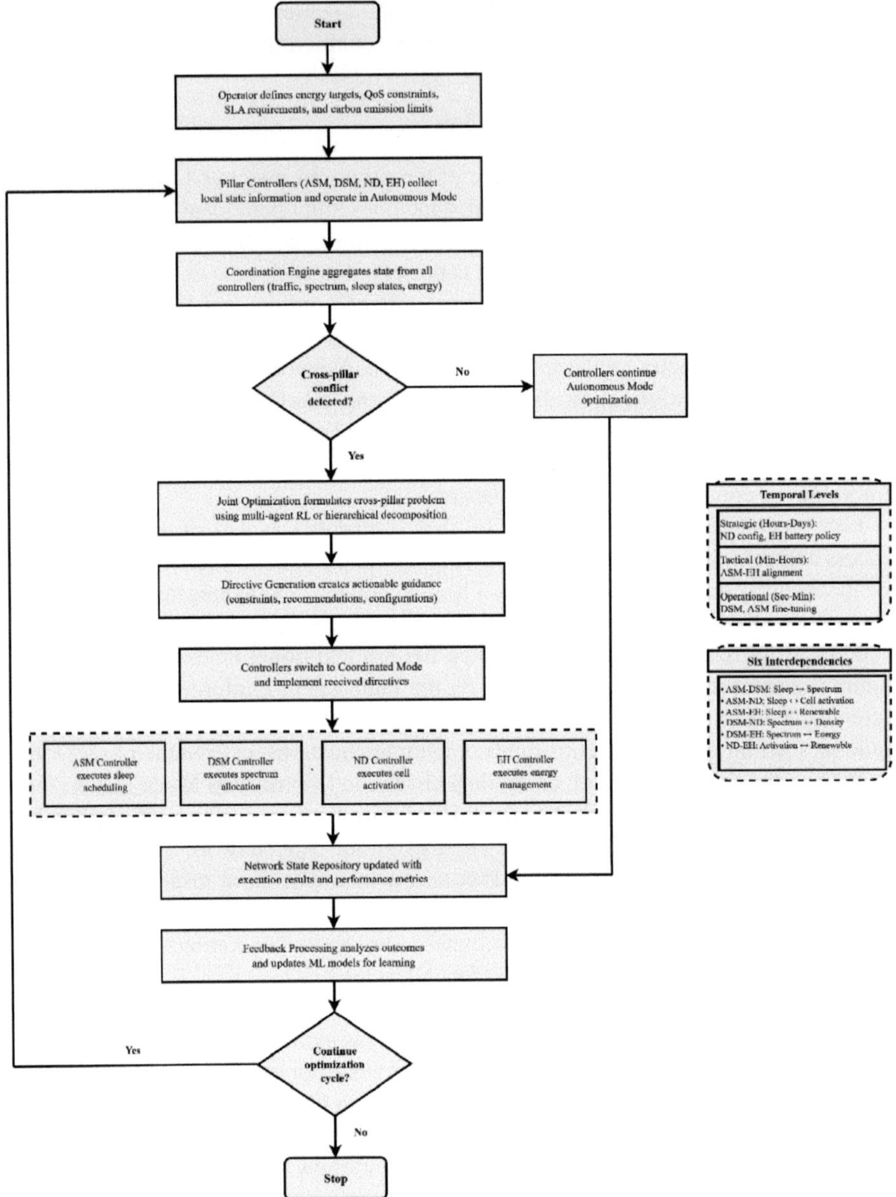

Fig. 1. Proposed Workflow Diagram

choices of pillars and computes joint rewards that promote cooperative behaviour such that a local optimal will also be global energy efficiency optimal.

3.2 Problem Formulation

The cross-pillar coordination problem can be modelled as a multi-agent MDP (S, A, P, R, γ), where:

State Space (S): The general state is the combination of local pillar states:

$$S = \{s_{ASM}, s_{DSM}, s_{ND}, s_{EH}\} \quad (1)$$

where:

- s_{ASM} = {sleeping Models, wake-up latencies, traffic Predictions}
- s_{DSM} = {spectrum allocation, channel quality indicators, interference levels}
- s_{ND} = {cell activation, load distribution, coverage maps}
- s_{EH} = {battery charge, renewable generation forecast, grid energy price}

Action Space (A): The joint action consists of the pillar-wise decisions:

$$A = \{a_{ASM}, a_{DSM}, a_{ND}, a_{EH}\} \quad (2)$$

where:

- $a_{ASM} \in$ {deep sleep, light sleep, active} for each cell
- $a_{DSM} \in$ {block assignments of spectrum blocks, power levels}
- $a_{ND} \in$ {activate, deactivate} for every small cell
- $a_{EH} \in$ {grid, battery, renewable} energy source choice

Reward Function (R): The global reward is a combination of several factors:

$$R = w_1 \cdot E_{saved} + w_2 \cdot QoS_{maintained} + w_3 \cdot RE_{utilized} - w_4 \cdot C_{penalty} \quad (3)$$

where:

- E_{saved} stands for energy savings as compared to the baseline operation.
- $QoS_{maintained}$ measures the compliance to service quality (throughput, latency, coverage).
- $RE_{utilized}$ represents renewable energy utilization ratio.
- $C_{penalty}$ penalizes cross-pillar conflicts identified by the coordination engine.
- w_1, w_2, w_3, w_4 are weight values defined in the Policy Management Layer

The goal of the framework is to maximize long-term discounted reward:

$$\max \ E\left[\Sigma_{t=0}^{\infty} \gamma^t R_t\right] \quad (4)$$

where $\gamma \in [0,1]$ is a discount factor between immediate and future reward.

The cross-pillar coordination performed by the coordination engine is shown in Algorithm 1. The proposed algorithm continues to cycle, acquiring the state of pillars and checking conflicts among the six interdependencies till it finds conflicting, at which time joint MARL optimization process starts. In the absence of the detected conflicts, pillars perform independently to reduce coordination overhead.

Algorithm 1: Cross-Pillar Coordination Process

```
Input:  S = {sASM, sDSM, sND, sEH} - Local pillar states
Output: D - Coordinated directives for all pillars
```

```
WHILE network operational DO
  // State Aggregation
  Collect local states S ← {sASM, sDSM, sND, sEH}
  Sglobal ← Aggregate(S)
  // Conflict Detection (6 interdependencies)
  conflict_set ← ∅
  FOR each pair (i, j) ∈ {ASM-DSM, ASM-ND, ASM-EH,
                          DSM-ND, DSM-EH, ND-EH} DO
    IF DetectConflict (si, sj) THEN
      conflict_set ← conflict_set ∪ {(i, j)}
    END IF
  END FOR
  // Joint Optimization or Autonomous Operation
  IF conflict_set ≠ ∅ THEN
    joint_action ← MARL_Optimize (Sglobal, conflict_set)
    D ← GenerateDirectives(joint_action)
    SetMode (COORDINATED) for affected pillars
  ELSE
    SetMode(AUTONOMOUS) for all pillars
  END IF
  // Execution and Feedback
  Execute pillar actions based on D or local decisions
  Update Network State Repository
  Compute reward R using Equation (3)
  Update MARL models via experience replay
END WHILE
```

3.3 Coordination Mechanisms

There are two modes for both of the controllers: Autonomous Mode, which allows them both to make decisions independently based on local information when there is no necessity of coordination and Coordinated Mode, which enables them to cooperate by sharing state information and commands. The modes are switched dynamically according to network environments.

The coordination engine addresses six pillar interdependencies: ASM-DSM (sleep influencing interference and spectrum selection), ASM-ND (overlapping cell activation at different time scales), ASM-EH (sleep aligning with renewable availability); DSM-ND (spectrum portioning due to cell density), DSM-EH (transmit aggressiveness conditioned by energy abundance) and ND-EH (cell activation in renewable rich regions).

In order to handle these complex dependencies, the engine defines four basic operations. State Aggregation creates a holistic network view among all the controllers. A conflict detection is employed to detect when separate decision making would lead to a less-than-optimal resolution. Joint optimization constructs cross-pillar problems with multi-agent reinforcement learning or hierarchical decomposition. Directives generation encoding results to constraints, recommendations or configurations for pillar controllers.

4 Expected Benefits and Challenges

The integrated framework has several strengths compared with single-pillar optimization. By removing between-pillar conflicts and utilizing synergies, the approach is expected to provide more energy savings than combined individualistic optimizations. For example, synchronizing sleep with renewal availability avoids cases where ASM turns off (deactivates) cells just as EH has excess green energy to work on. By matching cell activation and traffic routing with renewable predictions; this increases the fraction of traffic that is served by green energy, with positive environmental sustainability impacts; as well as OpEx benefits in time-of-use electricity pricing domains. The global visibility offered by the coordination engine allows for a joint energy optimization respecting QoS constraints, whereas independent controllers may make conflicting decisions leading to service degradation. Internally, it is complex to coordinate, but the approach simplifies the interface for energy management providing a uniform framework instead of needing integration by hand across multiple subsystems.

There are a number of practical issues which must be addressed. Cross-pillar optimization adds dimensionality to the problem substantially, and makes it difficult for real time run and may call for decomposition approaches and approximate methods. Coordination requires communication between the distributed devices and a central engine, which drives careful protocol design to trade timeliness against available bandwidth. The system should be scalable to networks with thousands of cells, which will probably have hierarchical decomposition and regional coordination in such a case. Lastly, operational deployment demands inter-working with legacy network management systems and adherence to 3GPP standardized SON interfaces - aspects supported in the framework via clearly defined APIs and support for incremental adoption.

5 Validation Approach

The proposed framework will be implemented and validated in simulation with MATALAB based network simulators. The simulation framework will capture a suboptimal operation of 5G cellular network with realistic traffic, energy profiles drawn from standard base station datasheets, renewable generation models based on past observed solar and wind profiles and representative urban deployment models.

Benchmarking will include the total energy consumption and the percentage of served traffic renewable energy, QoS metrics (throughput, latency and coverage probability) and compared with the introduced coordination overhead due to the proposed mechanisms. They will compare their performance with the baseline, including independent pillar optimization and static network configurations. As illustrated in Fig. 2, the validation framework is a MATLAB environment that models real-world 5G network scenarios (e.g., traffic demand, spectrum usage, energy availability) in combination with an ML-based coordination engine. The engine, based on multi-agent reinforcement learning, computes coordinated control actions for ASM, DSM, ND and EH in a heterogeneous network composed of macro and small cells that make use of renewable energy. We log key performance indicators to support closed-loop learning and contribute to fair comparison with baseline approaches under various traffic and energy conditions. This framework will be further extended to evaluate the sensitivity on more advanced large-scale networks with different densities and renewable availabilities.

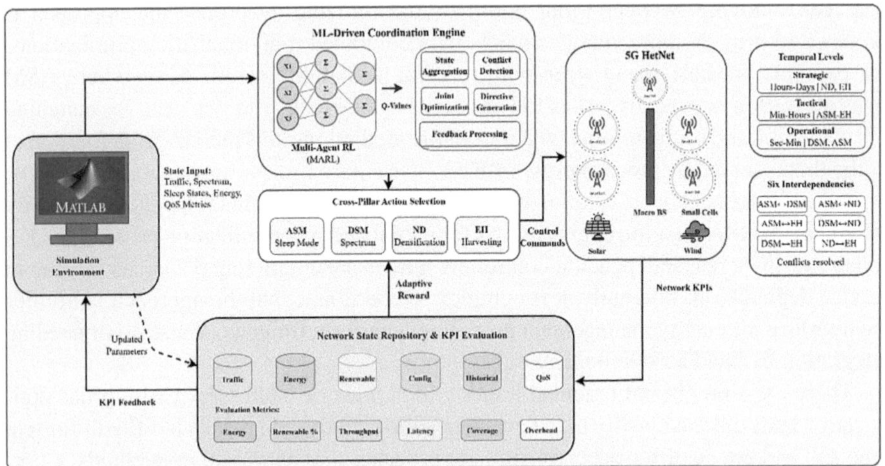

Fig. 2. Simulation using MATLAB with Realistic Deployment Scenarios

5.1 Complexity Analysis

The complexity of the coordination protocol is discussed in terms of four principal operations:

State Aggregation: O(P) is for P = 4 pillars. It has constant complexity and for some overall pillar count and width, linear with respect to the number of pillars.

Conflicts Detection: $O(P^2/2) = O(6)$ to compute pairwise interdependencies. Four sorts of couples, or six unique pairs (ASM-DSM, ASM-ND, ASM-EH, DSM-ND, DSM-EH and ND-EH), are tested at each coordination cycle. Every conflict test compares a pair of relevant state variables in time which is constant per pair.

Joint Optimization: O (|A|^P × T) in the worst case, where |A| is size of discretization A per pillar and T is the iterations of MARL training. For practical deployment, this is alleviated by: (1) discretization of action space to tame |A|, (2) experience replay to cut down necessary iterations T, and (3) hierarchical decomposition that confines joint optimization with conflicting pillar subsets.

Command Generation: Uniform in the number of pillars, O(P), for mapping cooperative actions to topical commands sent to individual controllers.

Total per-Cycle Complexity: The final complexity per coordination cycle overall is $O(P^2 + |A|\, k \times T)$ where $k \leq P$ is the number of pillars experiencing detected conflicts. In the absence of conflicts ($k = 0$) the scheme works in $O(P^2)$ with independent pillar operation. The higher order term remains manageable for the four-pillar arrangement, while increasing the number of pillars would necessitate hierarchical regional decomposition.

Table 1 compares the proposed unified framework with other existing optimization-based methods from theory point-of-view. The proposed approach has shown its efficacy in three aspects which are illustrated in Table 1: (i) coordinating across-pillars instead of separately acting; and (ii) resolving conflicts through six interrelated dependencies; and (iii) multi-agent learning that enables faster proactive adaptation at scale - composing these features.

Table 1. Theoretical Comparison of Energy Optimization Approaches

Criteria	Static	Single-Pillar	Proposed
Approach	Fixed thresholds	ML-based isolated	MARL coordinated
Coordination	None	None	Cross-pillar
Conflict Resolution	None	None	Intelligent
Adaptation	Reactive	Local only	Global proactive
Learning	None	Single-agent	Multi-agent
Synergy Capture	No	No	Yes
Expected EE Gain	Baseline	Moderate	Significant

6 Conclusion

In this paper, we have developed a unified conceptual framework to integrate the ML guided energy optimization for 5G SONs which fills an important gap that is left unmet in the literature where ASM, DSM, ND and EH are optimized independently. The overall hierarchical sensor architecture is hierarchically combining pillar specific agents with a cross-pillar coordination engine, which takes advantage of the dependencies between pillars and resolves potential conflict in several time scales.

The framework provides a structured discovery of six essential interdependencies of pillars abandoned in existing literature, a four-layer architecture based on clear coordination methods and multi-timescale optimization for aligning strategic, tactical and operational decisions. The future work is to implement the proposed scheme in a real and larger-scale system, design more ML algorithms for cross-pillar coordination formulation, and confirm the expected gain of energy in some complex environments via elaborate performance analysis.

With the development of mobile networks towards 6G and stricter sustainability requirements, integrated energy optimization approaches become indispensable aspects toward environmentally sustainable infrastructures.

References

1. Malmodin, J., Lövehagen, N., Bergmark, P., Lundén, D.: ICT sector electricity consumption and greenhouse gas emissions – 2020 outcome. Telecommun. Policy. **48**, 102701 (2024)
2. Freitag, C., Berners-Lee, M., Widdicks, K., et al.: The real climate and transformative impact of ICT: a critique of estimates, trends, and regulations. Patterns. **2** (2021)
3. Brandl, M., Kellner, K.: Performance evaluation of power-line communication systems for LIN-bus based data transmission. Electronics. **10**, 85 (2021). https://doi.org/10.3390/electronics10010085
4. Mughees, A., Tahir, M., Sheikh, M.A., Ahad, A.: Towards energy efficient 5G networks using machine learning: taxonomy, research challenges, and future research directions. IEEE Access. **8**, 187498–187522 (2020)
5. Lu, X., Wang, P., Niyato, D., et al.: Wireless networks with RF energy harvesting: a contemporary survey. IEEE Commun. Surv. Tutor. **17**, 757–789 (2015)
6. Salahdine, F., Opadere, J., Liu, Q., et al.: A survey on sleep mode techniques for ultra-dense networks in 5G and beyond. Comput. Netw. **201**, 108567 (2021)
7. López-Pérez, D., De Domenico, A., Piovesan, N., et al.: A survey on 5G radio access network energy efficiency: massive MIMO, lean carrier design, sleep modes, and machine learning. IEEE Commun. Surv. Tutor. **24**, 653–697 (2022)
8. Wu, J., Zhang, Y., Zukerman, M., Yung, E.K.-N.: Energy-Efficient Base-stations sleep-mode techniques in green cellular networks: a survey. IEEE Commun. Surv. Tutor. **17**, 803–826 (2015)
9. Sabaawi, A.M.A., Almasaoodi, M.R., El Gaily, S., Imre, S.: Energy efficiency optimisation in massive multiple-input, multiple-output network for 5G applications using new quantum genetic algorithm. IET Netw. **13**, 165–177 (2024)
10. Bhat, S.: Low-power 5G protocols for sustainable communication in rural and remote networks. Eur. J. Electr. Eng. Comput. Sci. **9**, 17–22 (2025)
11. Gong, J., Thompson, J.S., Zhou, S., Niu, Z.: Base Station sleeping and resource allocation in renewable energy powered cellular networks. IEEE Trans. Commun. **62**, 3801–3813 (2014)
12. Shabbir, A., Rizvi, S., Shirazi, M.F., et al.: Maximizing energy efficiency in HetNets through centralized and distributed sleep strategies under QoS constraint. Sci. Rep. **14**, 25839 (2024)
13. Ma, D., Lan, G., Hassan, M., et al.: Sensing, computing, and Communications for Energy Harvesting IoTs: a survey. IEEE Commun. Surv. Tutor. **22**, 1222–1250 (2020)
14. Wang, Z., Hu, J., Min, G., et al.: Spatial-temporal cellular traffic prediction for 5G and beyond: a graph neural networks-based approach. IEEE Trans. Ind. Inform. **19**, 5722–5731 (2023)
15. Kim, S., Son, J., Shim, B.: Energy-efficient ultra-dense network using LSTM-based deep neural networks. IEEE Trans. Wirel. Commun. **20**, 4702–4715 (2021)
16. Jang, G., Kim, N., Ha, T., et al.: Base Station switching and sleep mode optimization with LSTM-based user prediction. IEEE Access. **8**, 222711–222723 (2020)
17. El Amine, A., Iturralde, M., Al Haj Hassan, H., Nuaymi, L.: A distributed Q-learning approach for adaptive sleep modes in 5G networks. In: WCNC 2019 : IEEE Wireless Communications and Networking Conference. Marrakech, Morocco (2019)

18. Salem, F.E., Gati, A., Altman, Z., Chahed, T.: Advanced sleep modes and their impact on flow-level performance of 5G networks. In: 2017 IEEE 86th Vehicular Technology Conference (VTC-Fall), pp. 1–7 (2017)
19. Feng, M., Mao, S., Jiang, T.: Base Station ON-OFF switching in 5G wireless networks: approaches and challenges. IEEE Wirel. Commun. **24**, 46–54 (2017)
20. Peng, J., Xiong, Z., Tan, H., et al.: Boosting photon-efficient image reconstruction with a unified deep neural network. IEEE Trans. Pattern Anal. Mach. Intell. **45**, 4180–4197 (2023)
21. Giannopoulos, A., Spantideas, S., Kapsalis, N., et al.: Deep reinforcement learning for energy-efficient Multi-Channel transmissions in 5G cognitive HetNets: centralized, decentralized and transfer learning based solutions. IEEE Access. **9**, 129358–129374 (2021)
22. Tehrani, R.H., Vahid, S., Triantafyllopoulou, D., et al.: Licensed Spectrum sharing schemes for Mobile operators: a survey and outlook. IEEE Commun. Surv. Tutor. **18**, 2591–2623 (2016)
23. Yonis, A.Z., Nawaf, A.: Investigation of evolving multiple access technologies for 5G wireless system. In: 2022 8th International Engineering Conference on Sustainable Technology and Development (IEC), pp. 118–122 (2022). https://doi.org/10.1109/IEC54822.2022.9807471
24. Amine, A.E., Chaiban, J.-P., Hassan, H.A.H., et al.: Energy optimization with multi-sleeping control in 5G heterogeneous networks using reinforcement learning. IEEE Trans. Netw. Serv. Manag. **19**, 4310–4322 (2022)
25. Zhang, H., Zeng, K.: Pairwise Markov chain: a task scheduling strategy for privacy-preserving SIFT on edge. In: IEEE INFOCOM 2019 - IEEE Conference on Computer Communications, pp. 1432–1440 (2019)
26. Ge, X., Yang, J., Gharavi, H., Sun, Y.: Energy efficiency challenges of 5G small cell networks. IEEE Commun. Mag. IEEE Commun. Soc. **55**, 184–191 (2017)
27. Alamu, O., Gbenga-Ilori, A., Adelabu, M., et al.: Energy efficiency techniques in ultra-dense wireless heterogeneous networks: an overview and outlook. Eng Sci. Technol. Int. J. **23**, 1308–1326 (2020)
28. Tian, Z., Lu, H., Yang, B., et al.: Microfiber with methyl blue-functionalized reduced graphene oxide and violet light sensing. IEEE Photon. Technol. Lett. **27**, 798–801 (2015)
29. Chandran, Y., Noor, R.M., Munir, M.U., et al.: Machine learning-driven wireless optimization: improving network performance and ISP selection. In: Meng, L. (ed.) 3rd International Conference on Cloud Computing and Computer Networks, pp. 166–179. Springer Nature Switzerland, Cham (2025)
30. Kaur, P., Garg, R., Kukreja, V.: Energy-efficiency schemes for base stations in 5G heterogeneous networks: a systematic literature review. Telecommun. Syst. **84**, 115–151 (2023)
31. Sharmin, S., Ahmedy, I., Jabaraj, B.R.P., et al.: A comprehensive review of machine learning-driven self-organizing networks for 5G coverage and capacity. Telecommun. Syst. **89**, 44 (2026). https://doi.org/10.1007/s11235-026-01424-0
32. Sharmin, S., Ahmedy, I., Ismail, H.: Enhancing 5G/6G network efficiency and security via artificial intelligent-driven real-time analytics and resource optimization. In: 2025 IEEE International Conference on Artificial Intelligence in Engineering and Technology (IICAIET), pp. 13–16 (2025)

Simulation of Care Processes to Maximise Operational Efficiency Under Industry 5.0 Paradigms

Luis Nuela and John Reyes(✉)

Universidad Técnica de Ambato, Ambato, Ecuador
johnreyes@uta.edu.ec

Abstract. This study presents a methodology that integrates discrete event simulation (DES) with design of experiments (DoE) to maximise operational efficiency, with a focus on sustainability and resilience in the context of industry 5.0. Thus, a DES model is developed in FlexSim 2025 that represents the areas of interest. In addition, to parameterise the process times, 60 observations are recorded to adjust the probability distributions with the Fitted Distribution tool. Thus, a DoE based on a 2^3 full factorial design with three factors is implemented: doctors, nurses and two-level triage. Here, 80 replicates per scenario are run in the Experimenter tool to evaluate two demands of 10 and 40 patients per day. In this context, two metrics are analysed to evaluate the operational performance: the total patient time in the system (TTS) and the waiting time to enter the medical consultation. Consequently, the results of the experiment determine that the medical consultation is the bottleneck. Therefore, it is validated that the number of doctors is the determining factor in operational performance management. Thus, increasing the number of doctors from one to two reduces TTS by 20% with low demand and by 28% with high demand, the impact of which is reflected in the waiting time to enter the medical consultation, which is reduced by 91% with low demand and by 56% with high demand. Finally, this study can serve as a guide for other healthcare centres to anticipate bottlenecks and validate changes virtually, thus ensuring improvements with a guaranteed impact.

Keywords: industry 5.0 · simulation · operational efficiency · sustainability · resilience

1 Introduction

Industry 5.0 (I5.0) emphasises an evolutionary paradigm that encompasses the emerging technologies of industry 4.0 (I4.0) [1], for example, simulation [2]. So, the I5.0 integrates collaboration between machines and humans with advanced technology towards a human-centred approach, sustainability and resilience [1]. In this context, sustainability is defined as a balance in social and economic development, which seeks to facilitate social well-being through personalised services and strengthen resilience [3]. Furthermore, resilience encompasses the ability to face adversity, recover and adapt in changing environments [4].

N. A. Mohd Subha et al. (Eds.): SMS 2026, CCIS 2956, pp. 86–100, 2026.
https://doi.org/10.1007/978-981-92-0306-2_7

Healthcare 5.0 (H5.0) in the healthcare sector is aligned with I5.0, proposing that technology amplifies human capabilities and collaboration between doctor, patient and machine to improve care and experience [5]. However, in the healthcare sector, sustainability is geared towards patient well-being, integrating social and economic responsibilities. In this context, various I5.0 technologies are being used in the healthcare sector, for example, artificial intelligence (AI), cloud computing, digital twins and the Internet of Things (IoT) [6]. In addition, in practice, health services operate under uncertainty and may face disruptive conditions due to external causes or variability inherent in the system that compromise continuity and operational performance [7]. Here, discrete-event simulation (DES) is used to represent clinical and operational flows, quantify waiting times and resource utilisation in order to evaluate improvement alternatives through virtual experimentation before implementing changes in the real system [8]. Therefore, simulation is established as a key technology for I5.0, enabling low-risk virtual experimentation to identify operational bottlenecks and validate productivity improvements that facilitate the transition to more resilient healthcare systems [9]. Likewise, DES supports decision-making at the strategic, tactical and operational levels [7].

The latest research indicates that I5.0 is poised to be the next industrial evolution. However, its implementation in healthcare remains in its early stages, as the concepts of sustainability and resilience are not yet fully aligned with I4.0 technologies, and there are still significant gaps in the development of enabling technologies [1]. So, H5.0 proposes a human-centred, sustainable and resilient healthcare system, supported by smart technologies [5]. Therefore, simulation is a valuable technology for analysing healthcare systems; however, its effective adoption is still limited [8]. Nevertheless, implementing the simulation has the potential to reduce waiting times and enhance resource management. The main barrier to wider adoption is the low level of penetration in healthcare practice [8]. In contrast, DES allows experimental scenarios to be evaluated and the impact of interventions to be anticipated without altering the real system, because modelling is often time-intensive and data-dependent, requiring the strengthening of methodological frameworks that facilitate the adaptation of the model to changing conditions [8].

In order to address the identified gap and inconsistency in the literature, this research aims to implement a DES model using FlexSim 2025 Update 2.0 as simulation software. This will help improve decision-making related to the operational efficiency of healthcare processes under I5.0 paradigms. Furthermore, an experimental methodological approach is proposed that can be adapted to other service industries, with the aim of improving current knowledge. In this context, this research poses the following research questions (RQ):

RQ1: How is simulation integrated as an enabling technology for I5.0 in healthcare processes to maximise operational efficiency?

RQ2: How does DES in healthcare settings promote sustainability in health service management, favouring a balance between operational efficiency and responsible use of available resources?

RQ3: How does healthcare process simulation facilitate decision-making in disruptive scenarios, providing operational resilience and ensuring service continuity in healthcare systems?

This study adopts the I5.0 paradigm when evaluating healthcare performance beyond efficiency. Here, it addresses human-centredness by reducing waiting times and length of stay; resilience by testing performance under demanding conditions; and sustainability by using capacity management to support responsible resource use and long-term viability. Moreover, simulation provides quantitative evidence to support decisions aligned with I5.0 [4, 5].

The rest of the article is structured as follows. Section 2 presents a literature review linking the I5.0 paradigms with simulation. Section 3 details the method used to develop the simulation in the healthcare centre. Section 4 shows the results obtained from the simulation in the healthcare centre in the patient care process. Section 5 presents a discussion that delves deeper into the results obtained. Finally, Sect. 6 presents the conclusions and limitations of the study and future research.

2 Literature Review

I4.0 takes a paradigm shift with the advancement of new technologies that merge physical and digital systems, enabling the creation of intelligent production networks to transform the way we interact with the environment [1]. So, I5.0 extends this transformation by explicitly prioritising the human-centred approach and incorporating it with the paradigms of sustainability and resilience as fundamental objectives, rather than secondary outcomes [1, 3].

In this context, sustainability is understood as a comprehensive approach that aims to ensure the long-term well-being of society and the preservation of the planet, promoting a balance between human progress and care for the environment [4]. On the other hand, resilience is the ability to withstand, absorb and recover from disruptions, preserving operations and adapting to dynamic environments to ensure long-term sustainability [2]. However, I5.0 has driven the concept of H5.0, which adopts a patient-centred approach to improve diagnosis, treatment and the healthcare experience [5]. In addition, in H5.0, sustainability is conceived as an integral objective that requires balancing and including economic, environmental, and social pillars; within this framework, I5.0 seeks to promote goals such as socio-environmental sustainability, resilience, and the centrality of human beings [3]. Furthermore, resilience in hospital systems allows them to adapt to changes and challenges in order to maintain high-quality medical care, helping to deal with critical situations and ensuring a continuous response [7].

On the other hand, H5.0 transforms healthcare by integrating I5.0 technologies such as IoT, AI, and big data to drive a more sustainable, people-centred model with the potential to reduce costs and improve patient outcomes [5]. Moreover, the literature on I5.0 identifies enabling technologies such as simulation, cloud computing, and blockchain, among others, applicable to outpatient care settings, which favour more efficient and reliable digital infrastructures for clinical information management [1].

Therefore, computational simulation allows real systems to be represented using approaches such as DES, Monte Carlo and system dynamics; in particular, DES explicitly models variability in service times and arrival patterns, which is why it is widely used in healthcare to support decision-making in complex care settings [8]. Therefore, simulation supports decision-making by providing a quantitative and predictive basis that allows scenarios to be modelled, critical processes to be evaluated, and resources to be allocated through virtual experimentation before implementing changes. In the healthcare sector, this approach drives operational efficiency and cost reduction, while improving the patient experience and quality of care [10].

Furthermore, simulation supports coordinated responses to disruptive events by evaluating system performance and guiding investment decisions, staffing, and resilience strategies to mitigate risks and facilitate post-disaster adaptation [7]. In the context of software, various commercial platforms are currently used to perform simulations; some of the best known in the healthcare sector are FlexSim, Arena, Simul8, AnyLogic, and ProModel [11]. Here, FlexSim has gained attention for its 3D environment geared towards clinics and hospitals, with favourable reports on patient flow simulation thanks to its graphics capabilities and clinical libraries. It also integrates tools such as Experimenter and Fitted Distribution to analyse time study data, run scenarios, and optimise processes by managing queues, trajectories, and operating parameters to identify delays [12].

In this sense, DES has become a solid support for decision-making in healthcare organisations [10]. Moreover, the literature highlights that DES at a strategic level is used in the design of department distribution, room allocation and capacity policies, thus evaluating the physical distribution of the healthcare centre and the allocation rules with technologies such as simulation [12]. Furthermore, at the tactical level, DES is applied to determine staffing levels, schedule shifts, and adjust admission and prioritisation policies [10]. Finally, at the operational level, it is often integrated with optimisation and data mining to propose near-optimal solutions that reduce waiting and stay times, and to derive practical design and operation rules in services such as emergency rooms and outpatient clinics [13]. Finally, Table 1 summarises the reviewed literature, classifying it by simulation type, sustainability, resilience and level of decision-making in the healthcare field. This enables research gaps to be identified.

ENV: Environmental Sustainability; ECO: Economic Sustainability; SOC: Social Sustainability; HCR: Healthcare Resilience; DR: Disruptive Resilience; PR: Pandemic Resilience; SD: Strategic decisions; TD: Tactical decisions; OD: Operational decisions; BLANK: Not applicable or not mentioned.

3 Methodology

This study follows a descriptive-evaluative approach based on DES. Consequently, DES is used to develop a model to describe patient flow and resource interactions in the current medical centre. In addition, performance is modelled in order to evaluate alternative operational configurations through scenario-based experiments, in line with the use of DES to analyse and improve healthcare systems [14]. In this context, an experimental methodology is applied using FlexSim 2025 Update 2.0 software, modelling the main areas of the healthcare centre based on direct observation and operational records to estimate arrival rates and service times. Moreover, data collection is carried out through an observational study of times, recording service and waiting times in the main areas of the care flow. Here, the records are consolidated and refined to ensure consistency and representativeness of the observed trajectories. Subsequently, the input data is analysed in

Table 1. Literature review synthesis

Author	Simulation type	Type of sustainability	Type of resilience	Industry	Type of decision
[1]	DT	ENV		I5.0	
[2]			DR	I4.0	SD, TD
[3]		ENV, SOC	DR	I5.0	SD
[4]		ECO, SOC	DR	I5.0	SD
[5]		SOC		I5.0	SD
[7]			HCR, DR		SD, TD
[8]					SD
[10]					SD
[11]		SOC			SD
[12]				I4.0	TD, OD
[13]	DT, DES				TD, OD

FlexSim using the Fitted Distribution tool, with probability distributions for each process being adjusted accordingly. Furthermore, after parameterising these distributions, the result is a DES model that incorporates stochastic variability and reflects the complexity of a system with multiple interconnected decision factors. Consequently, the evaluation is structured using a design of experiments (DoE) approach, rather than varying one factor at a time. This is because the latter approach tends to omit interaction effects and lead to erroneous conclusions. In this context, 2^k factorial designs are a robust strategy for estimating main effects and interactions, providing a controlled number of replicates in simulation experiments [15]. Therefore, this study implements a complete 2^3 factorial experimental design to quantify the effect of doctors, nurses, and triage capacity on healthcare flow performance indicators. This identifies the most favourable configuration among the tested scenarios for each demand regime.

Finally, the simulation results are analysed in R using a script developed to process the outputs exported from FlexSim. So, first, an analysis of variance (ANOVA) is applied and compliance with its assumptions of normality of residuals and homogeneity of variances is verified using standard statistical tests. Furthermore, when the assumptions are not satisfied, a generalised linear model (GLM) is generally fitted to obtain robust estimates that are appropriate for the observed distribution of performance metrics [16].

The researcher confirms that generative AI technologies, including ChatGPT (version 5.0), are used for the sole purpose of improving the writing and clarity of the text. Moreover, its application is limited to refining language, rephrasing sentences, and ensuring terminological consistency, mainly in the Methodology and Results sections. The authors thoroughly review and edit all AI-generated content to ensure its accuracy and the academic integrity of the work.

3.1 Data Collection

To determine the parameterisation of the DES model, an observational study of process times is carried out at the healthcare centre, recording the duration from start to finish of each activity and its associated waiting times. Here, service times are measured at reception, triage and consultation rooms, as well as the waiting time between each process. In this context, 60 complete records are created through consecutive sampling over eight days, from 08:00 to 18:00, using direct timing and standardised recording. The data are consolidated and refined by checking for temporal consistency and outliers, before being used to adjust input distributions in FlexSim.

3.2 Analysis of Data

The DES model is parameterised using the Fitted Distribution feature in the FlexSim 2025 Update 2.0 software, which adjusts the probability distributions to the observed data. In addition, Fig. 1 presents the Fitted Distribution environment in FlexSim.

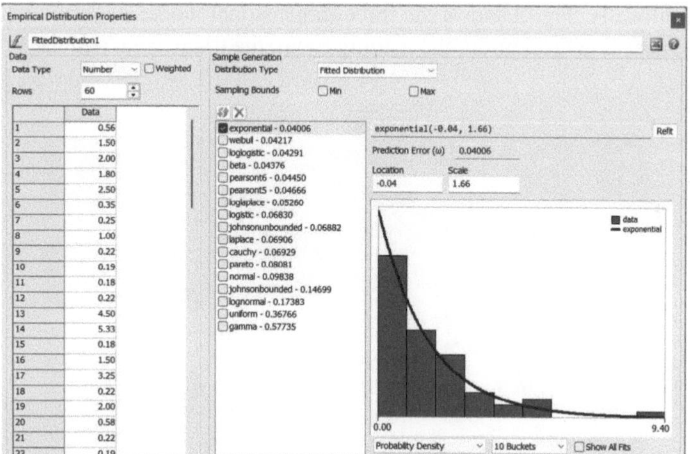

Fig. 1. Goodness of fit analysis of input data: histogram and exponential probability density function, from FlexSim.

Figure 1 shows the adjustment of patient discharge times to an exponential distribution using FlexSim Fitted Distribution, incorporated into the DES model to represent the stochastic variability of the flow. In this context, Table 2 presents the distributions selected for reception, waiting room, triage, and consultation rooms, used to calibrate service times and incorporate the variability of the care flow. These distributions were obtained using FlexSim Fitted Distribution tool, ensuring proper statistical fit to the observed operational data.

Table 2. Probability distributions adjusted by stage of the care process and estimated parameters.

Process	Distribution
Reception	Loglogistic (0.13, 0.19, 1.50, <stream>)
Waiting to enter triage	Beta (-0.29, 7.95, 1.24, 5.28, <stream>)
Triage	Cauchy (3.12, 0.31, <stream>)
Waiting to enter a medical consultation	Weibull (-0.03, 8.54, 0.82, <stream>)
Medical care	Loglogistic (5.64, 5.75, 1.46, <stream>)
Service payment	Exponential (-0.04, 1.66, <stream>)

3.3 Model Development

The simulation model is developed using FlexSim 2025 Update 2.0 for a healthcare facility and runs on an ASUS VivoBook X1605VA laptop with an Intel® Core™ i9-13900H processor and 32 GB of RAM. This allows multiple replicas and scenarios to run stably. Additionally, Fig. 2 shows the three-dimensional model, highlighting the main areas of the care flow: reception, the waiting room, the triage area, and the two medical offices.

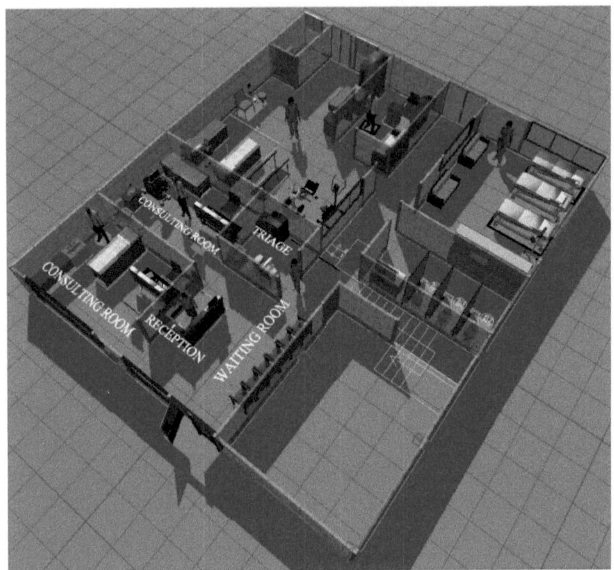

Fig. 2. Three-dimensional design of the healthcare centre process in FlexSim.

Furthermore, Fig. 2 shows the distribution and logical links between healthcare areas and the patient's journey through the medical centre. This 3D representation allows for realistic modelling of resource location, staff location, and potential congestion points in the DES model. Finally, patient flow is modelled by defining the resources of doctors, nurses and infrastructure. Service times are parameterised by applying the adjusted probability distributions obtained from the adjusted distribution analysis to each process.

3.4 Experimentation

In this section, the study implements a full factorial design 2^3 to evaluate three factors: doctors, nurses, and triage stations. Here, each factor is evaluated at two levels: low and high. This results in eight experimental scenarios, which are detailed in Table 3 and run in FlexSim 2025 using the Experimenter tool. The baseline scenario includes two doctors, one nurse, and one triage station, representing the current configuration of the medical centre. Finally, the remaining seven scenarios test alternative combinations of resources to evaluate performance and identify the most efficient and robust staffing combination [8, 11].

Table 3. Scenarios of the 2^3 full factorial experimental design implemented in FlexSim

Scenario ID	Scenario name	Doctors	Nurses	Triage
S1	Basic centre configuration	2	1	1
S2	Reduced doctors scenario	1	1	1
S3	Increased nurses scenario	2	2	1
S4	Low doctors high nurses scenario	1	2	1
S5	Expanded triage capacity	2	1	2
S6	Low staffing with extra triage	1	1	2
S7	High staffing across all resources	2	2	2
S8	High support staff with low doctors	1	2	2

To evaluate system performance, two metrics are defined: total time in the system (TTS), used as the main indicator of the healthcare centre's operational efficiency, and waiting time to access a medical consultation, used as an indicator of service sustainability, where less congestion reflects a more stable system. To ensure statistical validity, 80 replicates are performed for each experimental scenario, based on a pilot test and the application of the sample size formula to ensure a confidence level of 95%. Finally, resilience is analysed by subjecting each scenario to two demand conditions: low demand of 10 patients per day and high demand of 40 patients per day. This experimentation phase supports the descriptive and evaluative nature of the study. The DES model provides a representation of the current medical centre, and experimental runs generate comparable performance tests in different scenarios. Therefore, the analysis focuses on evaluating how alternative resource configurations affect operational performance under different demand regimes, using the selected indicators.

3.5 Model Validation

The model outputs are validated by comparing the simulated results with the actual trajectories observed in 60 patients. Therefore, under routine operating conditions, the empirical mean TTS is 31.49 min, with a 95% confidence interval (CI) between 27.78 and 35.20 min; and the empirical mean waiting time to enter the doctor's office is 9.44 min, with a 95% CI between 6.57 and 12.30 min. Here, the S1 reference scenario configuration is simulated with trace-guided arrivals, scheduling each arrival according to the timestamps observed during the working day. Thus, the resulting simulated averages are 33.59 min for TTS, with a 95% CI between 33.11 and 34.07 min, and 9.15 min for the waiting time to enter the doctor's office, with a 95% CI between 8.80 and 9.47 min. Therefore, relative errors of 6.67% and 3.07%, respectively, are generated compared to the empirical results. Consequently, the match is supported by the overlap of the confidence intervals, indicating adequate fidelity of the model for scenario-based experimentation and staff evaluation.

4 Results

Table 4 summarises the descriptive statistics of the TTS for the eight scenarios under a low demand condition. For each scenario, the sample mean, its 95% CI, and standard deviation are presented.

Table 4. Average time metrics under low demand conditions for TTS.

Scenario	Doctors	Nurses	Triage	Mean (95% CI)	Sample Std Dev
S1	2	1	1	36.44 ± 2.47	11.07
S2	1	1	1	45.36 ± 6.16	27.64
S3	2	2	1	36.44 ± 2.47	11.07
S4	1	2	1	45.36 ± 6.16	27.64
S5	2	1	2	35.68 ± 2.07	9.29
S6	1	1	2	44.61 ± 6.05	27.16
S7	2	2	2	35.68 ± 2.07	9.29
S8	1	2	2	44.61 ± 6.05	27.6

Under low demand scenarios with two doctors (S1, S3, S5, S7) maintain a TTS close to 36.4 min, while configurations with a single doctor (S2, S4, S6, S8) increase to approximately 45.4 min. Consequently, it is evident that greater medical capacity improves efficiency by reducing the total time the patient remains in the system. Finally, a statistical analysis is performed using a GLM after verifying the non-normality of the ANOVA residuals, confirming that the number of doctors is the only statistically significant factor ($p < 0.001$) affecting TTS. Therefore, increasing from one to two doctors improves TTS by 20%.

On the other hand, Table 5 summarises the descriptive statistics of the TTS for the eight scenarios under a high demand condition.

Table 5. Average time metrics under high TTS demand conditions.

Scenario	Doctors	Nurses	Triage	Mean (95% CI)	Sample Std Dev
S1	2	1	1	41.79 ± 2.01	9.02
S2	1	1	1	57.65 ± 4.07	18.27
S3	2	2	1	41.79 ± 2.01	9.02
S4	1	2	1	57.65 ± 4.07	18.27
S5	2	1	2	41.13 ± 2.25	10.08
S6	1	1	2	55.32 ± 4.32	19.41
S7	2	2	2	41.13 ± 2.25	10.08
S8	1	2	2	55.32 ± 4.32	19.41

In situations of high demand, the resilience of the system becomes evident. Thus, a single doctor configuration is not viable leading to a deterioration in operational performance with an average TTS of 56.5 min. In contrast, the addition of a second doctor stabilises the flow and reduces the average TTS to 41.5 min, a reduction of 28%. The GLM results are consistent with the low demand condition, again identifying doctors as the only significant factor.

The waiting time to access medical consultation is used as an indicator of service saturation. The GLM indicates the same dominant effect without the other factors showing a statistically significant contribution. Furthermore, within the tested ranges, changes in nursing and triage capacity do not produce statistically significant improvements. Although physician capacity is an intuitively plausible limitation, the contribution here is to quantify the magnitude of its effect across all demand regimes. Finally, as demand increases, waiting times increase in all scenarios, but the single-physician configuration shows a disproportionate increase. As shown in Table 6, at low demand, going from one to two doctors reduces the average waiting time from 3.7 to 0.3 min, equivalent to a 91% reduction. At high demand, the same change decreases waiting time from 15.0 to 6.8 min, representing a 56% improvement.

Table 6. The waiting time for a medical consultation under conditions of low and high demand.

Scenario	Doctors	Nurses	Triage	Demand	Mean (95% CI)	Sample Std Dev
S1	2	1	1	Low	0.346 ± 0.211	0.946
S1	2	1	1	High	6.71 ± 1.35	6.07
S2	1	1	1	Low	3.737 ± 2.911	13.072
S2	1	1	1	High	15.26 ± 2.50	6.07
S3	2	2	1	Low	0.346 ± 0.211	0.946
S3	2	2	1	High	6.71 ± 1.35	6.07
S4	1	2	1	Low	3.737 ± 2.911	13.072

(continued)

Table 6. (*continued*)

Scenario	Doctors	Nurses	Triage	Demand	Mean (95% CI)	Sample Std Dev
S4	1	2	1	High	15.26 ± 2.50	6.07
S5	2	1	2	Low	0.339 ± 0.211	0.947
S5	2	1	2	High	7.12 ± 1.41	11.28
S6	1	1	2	Low	3.733 ± 2.911	13.072
S6	1	1	2	High	14.64 ± 2.51	11.28
S7	2	2	2	Low	0.339 ± 0.211	0.947
S7	2	2	2	High	7.12 ± 1.41	6.33
S8	1	2	2	Low	3.733 ± 2.911	13.072
S8	1	2	2	High	14.64 ± 2.51	11.28

Figure 3 shows that, during periods of high demand, scenarios with one doctor (S2, S4, S6, S8) have longer waiting times and greater dispersion, while with two doctors (S1, S3, S5, S7) the average waiting times are considerably shorter.

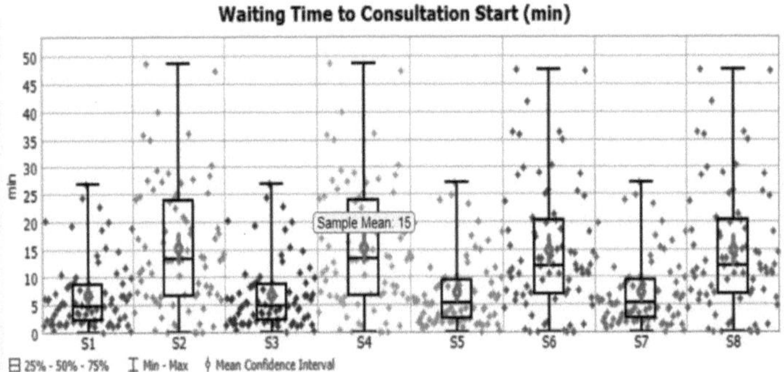

Fig. 3. Distribution of waiting time to enter a doctor's appointment, from FlexSim.

Furthermore, in Fig. 3, each scenario described in Table 6 includes 80 replicates, represented as coloured dots to show variability. Here, overlaying the boxplots makes it easier to identify concentrations and outliers, and supports the stability of the mean and its confidence intervals.

5 Discussion

The results consistently indicate that physician capacity is the determining factor in congestion as the doctor's office acts as the bottleneck of the system in all demand regimes. This pattern is consistent with previous DES evidence in outpatient and clinical services, where physician availability is explicitly modelled as the main limited resource and directly conditions queue growth and waiting time propagation [12]. The reductions in TTS reported in Sect. 4 under both demand conditions support the interpretation that additional physician capacity is the most effective lever for improving system performance.

Therefore, this study contributes to quantifying the buffering capacity needed to absorb fluctuations in demand without affecting the patient experience. Increasing the number of doctors acts as a buffer against variability stabilising patient flow in stressful situations and promoting service continuity, operational sustainability and patient experience within the framework of I5.0. Overall, these results indicate that consultation capacity is the main lever for improving efficiency in the face of demand variability, increasing resilience by absorbing flow variations without causing congestion, and promoting sustainability by reducing waiting times and the operational burden on staff, in line with the I5.0 approach to service continuity and patient experience.

On the other hand, although factors related to nursing and triage are relevant from an operational point of view, statistical analysis shows that they did not have a significant effect on TTS, either individually or in their interactions. This is consistent with the fact that consultation is the main limitation. Therefore, it is unlikely that investing resources in stages that are not bottlenecks will produce significant improvements in the overall performance of the system. This is consistent with the literature on simulation, which indicates that improvements outside the bottleneck have a marginal impact [8].

Increased demand acts as a stress test for operational robustness. In high-demand situations, the marked deterioration observed in single-doctor scenarios supports the interpretation that the capacity of a single doctor is insufficient to absorb peaks in demand, while two-doctor configurations provide a stabilising buffer and reduce dispersion. This behaviour is consistent with analyses based on simulations of service systems under conditions of variability, in which capacity buffers in the limiting resource improve the ability to absorb peaks and stabilise the dynamics of waiting times [12].

Furthermore, given the stochastic nature of the system and the possible interactions between decision factors, the 2^3 factorial design allows for a rigorous evaluation of the main effects and interactions between medical factors, nursing staff, and triage. Therefore, the integration of DES and DoE to analyse multiple configurations simultaneously represents a recommended practice in recent literature, as it provides a solid empirical basis for decision-making regarding service capacity [17].

From a decision-making perspective, the results provide clear prioritisation at all levels. Strategically, increasing medical capacity is the most defensible intervention for improving efficiency and maintaining the viability of the service in the face of growing demand. Therefore, these findings provide strong evidence for strategic decision-making: hiring a second doctor is not only an operational improvement but also a crucial investment to ensure the sustainability and viability of the service. From a tactical standpoint, changes in nursing or triage should be considered secondary until consultation capacity is addressed. From an operational standpoint, day-to-day management should focus on mitigating queues before the medical consultation and maintaining stable utilisation of doctors. This interpretation is consistent with the literature describing DES as a decision-support tool capable of linking experimental evidence to capacity and resource allocation policies in healthcare operations [11].

6 Conclusions

This study develops and applies a DES model in FlexSim for a healthcare centre, using a complete 2^3 factorial design and GLM models for statistical analysis of the results with two daily patient demands. Moreover, key metrics such as TTS and waiting time to enter a medical consultation are analysed. Therefore, simulation allows bottlenecks to be identified, the effect of different allocations to be quantified, and alternative configurations to be evaluated before implementation. In this context, the results position DES as an enabling technology for I5.0 to improve operational efficiency and support decisions oriented towards sustainability and resilience paradigms in the face of demand variations. Thus, the questions posed for this research are answered.

Furthermore, this study demonstrates that the usefulness of simulation in I5.0 lies not only in the detection of scarce resources, but also in the characterisation of structural robustness thresholds. It concludes that the number of doctors should not be calculated solely to meet average demand, but as a resilience factor that guarantees the stability of the system in the face of uncertainty. The model provides a methodology for determining the 'safety margin' necessary to protect human-centricity (waiting times) and operational sustainability, and transforms bottleneck management into a strategic design of healthcare systems capable of absorbing care stress without collapsing. Beyond identifying the bottleneck, the study quantifies the size of the effect across all demand regimes and demonstrates the capacity margin needed to preserve human-centred performance under conditions of variability.

RQ1: How is simulation integrated as an enabling technology for I5.0 in healthcare processes to maximise operational efficiency? DES is integrated as an enabling technology for I5.0 by creating a model for healthcare centres that improves efficiency with an approach that prioritises human well-being. In this context, this study addresses human-centred care by focusing on metrics that directly impact the patient experience, reducing TTS by up to 28% and proposing solutions that prevent medical staff overload. So, DES is establishing itself as a technology that not only models a process, but also improves it by balancing operational efficiency with human well-being and system stability.

RQ2: How does DES in healthcare settings promote sustainability in health service management, favouring a balance between operational efficiency and responsible use of available resources? DES promotes sustainability by transforming decision-making from an intuitive exercise to one based on quantitative evidence. In this study, DES demonstrates how to strike a balance between efficiency and resource use in two key ways. Firstly, promoting economic sustainability by revealing that only the number of doctors is a statistically significant factor ($p < 0.001$), which justifies investing solely in the resource with the greatest impact and avoiding expenditure on nursing or triage. Secondly, social sustainability is ensured by quantifying how a system with a single doctor collapse under high demand with a TTS of 56.5 min, while the addition of a second doctor reduces the TTS by 28%, thus guaranteeing the long-term stability and viability of the service. In this context, DES is consolidated as a strategic decision-making tool that enables the design of efficient healthcare systems without compromising their economic stability.

RQ3: How does healthcare process simulation facilitate decision-making in disruptive scenarios, providing operational resilience and ensuring service continuity in healthcare systems? DES facilitates decision-making for disruptive scenarios ranging from natural disasters to pandemics such as COVID-19 by quantifying operational resilience through the assessment of conditions in a virtual environment. This is aligned with other studies [4, 7]. Moreover, in this study, the high demand scenario acts as such a test, demonstrating that the single doctor configuration is fragile, with a collapse in operational performance. In contrast, the system with two doctors proves to be resilient, absorbing the stress of peaks in demand and managing to reduce TTS. However, this evidence allows us to establish a proactive contingency rule: the capacity of two doctors is essential to guarantee continuity of service during peaks in demand at the healthcare centre. Here, DES enables the design of robust, data-driven systems, rather than reacting to unfolding crises.

This study has several limitations, given that the model is based on a single medical centre, meaning that generalisation to other contexts may involve other processes that are not analysed in this research. Furthermore, the experimental design is restricted to three factors doctors, nurses, and two-level triage in addition to two patient demands per day, without exploring other relevant determinants of operational performance.

As future lines of research, it is proposed to evolve the DES model towards a digital twin approach that incorporates system status updates to enable decisions closer to real time, in line with recent frameworks that integrate DES with digital twin architecture to optimise clinical processes, highlighting efficiency and sustainability [6]. This extension would enable near real time, feedback-driven operational decisions, incorporating adaptive mechanisms such as queue length triggered capacity activation and dynamic triage rules. Therefore, it is recommended to expand the decision space from static allocation to programming, prioritisation, and sequencing policies, given that programming strategies evaluated with DES have shown potential to reduce congestion and improve performance in emergency services [17]. Moreover, to strengthen the environmental sustainability pillar with more explicit metrics, it is advisable to integrate indicators of waste and operational inefficiencies, workloads and utilisation under a combined approach of continuous improvement and simulation, consistent with recent evidence that integrates lean healthcare to support redesign and capacity decisions with quantifiable improvements [12]. From a resilience perspective, it is also pertinent to incorporate additional disruptive scenarios, such as more severe stochastic peaks, staff absences, patient variations and operational failures. It is then possible to evaluate robustness through frameworks geared towards responding to high pressure in the healthcare sector [13].

References

1. Maddikunta, P.K.R., et al.: Industry 5.0: a survey on enabling technologies and potential applications. J. Ind. Inf. Integr. **26**, 100257 (2022). https://doi.org/10.1016/j.jii.2021.100257
2. Ivanov, D., Dolgui, A., Sokolov, B.: The impact of digital technology and industry 4.0 on the ripple effect and supply chain risk analytics. Int. J. Prod. Res. **57**, 1–18 (2018). https://doi.org/10.1080/00207543.2018.1488086
3. Ghobakhloo, M., Iranmanesh, M., Mubarak, M.F., Mubarik, M., Rejeb, A., Nilashi, M.: Identifying industry 5.0 contributions to sustainable development: a strategy roadmap for

delivering sustainability values. Sustainable Prod. Consumption. **33**, 716–737 (2022). https://doi.org/10.1016/j.spc.2022.08.003

4. Ivanov, D.: The industry 5.0 framework: viability-based integration of the resilience, sustainability, and human-centricity perspectives. Int. J. Prod. Res. **61**, 1683–1695 (2023). https://doi.org/10.1080/00207543.2022.2118892
5. Basulo-Ribeiro, J., Teixeira, L.: The future of healthcare with industry 5.0: preliminary interview-based qualitative analysis. Future Internet. **16**(3), 68 (2024). https://doi.org/10.3390/fi16030068
6. Kuruppu Appuhamilage, G.D.K., Hussain, M., Zaman, M., Khan, A.: W.: a health digital twin framework for discrete event simulation based optimised critical care workflows. NPJ Digit. Med. **8**, 376 (2025). https://doi.org/10.1038/s41746-025-01738-4
7. Mahmoud, H., Kirsch, T., O'Neil, D., Anderson, S.: The resilience of healthcare systems following major disruptive events: current practice and a path forward. Reliab. Eng. Syst. Saf. **235**, 109264 (2023). https://doi.org/10.1016/j.ress.2023.109264
8. Vázquez-Serrano, J.I., Peimbert-García, R.E., Cárdenas-Barrón, L.E.: Discrete-event simulation modeling in healthcare: a comprehensive review. Int. J. Environ. Res. Public Health. **18**(22), 12262 (2021). https://doi.org/10.3390/ijerph182212262
9. Reyes, J., et al.: A study on modeling and simulation of automobile painting process based on flex Sim. Adv. Intell. Syst. Comput. **1078**, 260–273 (2020). https://doi.org/10.1007/978-3-030-33614-1_18
10. Ruiz, M., Orta, E., Sánchez, J.: A simulation-based approach for decision-support in healthcare processes. Simul. Model. Pract. Theory. **136**, 102983 (2024). https://doi.org/10.1016/j.simpat.2024.102983
11. Vecillas Martin, D., Berruezo Fernández, C., Gento Municio, A.M.: Systematic review of discrete event simulation in healthcare and statistics distributions. Appl. Sci. **15**(4), 1861 (2025). https://doi.org/10.3390/app15041861
12. Osorio-Canchig, K., Reyes-Vásquez, J., Aldás-Salazar, D.: Improvement of the efficiency of hospital care: a simulation-based approach. Dyn. Med. **92**(239), 101–110 (2025). https://doi.org/10.15446/dyna.v92n239.120207
13. Ouda, E., Sleptchenko, A., Simsekler, M.C.E.: Optimizing emergency department operations: a simulation framework for managerial decision-making. J. Nurs. Manag. **2025**, 8874680 (2025). https://doi.org/10.1155/jonm/8874680
14. Günal, M.M., Pidd, M.: Discrete event simulation for performance modelling in health care: a review of the literature. J. Simul. **4**, 42–51 (2010). https://doi.org/10.1057/jos.2009.25
15. Sanchez, S.M., Sanchez, P.J., Wan, H.: Work smarter, not harder: a tutorial on designing and conducting simulation experiments. In: In: 2020 Winter Simulation Conference (WSC), pp. 1128–1142 (2020). https://doi.org/10.1109/WSC48552.2020.9384057
16. Provencher St-Pierre, A., Pemberton-Renaud, V., Schneider, D.: Count data in biology data transformation or model reformation? Ecol. Evol. **8**(6), 3077–3085 (2018). https://doi.org/10.1002/ece3.3807
17. Lv, W., Liu, R., Yan, F., Wang, Y.: Discrete event simulation-based analysis and optimization of emergency patient scheduling strategies. Health. **14**(1), 99 (2026). https://doi.org/10.3390/healthcare14010099

Modelling the Impact of Sugary Food Consumption on Diabetes Dynamics

Norazaliza Mohd Jamil[1(✉)], Chai Jian Tay[1], Nadirah Mohd Nasir[1], Balvinder Singh Gill[2], Sharmeen Izat Hassan[3], and Salisu Ibrahim[4]

[1] Centre for Mathematical Sciences, Universiti Malaysia Pahang Al-Sultan Abdullah, Lebuh Persiaran Tun Khalil Yaakob, 26300 Gambang, Pahang, Malaysia
norazaliza@umpsa.edu.my

[2] Institute for Medical Research (IMR), Ministry of Health Malaysia, 50588 Kuala Lumpur, Malaysia

[3] Department of Mathematics Education, Faculty of Education, Tishk International University, Erbil, Iraq

[4] Department of Information Technology, Faculty of Applied Sciences, Tishk International University, Erbil, Iraq

Abstract. The global prevalence of diabetes is increasing, partly due to lifestyle factors such as high sugar consumption. While mathematical models have often been used to study how diabetes develops, the direct effect of eating sugary foods on progression between disease stages has not been well studied. This study presents a compartmental ordinary differential equation model that characterizes transitions among susceptible, prediabetic, and diabetic populations, with sugar consumption intensity incorporated as a key driver of disease progression. The model is studied using equilibrium analysis, nullcline structure, and phase-plane dynamics within the prediabetes to diabetes state space. Analytical results identify threshold conditions that determine whether the system exhibits prediabetes-dominant or diabetes-dominant equilibria. Increased sugar intake shifts the system toward persistent diabetes. Phase-plane simulations indicate that increased sugar consumption qualitatively modifies long-term system dynamics, leading to higher diabetes prevalence. The results underscore the significant influence of dietary behaviour on diabetes progression and establish a mathematical framework for evaluating the long-term effects of sugar-reduction interventions.

Keywords: diabetes · prediabetes · dietary sugar · ordinary differential equations · phase-plane analysis

1 Introduction

Type 2 diabetes (T2D) is surging across the globe, bringing with it serious health risks and high financial costs [1]. The transition from prediabetes to diabetes is a critical window for intervention. Epidemiological studies show that sugar-sweetened beverages and high dietary sugar intake are linked to increased risk of T2D and cardiometabolic disease. Dietary patterns and lifestyle factors, especially high sugar consumption, contribute to

N. A. Mohd Subha et al. (Eds.): SMS 2026, CCIS 2956, pp. 101–113, 2026.
https://doi.org/10.1007/978-981-92-0306-2_8

the progression from prediabetes to diabetes [1]. Some epidemiological evidence linking lifestyle or sugar to diabetes is as follows: Corte et al. [1] presented systematic evidence that highlighted that sugar intake is a risk factor for having T2D. There is an association between total sugar intake and diabetes prevalence, and the mediating role of factors such as body mass index [2]. Global burden of disease attributable to sugar-sweetened beverages, reinforcing public health relevance [3].

Mathematical modelling is applied across various scientific disciplines, as in [4–7], offering essential insights into systems that cannot be fully understood through empirical observation alone. Various mathematical models of diabetes dynamics have been proposed. For example, Ferdous [8] proposed an ordinary differential equation model incorporating lifestyle intervention to assess its effect on diabetes prevalence. In another study, Yao et al. [9] developed an ODE model to investigate the dynamics of type 2 diabetes and the impact of risk factors and interventions.

The existing mathematical model in diabetes, including those in [8], often focuses on physiological glucose-insulin dynamics or population projections without explicitly modelling dietary sugar intake as a dynamic driver of transitions between susceptible, prediabetic, and diabetic states. Hence, there is a need for dynamical system frameworks that directly incorporate the intensity of sugar intake to understand qualitative shifts in disease progression.

The aim of this paper is to develop a compartmental model comprising susceptible, prediabetic, and diabetic classes, formulated as ordinary differential equations, to investigate the effect of levels of sugary food consumption on the dynamics. Secondly, the aim is to analyse equilibria and phase-plane dynamics to reveal conditions under which prediabetes or diabetes states dominate. These frameworks can generate insights that inform preventive strategies.

2 Model Formulation

Mathematical modelling has proven to be a tool for anticipating national health crises in Malaysia. Previous work [10–13] demonstrated the utility of a compartmental framework in projecting the transmission dynamics of COVID-19. Building upon that methodological foundation, this study applies a deterministic compartmental model to diabetes population projection based on ordinary differential equations.

In this study, three epidemiologically relevant states representing the progression from health to diabetic disease under the influence of dietary behaviour are considered: Susceptible (S), Prediabetic (P), and Diabetic (D), and their description are in Table 1. Figure 1 shows the schematic diagram of the compartmental model illustrating the progression from susceptible to diabetic states influenced by sugary food consumption. Prediabetes is a distinct stage in diabetes. It is an intermediate stage between normal glycemia and diabetes and an important target for preventive intervention [14].

The proposed model is governed by a system of ordinary differential equations for $S(t)$, $P(t)$, and $D(t)$. Population enters the susceptible class at a constant rate b, and all compartments are subject to a removal rate at rate μ, representing the natural mortality rate. The resulting system is:

$$\frac{dS}{dt} = b - (\varepsilon_0 + \varepsilon_1 F)S - \mu S \tag{1}$$

Table 1. Compartmental model classes.

Variable	Class name	Description
$S(t)$	Susceptible	Non-Diabetic population (individuals without impaired glucose regulation)
$P(t)$	Prediabetic	At-risk population (individuals with elevated blood glucose indicative of high risk for type 2 diabetes)
$D(t)$	Diabetic	Diabetic population (individuals meeting clinical criteria for diabetes)

$$\frac{dP}{dt} = (\varepsilon_0 + \varepsilon_1 F)S - (\tau_0 + \tau_1 F)P - \mu P \tag{2}$$

$$\frac{dD}{dt} = (\tau_0 + \tau_1 F)P - \mu D \tag{3}$$

Let F be sugary food consumption intensity at the population level, directly affects the transition rates from S to P and from P to D. Empirical evidence indicates that increased intake of sugar-rich foods and beverages elevates the risk of impaired glucose regulation and progression to diabetes. The parameter F is treated as a constant and dimensionless value within the model framework. Low F means sugary food is rarely consumed and high F means sugary food is frequently consumed. If $F = 0$ means no dietary effect and if F incresease, the progression from susceptible to prediabetic is increased.

The parameter ε_0 is the baseline transition from S to P due to the factors of genetic, age and environment and $\varepsilon_1 F$ is the additional effect of sugar consumption on S to P. Similarly, τ_0 and $\tau_1 F$ play the same role for P to D, respectively. These parameter setups add controllable, interpretable driver representing sugary food intensity, which is distinct from demographic transitions or physiological processes. All parameter values are recorded in Table 2. These parameters are assumed positive and constant for analytical tractability.

The total population, N is assumed to be constant, where $N(t) = S(t) + P(t) + D(t)$ and $\frac{dN}{dt} = 0$. Thus, $b = \mu N$ to ensure that the total population remains constant.

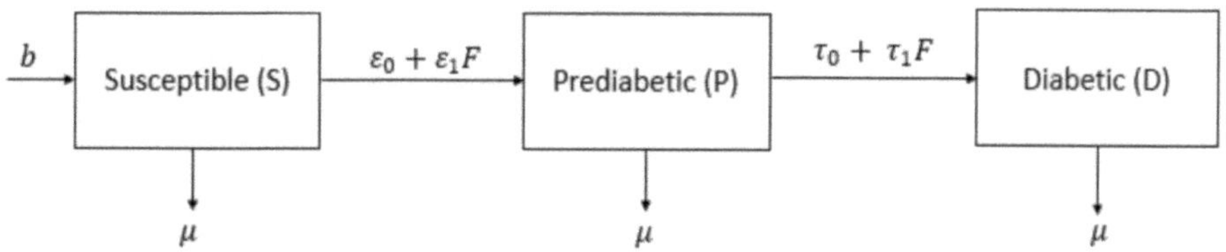

Fig. 1. Compartmental model diagram illustrating transitions between susceptible, prediabetes and diabetes under sugar consumption intensity

The model assumptions are as follows:

(i) Close population system: The model assumes a constant population $N = 34{,}300{,}000$ [15].
(ii) Linear transition dynamics: The transition rate from S to P and from P to D is assumed to be a linear function of sugary food consumption intensity (F).
(iii) Unidirectional disease pathway: The model assumes a move from S to P and P to D without returning to S or P. This represents the worst-case scenario projection for healthcare resource planning.

To ensure the model represents the current epidemiological status in Malaysia, initial conditions were calibrated using the National Health and Morbidity Survey (NHMS) 2023 [16]. The prevalence of diabetes and prediabetes among Malaysian adults was recorded at 15.6% and 11.6%, respectively. Based on reports from the Malaysian Medical Association [17] and The Star [18], the diabetes prevalence rate in Malaysia has surged to 21%, ranking the nation 13th globally and the highest in Southeast Asia. This aligns with the International Diabetes Federation 2025 estimates [19], which place the absolute number of diabetic adults in Malaysia at approximately 4.75 million. These figures represent a critical increase from the 15.6% recorded in the NHMS 2023 report, necessitating a recalibration or the initial state $D(0) = 0.21N$ to reflect the current national health crisis. To translate these proportions into population counts for the compartments $D(0)$ and $P(0)$, the 2025 national population estimates of $N = 34{,}300{,}000$ provided by the Department of Statistics Malaysia [15] was utilized. Based on these reports, the initial conditions for all simulation study in this paper are set to $P(0) = 0.116N$, $D(0) = 0.21N$, $S(0) = N - P(0) - D(0)$. Based on demographic data from the Department of Statistics Malaysia [15], the natural mortality rate, denoted by the parameter μ is set at 0.0058. This value is utilised throughout the simulation.

Table 2. Parameter description.

Parameter	Description
b	Birth rate in Malaysia
μ	Natural mortality rate in Malaysia
ε_0	Baseline progression from susceptible to prediabetic population
ε_1	Strength of sugary food effect on progression
τ_0	Baseline progression from prediabetic to diabetic population
τ_1	Strength of sugary food effect on progression
F	Sugary food consumption intensity at the population level

3 Equilibrium Analysis

Equilibrium analysis provides a qualitative understanding of the model's long-term behaviour. Identifying steady states and their dependence on key parameters, especially sugar consumption intensity (F), enables determination of conditions under which prediabetes or diabetes will persist or when susceptibility will dominate.

3.1 Equilibria Identification

By setting all derivatives in Eq. (1), (2) and (3) to zero, the unique equilibrium solution corresponding to a steady distribution of susceptible, prediabetic, and diabetic individuals $E^* = (S^*, P^*, D^*)$ is given by:

$$S^* = \frac{b}{\varepsilon_0 + \varepsilon_1 F + \mu}, P^* = \frac{\varepsilon_0 + \varepsilon_1 F}{\tau_0 + \tau_1 F + \mu} S^*, D^* = \frac{\tau_0 + \tau_1 F}{\mu} P^* \tag{4}$$

There is only one equilibrium as is known as diet-driven endemic equilibrium because diabetes is non-infectious. There is no disease-free equilibrium. There is monotonic dependence on sugary food intensity. S^* decreases with increasing F. P^* and D^* increase with increasing F. These expressions indicate that increased sugary food consumption intensity F reduces the steady-state susceptible population while increasing both prediabetic and diabetic population. The interpretation is that sugary food consumption shifts population mass downstream along the disease pathway.

The equilibrium structure is preserved. There exists a unique equilibrium for all $F \geq 0$. No equilibrium disappearance or bifurcation. The interpretation is that dietary behavior modifies equilibrium levels without altering the system structure.

3.2 Nullclines

Without loss of generality, let $\varepsilon = \varepsilon_0 + \varepsilon_1 F$ and $\tau = \tau_0 + \tau_1 F$. The P-nullcline, $\frac{dP}{dt} = 0$ gives $\varepsilon S^* - \tau P - \mu P = 0$ implies $P = \frac{\varepsilon S^*}{\tau + \mu}$. Since $S^* = \frac{\mu N}{\varepsilon + \mu}$, the P-nullcline is a vertical line in the P-D plane. The D-nullcline, $\frac{dD}{dt} = 0$ gives $\tau P - \mu D = 0$ implies $D = \frac{\tau}{\mu} P$. This is a straight line through the origin in the P-D plane. Both P-nullcline and D-nullcline are shown in Fig. 2.

The reduced Jacobian in (P, D) is $J = \begin{pmatrix} -(\tau + \mu) & 0 \\ \tau & -\mu \end{pmatrix}$, and the eigenvalues are $\lambda_1 = -\tau + \mu$ and $\lambda_2 = -\mu$. Both eigenvalues are negative indicate globally asymptotically stable node. The equilibrium point is located at the intersection of the system's nullclines, which are defined as the lines where $\dot{P} = 0$ or $\dot{D} = 0$. Analysing the nullclines is therefore very helpful for understanding the system.

3.3 Conditions for Prediabetes Dominance

The condition for prediabetes dominance is when the equilibrium (steady state) satisfies $P^* > D^*$. Hence,

$$S^* = \frac{b}{\varepsilon + \mu}, P^* = \frac{\varepsilon}{\tau + \mu} S^*, D^* = \frac{\tau}{\mu} P^* \tag{5}$$

For prediabetes dominance which is defined as $P^* > D^*$, substituting D^* yields $P^* > \frac{\tau}{\mu} P^*$. Thus, $1 > \frac{\tau}{\mu}$ i.e. $\tau < \mu$. Therefore, if $\tau < \mu$, prediabetes dominance will occur regardless of initial conditions and if $\tau > \mu$, diabetes dominance will be observed at steady states. It demonstrates that prediabetic dominance at equilibrium occurs when progression to diabetes is slower than population natural mortality. The prediabetic population exceeds the diabetic population when the progression rate to diabetes is sufficiently slow relative to population natural mortality. This reflects realistic epidemiological conditions where lifestyle modification and early detection delay disease progression.

The qualitative behaviour of the system near equilibrium is characterized by examining the local stability properties of the steady states. Linearisation of the reduced system about an equilibrium point yields a Jacobian matrix whose eigenvalues govern the local dynamics. Rather than pursuing explicit eigenvalue expressions, stability is inferred through the structure of the nullclines and the relative magnitudes of the transition and removal rates.

In particular, the balance between the progression rate from prediabetes to diabetes, $(\tau_0 + \tau_1 F)$, and the natural removal rate μ plays a dominant role. When $(\tau_0 + \tau_1 F) > \mu$, trajectories in the neighborhood of the equilibrium are directed towards the diabetes-dominant state, indicating local asymptotic stability. Conversely, if $(\tau_0 + \tau_1 F) < \mu$, the system favours a prediabetes-dominant equilibrium, with trajectories repelled from the diabetes state. These stability characteristics are consistent with the nullcline geometry and are further illustrated through phase-plane simulations in the subsequent section.

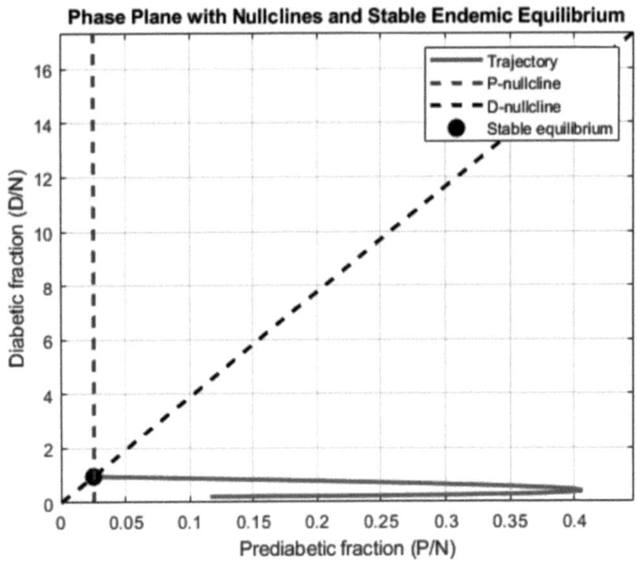

Fig. 2. Phase-plane trajectory and system nullclines at a moderate sugar consumption intensity ($F = 0.5$). The intersection of the vertical P-nullcline and the linear D-nullcline identifies the unique, globally asymptotically stable equilibrium.

4 Phase-Plane Analysis

A critical tool for investigating a dynamical system is the phase plane. The phase plane illustrates the behaviour of a dynamical system by plotting the two states against each other, and showing the possible trajectories a system might take. The phase plane is helpful to see the qualitative behaviour of the system.

The phase plane for P versus D, as well as its corresponding solution plot are shown in Fig. 3 with initial condition $P(0) = 0.116N$, $D(0) = 0.21N$, $S(0) = N - P(0) - D(0)$, $N = 34{,}300{,}000$ for parameters $\mu = 0.0058$, $\tau_0 = 0.4877$, $\tau_1 = 0.4877$. The system starts with non-zero prediabetic and non-zero diabetic population. This is the case for $\tau_0 > \mu$ and $\tau_1 > \mu$ meaning the progression rate from prediabetic to diabetic population and the strength of sugary food effect on progression are higher than natural mortality rate. The initial point is marked in red to indicate starting population. The red trajectory showing how the number of prediabetic and diabetic people will vary over time, for a specific initial condition. It shows the diabetic dominance occur in the end of the trajectory. A diabetes-dominant situation shows that natural mortality by itself cannot reduce the number of diabetes cases when unhealthy diets are common. This means that without early action to change eating habits, the risk of chronic diseases increases sharply.

Figure 4 shows the phase plane and its corresponding plot for parameters $\tau_0 < \mu$ and $\tau_1 < \mu$. Numerical simulations indicate that under Malaysia-like epidemiological conditions, the prediabetic compartment is more dominant over the diabetic population, reflecting delayed disease progression. The initial red point shows progression from diabetic dominance to prediabetic dominance. Even though the initial condition $P(0) < D(0)$, the prediabetic dominance occur in the end of the trajectory. The prediabetic population remains above diabetic throughout. Slow transition prevents rapid diabetic accumulation. System approached endemic equilibrium with $P > D$.

Figure 5 shows phase-plane trajectories for multiple curves at different levels of sugary food intensity, F indicating how sugary food alters the dynamics for parameters $\mu = 0.0058$, $\varepsilon_0 = 0.18$, $\varepsilon_1 = 0.60$, $\tau_0 = 0.2$, $\tau_1 = 0.05$. The chosen values for F are 0.1 for low consumption, 0.5 for moderate consumption, and 1 for high consumption of sugary foods. The system starts with a non-zero population of prediabetic and diabetic individuals. Dynamics show how prediabetes evolves into diabetes under different levels of sugary foods. All trajectories emanate from the same epidemiological baseline, ensuring a fair comparison across F.

The arrows along the curve indicate the direction of progression over time. Flow direction consistently points from P to D. Vector arrows indicate greater flow magnitude for higher F. Across all scenarios, trajectories converge toward a common attracting region, indicating consistent long-term disease dynamics. However, higher values of F produce steeper trajectories and faster progression toward elevated diabetic prevalence, highlighting the role of dietary behavior in accelerating disease evolution. Increasing F will steepen the trajectory slope, accelerate movement toward higher D, and reduce residence time in the prediabetic state. It indicates faster transition from prediabetes to diabetes and shorter intervention window.

The numerical results presented in Figs. 2, 3, 4 and 5 are simulated using a combination of empirical demographic constant values from the literature and assumed progression parameters used for exploratory simulation to examine the qualitative behaviour of

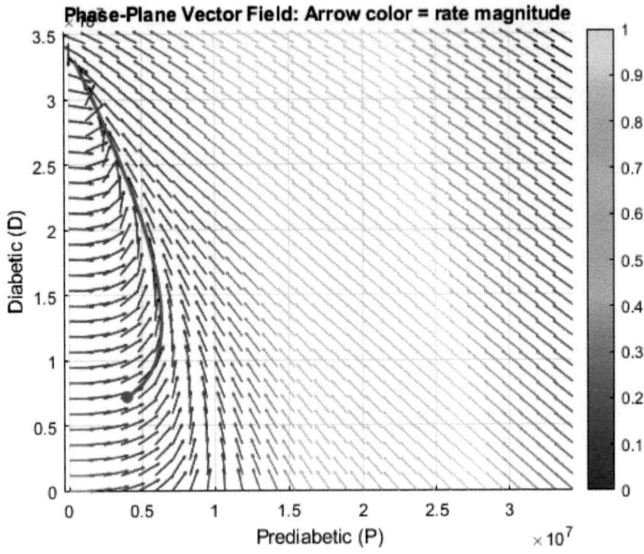

(a) Phase plane for parameters $\mu = 0.0058, \tau_0 = 0.4877, \tau_1 = 0.4877$

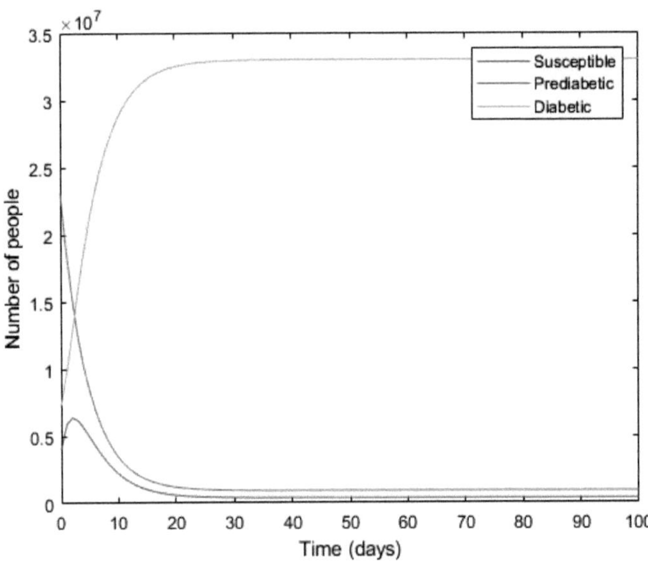

(b) Solution plot for parameters $\mu = 0.0058, \tau_0 = 0.4877, \tau_1 = 0.4877$

Fig. 3. Phase-plane trajectory and corresponding temporal plot for the diabetes-dominant regime ($\tau_0, \tau_1 > \mu$). The vector filed illustrates the global attraction toward a stable endemic equilibrium where the diabetic population exceeds the prediabetic reservoir.

the model. The initial conditions $P(0) = 0.116N, D(0) = 0.21N, S(0) = N - P(0) - D(0)$ and $N = 34{,}300{,}000$ are calibrated using the prevalence data by the Ministry of Health

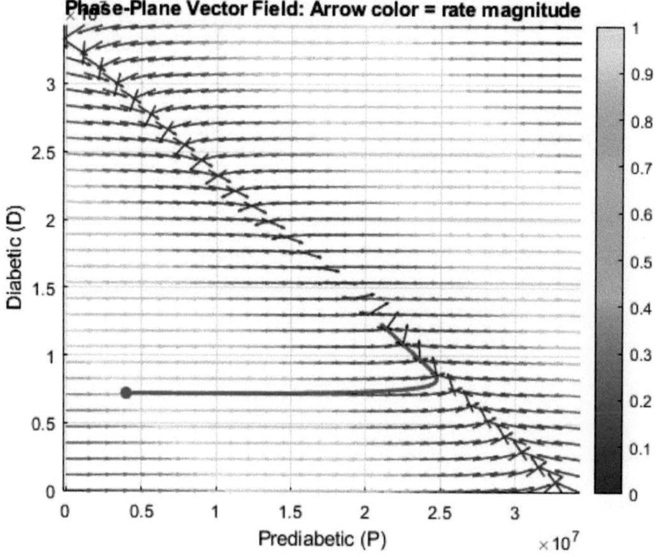

(a) Phase plane for parameters $\mu = 0.0058, \tau_0 = 0.004877, \tau_1 = 0.004877$

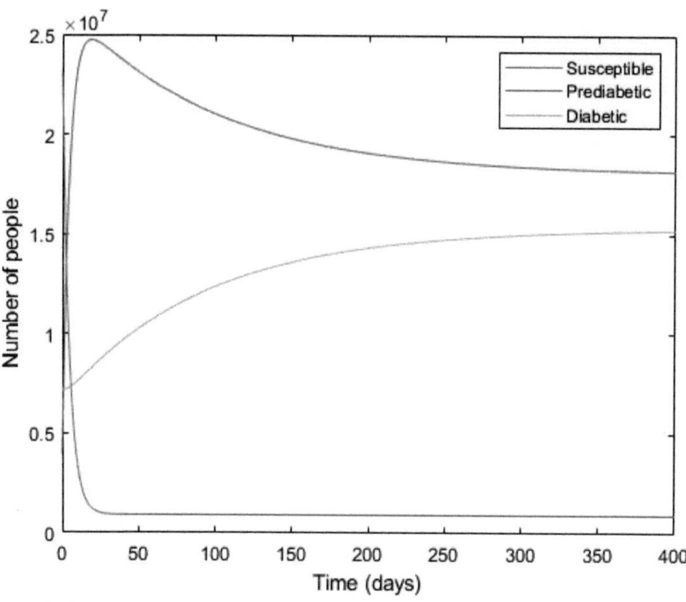

(b) Solution plot for parameters $\mu = 0.0058, \tau_0 = 0.004877, \tau_1 = 0.004877$

Fig. 4. Phase-plane trajectory and corresponding temporal plot for the prediabetes-dominant regime (τ_0, $\tau_1 < \mu$). The vector filed illustrates the global attraction toward a stable endemic equilibrium where the prediabetic population exceeds the diabetic reservoir.

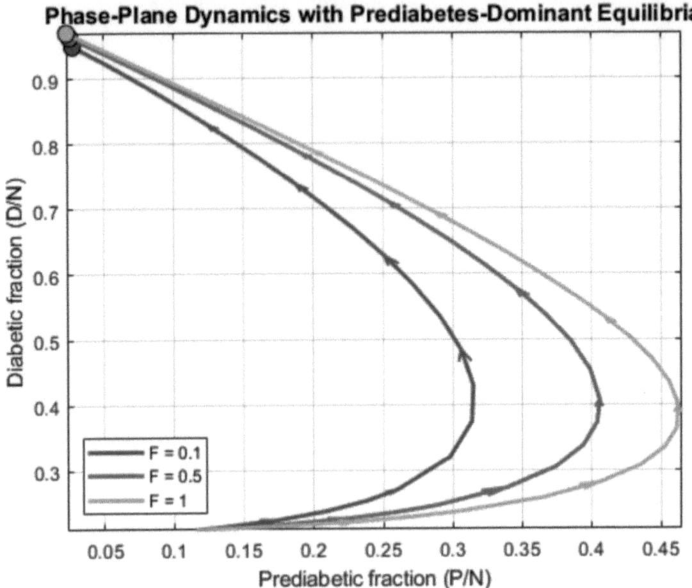

Fig. 5. Phase-plane trajectories illustrating the influence of sugar consumption intensity, F on diabetes progression dynamics. Increasing F steepens the trajectories and accelerates the transition from prediabetes to diabetes, indicating a faster disease progression rate.

Malaysia, whereas the natural mortality rate, $\mu = 0.0058$, is prescribed based on national statistics [15]. To explore the qualitative behaviour of the model, the transition parameters τ_0 and τ_1 are strategically varied to reflect different case of $\tau > \mu$ and $\tau < \mu$. These two cases illustrated different behaviours: diabetes dominant and prediabetes-dominant, as shown in Figs. 3 and 4, respectively.

5 Discussion

The analysis in this paper is comparative and illustrative. This paper aims to answer the question of how consumption of sugary foods influences the dynamics of diabetes progression. It is observed that the equilibrium point is consistent with phase-plane dynamics. Phase-plane trajectories converge toward the equilibrium point. Stability structure is preserved: The system retains the same qualitative dynamics, with no change in the number of equilibria.

The phase-plane analysis demonstrates that increased consumption of sugary foods primarily influences the rate of progression from prediabetes to diabetes. While the equilibrium behavior remains unchanged, higher sugary dietary intensity compresses the prediabetic phase, thereby reducing the effective window for preventive intervention. These findings suggest that dietary moderation can play a critical role in delaying disease progression even when long-term prevalence remains governed by systemic factors. Dietary modifications alter the rate of disease progression rather than the underlying disease structure.

Low F means slow drift and prolonged prediabetic phase. High F indicates rapid drift and early accumulation in the diabetic compartment. Higher sugar intake compresses the time available for preventive intervention. F functions analogously to a control knob. Increasing F results in greater consumption of sugary foods and accelerates the progression to diabetes. Conversely, decreasing F promotes a healthier diet and slows disease progression.

The proposed model provides a robust baseline for the natural progression of diabetes in Malaysia. However, certain structural simplifications present opportunities for future investigation. First, the proposed model assumes that transition rates increase linearly with F to establish a baseline mathematical framework that demonstrates the first-order impact of sugar intake on disease progression. Future research may replace this linear formulation with a saturating functional response, such as Michaelis-Menten kinetics, to reflect the idea that after a certain amount of sugar, the transition rates should plateau. This approach prevents the model from predicting infinitely fast disease progression at extreme levels of F. From a biological perspective, increasing sugar intake to high levels does not result in a proportional doubling of damage to pancreatic beta cells.

Second, this study models a unidirectional disease pathway to estimate the maximum potential burden on the healthcare system. The primary objective is to represent the natural progression and projected burden under current population-wide conditions. In this framework, diabetes is considered an inevitable outcome rather than a manageable condition. Future research could incorporate reversible pathways, such as transitions from prediabetes to a healthy state and from diabetes to prediabetes through lifestyle interventions leading to remission. Including a recovery rate would enhance the model's capacity to evaluate health policy interventions.

The model's primary significance lies in its ability to simulate the long-term impacts of policy interventions, including government fiscal measures such as sugar taxation and individual-level behavioural change. The model can be used to forecast how the Sugar-Sweetened Beverage (SSB) tax, implemented in 2019 and expanded in 2024/2025, influences the transition from susceptible to prediabetic states [20]. The proposed model offers a way to test the Malaysian Medical Association's plan [17] to cut added sugar in manufactured foods by 20% to 30% over the next three to five years [18]. By changing the F parameter, which represents sugar intake, policymakers can see how quickly the S to P transition could strain the healthcare system if current eating habits continue.

6 Conclusion

This paper presented a mathematical model to explore the impact of sugary food consumption on diabetes progression. Phase-plane analysis shows that increased sugary intensity accelerates transitions from prediabetes to diabetes without altering the qualitative structure of the system. This paper demonstrates that higher sugary intake compresses the prediabetic phase and accelerates accumulation in the diabetic compartment, highlighting the critical role of eating behaviour in shaping disease progression. The results highlight the value of the model for visualising lifestyle-driven disease dynamics and underscore the potential of dietary moderation in delaying disease progression. The framework supports early-stage analysis of behavioural interventions. Extensions incorporating clinical follow-up and sensitivity analysis are left for future work.

Acknowledgments. The authors would like to express sincere gratitude to the Universiti Malaysia Pahang Al-Sultan Abdullah, Internal Grant RDU240327 (Ref: UMP.05/26.10/03/RDU240327) for funding this research.

References

1. Della Corte, K.A., et al.: Dietary sugar intake and incident type 2 diabetes risk: a systematic review and dose-response meta-analysis of prospective cohort studies. Adv. Nutr. 100413 (2025)
2. Liu, Y., Cheng, J., Wan, L., Chen, W.: Associations between total and added sugar intake and diabetes among chinese adults: the role of body mass index. Nutrients. **15**(14), 3274 (2023)
3. Lara-Castor, L., et al.: Burdens of type 2 diabetes and cardiovascular disease attributable to sugar-sweetened beverages in 184 countries. Nat. Med. **31**(2), 552–564 (2025)
4. Jamil, N.M., Khan, I.: Unravelling the influence of advection and diffusion on the spatial and temporal evolution of pollutant concentration. J. Adv. Res. Fluid Mech. Therm. Sci. **119**(2), 1–12 (2024)
5. Azimi, A.I.F., Jamil, N.M., Rahim, M.H.A.: Neural ordinary differential equations for solving nonlinear system of ethanol fermentation in bioreactor. In: AIP Conference Proceedings, vol. 2895, p. 020012. AIP Publishing LLC (2024)
6. Othman, S.N., Jamil, N.M.: A graphical user interface for predicting quality parameters of deep-fried foods. J. Food Process Eng. **47**(9), e14737 (2024)
7. Othman, S.N., Jamil, N.M.: A modified predictive model for colour changes in French fries during frying. J. Phys. Conf. Ser. **1988**(1), 012030 (2021)
8. Ferdous, A.: An ordinary differential equation model for assessing the impact of lifestyle intervention on type 2 diabetes epidemic. Healthc. Anal. **4**, 100271 (2023)
9. Yao, S., et al.: Modeling diabetes progression with risk factors: a case study in China. Comput. Biol. Med. **186**, 109643 (2025)
10. Jamil, N.M., Gill, B.S.: Estimation of public compliance with COVID-19 prevention standard operating procedures through a mathematical model. In: Proceedings of 2nd International Conference on Mathematical Modeling and Computational Science: ICMMCS 2021, pp. 49–58. Springer Nature Singapore (2022)
11. Jamil, N.M.: Understanding Pandemic Challenges: Data-Driven and Computational Approaches, p. 89. UUM Press (2024)
12. Husnin, S.A.M., Mohd, N.: Campus reopening during the COVID-19: ODE-SIRD model. J. Eng. Sci. Technol. **18**(3), 1650–1669 (2023)
13. Mohd Jamil, N., Rosli, N., Muhammad, N.: Simulation of COVID-19 outbreaks via graphical user interface (GUI). J. Public Health Res. **11**(1), jphr-2021 (2022)
14. Echouffo-Tcheugui, J.B., Selvin, E.: Prediabetes and what it means: the epidemiological evidence. Annu. Rev. Public Health. **42**(1), 59–77 (2021)
15. Department of Statistics Malaysia: Demographic Statistics, Third Quarter 2025. DOSM Official Portal (2025)
16. Institute for Public Health: National Health and Morbidity Survey (NHMS) 2023: Non-communicable diseases and healthcare demand—Key Findings. Ministry of Health Malaysia (2024)
17. Malaysian Medical Association: MMA: Drastic action needed as one in five Malaysians now diabetic. https://codeblue.galencentre.org/2025/11/mma-drastic-action-needed-as-one-in-five-malaysians-now-diabetic/. Accessed 20 Feb 2026
18. The Star: MMA seeks tougher sugar control measures as diabetes cases rise. https://www.thestar.com.my/news/nation/2025/11/14/mma-seeks-tougher-sugar-control-measures-as-diabetes-cases-rise. Accessed 20 Feb 2026

19. International Diabetes Federation: IDF Diabetes Atlas 11th edition. Malaysia country report (2025). https://diabetesatlas.org/data-by-location/country/malaysia/. Accessed 20 Feb 2026
20. Ministry of Health Malaysia: The impact and effectiveness of sugar-sweetened beverage (SSB) taxation in Malaysia: Technical report. https://hq.moh.gov.my/nutrition/wp-content/uploads/2025/10/Ministry-of-Health-2024-Technical-Report-on-the-Impact-Effectiveness-of-SSB-Taxation-in-Malaysia.pdf (2024)

Effect of Hydrogen Bonding on Physicochemical Properties of Natural Deep Eutectic Solvents (NaDES)

Joanne Liew Xin Hui[1] and Azizul Azri Mustaffa[1,2](✉)

[1] Faculty of Chemical and Energy Engineering, Universiti Teknologi Malaysia, UTM, 81310 Johor Bahru, Johor, Malaysia
azizulazri@utm.my

[2] Centre of Lipids Engineering and Applied Research (CLEAR), Ibnu-Sina Institute for Scientific and Industrial Research (ISI-SIR), Universiti Teknologi Malaysia, UTM, 81310 Johor Bahru, Johor, Malaysia

Abstract. Natural Deep Eutectic Solvents (NADES) are emerging as sustainable alternatives to conventional organic solvents and ionic liquids, with promising applications in extraction, biocatalysis, and pharmaceutical formulation. However, the rational design of these solvents remains challenging due to the complex role that hydrogen bonding plays in determining their physicochemical properties. This study uses molecular dynamics simulations to obtain and correlate hydrogen bonding metrics—total hydrogen count, donor-acceptor distance, and bond angle—with five key properties in Type V NADES: solubility, viscosity, density, melting point and boiling point. Seven thymol–menthol (Thy-Men) and menthol–lauric acid systems (Men-Lau) were simulated using GROMACS at various molar ratios. Results show that stronger hydrogen bonding increases solubility, density, and boiling point while depressing viscosity and melting point. Highest number of hydrogen bond, N_{HB} were observed in the Men-Lau (0.5:1) and Thy-Men (2:1) systems, reaching 124.921 and 114.178, respectively. This work establishes hydrogen bonding as a predictive design parameter, enabling the tailored formulation of NADES for green chemistry applications.

Keywords: Natural Deep Eutectic Solvents · Hydrogen Bonding · Molecular Dynamics Simulation · Physicochemical Properties · GROMACS

1 Introduction

In recent years, the search for sustainable and environmentally friendly solvents has led to the development of Deep Eutectic Solvents (DES). When DES are formulated using natural acids and alcohol, they are referred to as Natural Deep Eutectic Solvents (NADES). These solvents have gained popularity due to their biodegradability and low toxicity, aligning with green chemistry principles [1].

However, the molecular mechanisms, particularly the role of hydrogen bonding, that govern the behaviour of DES remain inadequately understood [2]. Hydrogen bonding significantly influences key properties such as viscosity, density, and solubility,

N. A. Mohd Subha et al. (Eds.): SMS 2026, CCIS 2956, pp. 114–125, 2026.
https://doi.org/10.1007/978-981-92-0306-2_9

which are crucial for evaluating solvent performance. Despite this, existing research often focuses on specific HBD-HBA combinations which limit the exploration of other potential formulations, especially the newly identified type V NADES [3].

This study seeks to address these limitations by investigating the influence of hydrogen bonding on the properties of NADES. The ultimate objective is to gain insights that enhance the formulation and utility of NADES using combinations of natural acids and alcohols.

2 Methods

2.1 Molecular Model and Simulation Details

This study uses molecular simulation and analysis tools to explore the effect of hydrogen bonding on the physicochemical properties of NADES. The simulation workflow begins with the selection of NADES components, followed by molecular structure preparation, construction of the initial simulation box, generation of topology files and force field parameters, energy minimisation and equilibration, molecular dynamics (MD) simulation, and finally hydrogen bond analysis and visualisation.

2.2 Simulation Details

The study employs several computational tools: PubChem was used to obtain SMILES notations; RDKit in Python to build and optimize 3D molecular structures; LigParGen to generate OPLS-AA force field topology files; PACKMOL to pack molecules into a 5 × 5 × 5 nm3 simulation box; GROMACS for all MD simulations (energy minimisation, NVT/NPT equilibration at 298 K and 1 bar, and 1 ns production run); MDAnalysis and MDTraj in Python for hydrogen bond trajectory analysis; and OVITO for visualisation.

2.3 Systems Studied

The study focuses on seven (7) NADES systems (see Table 1): two thymol-menthol (Thy-Men) systems at 1:1 and 2:1 molar ratios, and five menthol-lauric acid (Men-Lau) systems at molar ratios from 0.5:1 to 2.5:1. Each system contained a total of 200 molecules.

2.4 Hydrogen Bond and Property Analysis

Hydrogen bond analysis was performed using custom Python scripts with MDAnalysis and MDTraj libraries, applying geometric criteria: donor-acceptor distance $\leq$3.5 Å and donor-hydrogen-acceptor angle $\geq$150°. Key physicochemical properties such as solubility, viscosity, density, melting point, and boiling point were predicted using established computational models (e.g., COSMO-RS, thermodynamic depression models) based on the simulation outputs and system composition.

Table 1. Components, molar ratio, and number of molecules in the 7 NADES systems

System label	Component 1	Component 2	Molar ratio	Number of molecules
Thy-Men11	Thymol	Menthol	1:1	200
Thy-Men21	Thymol	Menthol	2:1	200
Men-Lau0.51	Menthol	Lauric Acid	0.5:1	200
Men-Lau11	Menthol	Lauric Acid	1:1	200
Men-Lau1.51	Menthol	Lauric Acid	1.5:1	200
Men-Lau21	Menthol	Lauric Acid	2:1	200
Men-Lau2.51	Menthol	Lauric Acid	2.5:1	200

3 Results and Discussion

3.1 Properties and Hydrogen Bonding Metrics for NADES

The molecular structures and key descriptors of thymol, menthol, and lauric acid confirm their hydrogen bonding capacity (see Table 2 and Fig. 1). Lauric acid exhibits enhanced hydrogen bond acceptor functionality due to the presence of two oxygen atoms within its carboxyl group.

Table 2. SMILES Notation and its properties of the NADES components

Components	SMILES Notation	MW (g/mol)	T_m (°C)	T_b (°C)
Thymol	CC(C)C1 = CC = C(C = C1)O	150.22	50	232
Menthol	CC1CCC(C(C1)O)C(C)C	156.27	42	213
Lauric Acid	CCCCCCCCCCCC(=O)O	200.32	43	298

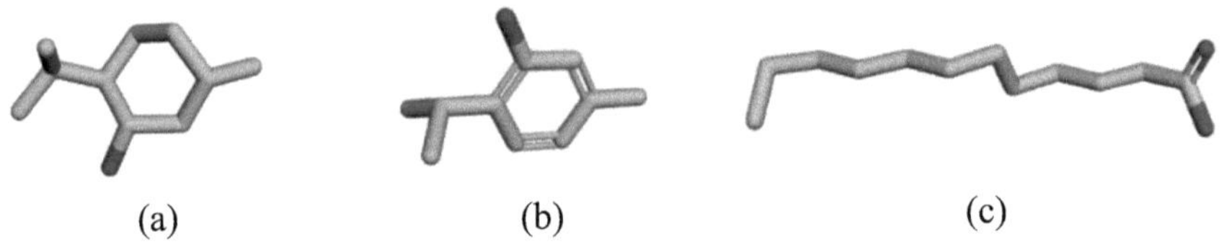

Fig. 1. 3D Molecular Structures of the Components Formed (a) Thymol, (b) Menthol, (c) Lauric Acid

A comprehensive summary of hydrogen bonding metrics and their physicochemical properties for all seven (7) systems is presented in Table 3 and 4. For the thymol-menthol systems, the total hydrogen bond count (N_{HB}) was higher for Thy-Men21 (114.2) than for Thy-Men11 (108.4). For the menthol-lauric acid system, N_{HB} decreased from 124.9

(Men-Lau0.51) to 62.3 (Men-Lau2.51) as the proportion of menthol increased. Concurrently, the average donor-acceptor distance was shortest for systems with the highest N_{HB} (e.g., 2.725 Å for Men-Lau0.51), and the donor-hydrogen-acceptor angle was more linear.

Table 3. Hydrogen Bonding Metrics for All Systems

System label	Number of hydrogen bond (N_{hb})	Average DA displacement (Å)	Average DHA angle (°)
Thy-Men11	108.356	2.887	163.192
Thy-Men21	114.178	2.874	163.216
Men-Lau0.51	124.921	2.725	164.964
Men-Lau11	99.02	2.753	164.84
Men-Lau1.51	81.188	2.775	164.453
Men-Lau21	71.337	2.804	164.317
Men-Lau2.51	62.257	2.822	164.046

Table 4. Physicochemical Properties for All Systems

System label	Solubility (mol/L)	Viscosity (mPa.s)	Density (kg/m^3)	Melting point (°C)	Boiling point (°C)
Thy-Men11	0.3097	28.48	929.4610	22.2	332.419
Thy-Men21	0.4510	25.29	931.4006	20.4	346.053
Men-Lau0.51	0.3100	13.87	908.8802	20.4	367.161
Men-Lau11	0.2328	18.16	907.4554	21.8	344.046
Men-Lau1.51	0.1985	21.34	903.8418	24.2	333.35
Men-Lau21	0.1823	23.73	901.2989	27.4	327.144
Men-Lau2.51	0.1710	25.64	900.2070	30.0	323.084

3.2 Correlation Between Hydrogen Bonding and Physicochemical Properties

Relationship Between Hydrogen Bonding and Solubility.
The relationship between hydrogen bonding and the solubility of lutein was assessed for each system. Figure 2 (a) demonstrates a positive relationship between the total hydrogen count and solubility at different molar ratios. This trend indicates that a stronger, more extensive hydrogen-bond network generally enhances solvent capacity, likely by facilitating solute–solvent interactions and stabilising the dissolved state of lutein.

Figure 2(b) shows an inverse correlation between the average donor–acceptor (DA) distance and solubility, where longer DA distances correspond to weaker, less favourable hydrogen bonds and lower predicted solubility. Conversely, Fig. 2 (c) reveals that solubility increases as the donor–hydrogen–acceptor (DHA) angle approaches linearity, indicating that stronger, more directional hydrogen bonds contribute positively to solvation capability.

Comparing across systems, Thy-Men21 exhibits the highest solubility (0.4510 mol/L), while Thy-Men11 shows the lowest (0.3097 mol/L) within the thymol–menthol series. In the menthol–lauric acid series, solubility is highest for Men-Lau0.51 (0.3100 mol/L) and decreases progressively to the lowest value for Men-Lau2.51 (0.1710 mol/L). Overall, the thymol–menthol systems demonstrate superior solubility performance compared to the menthol–lauric acid systems. This can be attributed to the more favourable balance of hydrogen bond strength and network geometry in the Thy-Men series, which appears to create a solvent environment more conducive to dissolving lutein [4].

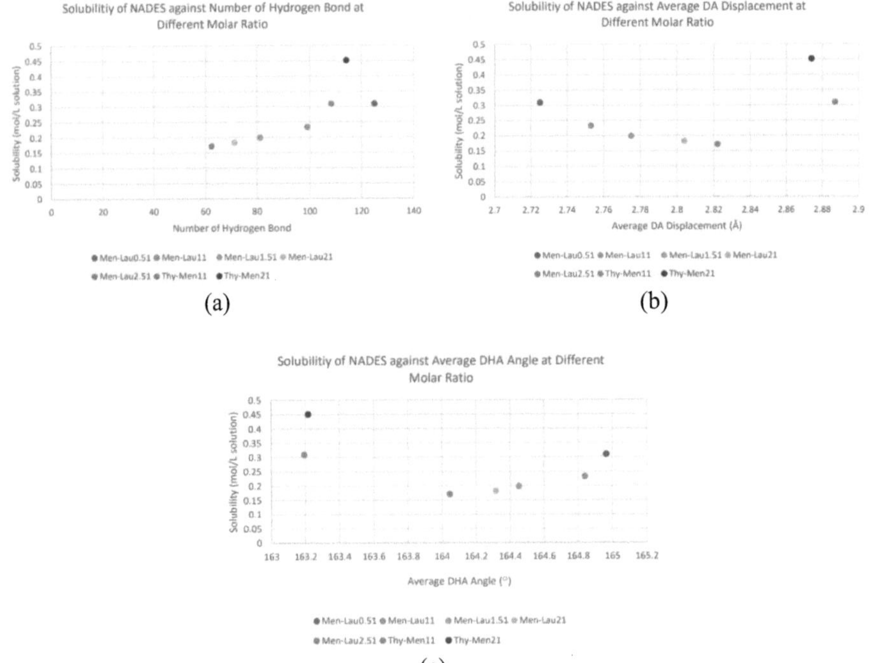

Fig. 2. Solubility of 7 different NADES systems against different hydrogen bond metrics, (a) N_{hb}, (b) Average DA displacement, (c) Average DHA angle

Relationship Between Hydrogen Bonding and Viscosity.

The correlation between hydrogen bonding metrics and predicted viscosity is illustrated in Fig. 3. Figure 3(a) demonstrates a positive relationship between the total hydrogen count and viscosity with different molar ratios. The trend indicates that systems with a higher total hydrogen bond count tend to exhibit lower predicted viscosity. This inverse correlation can be explained by considering the nature of the hydrogen bond network; while strong hydrogen bonds can restrict flow, a well-connected, extensive network may facilitate more cooperative molecular motion, reducing the effective resistance to shear and thus lowering viscosity in certain dynamic regimes [5].

Figure 3(b) shows a positive correlation between the average DA distance and viscosity. The trend indicates that systems with a longer DA distance, indicative of weaker hydrogen bond, and tend to exhibit higher viscosity. This suggests that weaker bonds allow for more transient and less structured interactions, which may impede collective molecular flow. Conversely, Fig. 3(c) demonstrates that viscosity decreases as the D-H-A angle approaches linearity. A more linear geometry signifies a stronger, more directional hydrogen bond, which can lead to a more organized and possibly more easily sheared network, thereby lowering viscosity.

Across the systems, Thy-Men21 demonstrates the lowest predicted viscosity (25.29 mPa·s), while Thy-Men11 shows the highest (28.48 mPa·s) in its series. In the menthol-lauric acid systems, Men-Lau0.51 exhibits the lowest viscosity (13.87 mPa·s), with values progressively increasing to the highest for Men-Lau2.51 (25.64 mPa·s). This is mainly due to the composed of similarly sized phenolic/alcoholic components in Thy-Men systems will form a more compact and rigid hydrogen-bonded matrix, whereas the Men-Lau systems incorporate a long-chain fatty acid (lauric acid), which introduces greater molecular flexibility and may disrupt the network's rigidity, leading to an overall lower viscosity baseline.

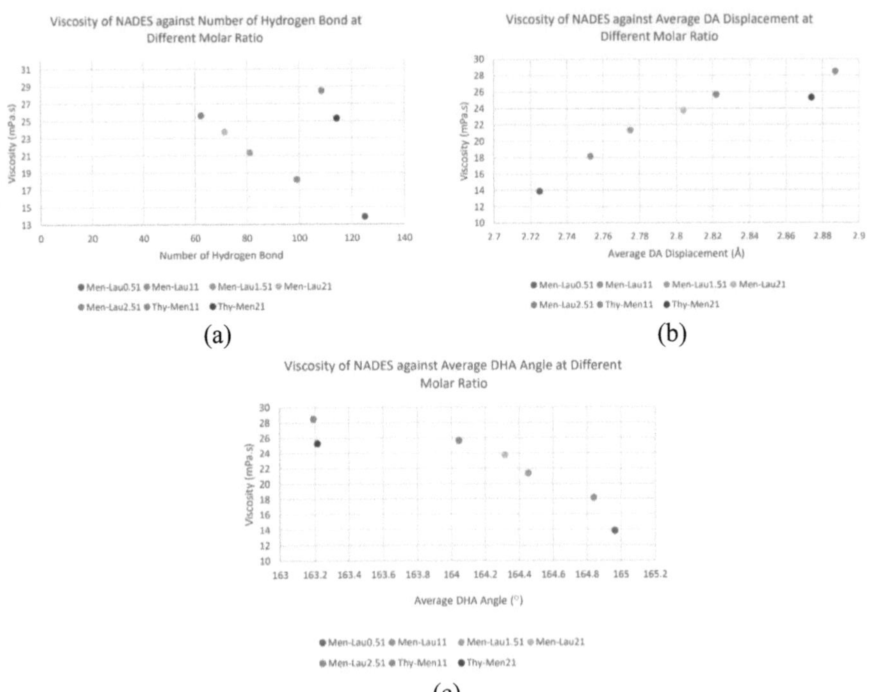

Fig. 3. Viscosity of 7 different NADES systems against different hydrogen bond metrics, (a) N_{hb}, (b) Average DA displacement, (c) Average DHA angle

Relationship Between Hydrogen Bonding and Density.
The correlation between hydrogen bonding metrics and predicted density is illustrated in Fig. 4. Figure 4(a) demonstrates a positive relationship between the total hydrogen count and density. This trend aligns with the expected physical behaviour, as stronger and more extensive hydrogen-bond networks promote closer intermolecular packing, reduce free volume, and thereby increase the overall mass per unit volume [6].

Figure 4(b) shows a positive correlation between the average DA distance and viscosity. It reveals that longer average donor-acceptor (DA) distances correlate with lower density. Since longer DA distances indicate weaker hydrogen bonds, this supports the notion that less cohesive intermolecular interactions result in looser molecular packing and reduced density.

Conversely, Fig. 4(c) demonstrates that density increases as the donor-hydrogen-acceptor (DHA) angle approaches 180°. More linear hydrogen bonds are generally stronger and more directional, facilitating efficient, ordered molecular arrangement and thus higher density. Between the two series, Thy-Men21 exhibits the highest density (931.401 kg/m^3), while Thy-Men11 shows the lowest (929.461 kg/m^3) among the thymol-menthol systems. In the menthol-lauric acid series, Men-Lau0.51 displays the highest density (908.880 kg/m^3), and density systematically decreases to the lowest value for Men-Lau2.51 (900.207 kg/m^3). Although the thymol-menthol systems consistently show higher absolute density than the menthol-lauric acid systems, this is primarily due to the higher molecular weight and more compact aromatic structure of thymol compared to the long aliphatic chain of lauric acid.

Relationship Between Hydrogen Bonding and Melting Point.
The correlation between hydrogen bonding metrics and predicted viscosity is illustrated in Fig. 5. Figure 5(a) demonstrates an inverse relationship between the total number of hydrogen bonds and the predicted melting point. It reflects the fundamental eutectic principle: the formation of a new, extensive inter-component hydrogen bond network disrupts the original, highly ordered crystalline lattices of the pure solids. This newly formed network is less stable and more disordered, requiring less thermal energy to melt. Thus, systems with stronger inter-component hydrogen bonding exhibit more significant melting point depression [7]. Figure 5 (b) shows a positive correlation between the average donor-acceptor (DA) distance and melting point. Longer DA distances indicate weaker hydrogen bonds; consequently, a weaker inter-component network causes less disruption to the components' inherent crystalline order, resulting in a less depressed, and therefore higher, melting point.

Conversely, Fig. 5(c) reveals that melting point decreases as the donor-hydrogen-acceptor (DHA) angle becomes more linear. A linear DHA angle signifies a stronger, more optimal hydrogen bond, which enhances network formation and promotes greater lattice disruption, thereby lowering the melting point more effectively. Across the systems, Thy-Men21 exhibits the lowest predicted melting point (20.4 °C), while Thy-Men11 shows the highest (22.2 °C) in its series.

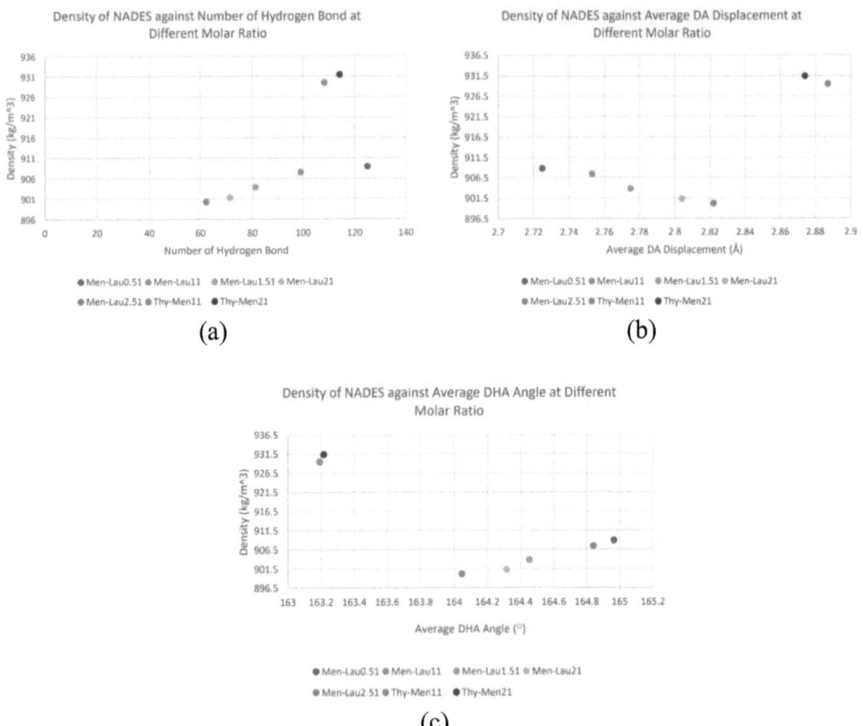

Fig. 4. Density of 7 different NADES systems against different hydrogen bond metrics, (a) N_{hb}, (b) Average DA displacement, (c) Average DHA angle

In the menthol-lauric acid series, Men-Lau0.51 has the lowest melting point (20.4 °C), with values progressively increasing to the highest for Men-Lau2.51 (30.0 °C). Overall, the thymol-menthol systems generally demonstrate lower melting points than the menthol-lauric acid systems. This can be attributed to the more effective disruption of crystalline order by the complementary phenolic and alcoholic groups in the Thy-Men series, compared to the interaction between menthol and the long-chain lauric acid, which may allow for more residual crystalline domains or a less effective charge delocalization within the network.

Relationship Between Hydrogen Bonding and Boiling Point.
The correlation between hydrogen bonding metrics and predicted boiling point is illustrated in Fig. 6. Figure 6(a) demonstrates a positive correlation between the total number of hydrogen bonds and the predicted boiling point. This is consistent with established physical chemistry principles: stronger and more extensive intermolecular hydrogen bonding increases the cohesive energy of the liquid. Consequently, more thermal energy is required to overcome these interactions during vaporization, resulting in a higher boiling point [7]. Figure 6(b) reveals an inverse relationship between the average donor-acceptor (D-A) distance and boiling point; longer D-A distances correspond to weaker hydrogen bonds and thus lower boiling points.

Conversely, Fig. 6(c) shows that boiling point increases as the donor-hydrogen-acceptor (D-H-A) angle becomes more linear, as optimal linear geometry signifies stronger, more effective hydrogen bonds that enhance the liquid's cohesive strength. Among the systems, Thy-Men21 exhibits the highest boiling point (346.053 °C) within the thymol-menthol series, while Thy-Men11 shows the lowest (332.419 °C). In the menthol-lauric acid series, Men-Lau0.51 possesses the highest boiling point (367.161 °C), with values progressively decreasing to the lowest for Men-Lau2.51 (323.084 °C). While the thymol-menthol and menthol-lauric acid series exhibit overlapping boiling point ranges, Men-Lau0.51 stands out with the highest overall value. This can be attributed to its unique combination of a high hydrogen bond count (124.921), short D-A distance (2.725 Å), and highly linear D-H-A angle (164.964°), indicating the formation of an exceptionally strong and cohesive hydrogen-bonded network that maximizes the energy required for vaporization.

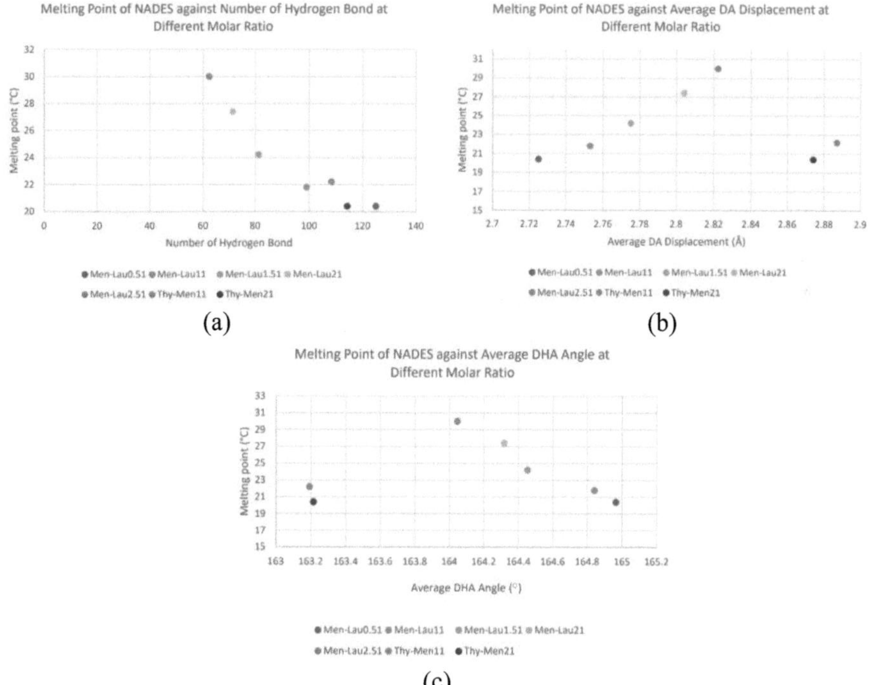

Fig. 5. Melting Point of 7 different NADES systems against different hydrogen bond metrics, (a) N_{hb}, (b) Average DA displacement, (c) Average DHA angle

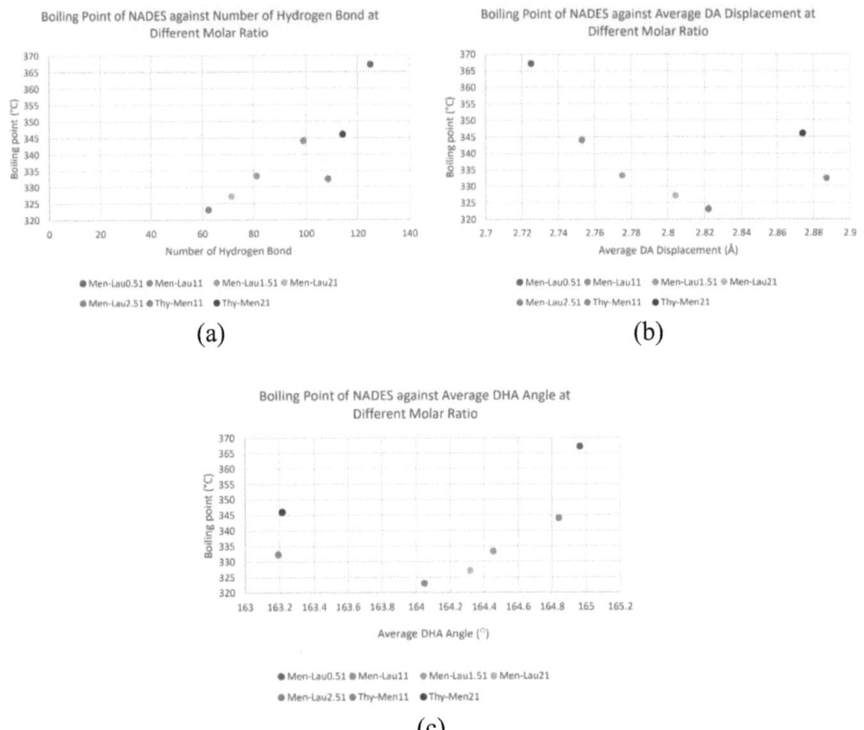

Fig. 6. Boiling Point of 7 different NADES systems against different hydrogen bond metrics, (a) N_{hb}, (b) Average DA displacement, (c) Average DHA angle

4 Conclusion

This study employed molecular dynamics (MD) simulations to investigate the role of hydrogen bonding in governing the physicochemical properties of seven Natural Deep Eutectic Solvent (NADES) systems formulated from thymol, menthol, and lauric acid.

The results systematically demonstrate that hydrogen bonding metrics—specifically the total bond count (N_{hb}), donor-acceptor (D-A) distance, and donor-hydrogen-acceptor (D-H-A) angle—are fundamental determinants of macroscopic solvent behaviour. A strong, extensive hydrogen-bond network, characterised by a high N_{hb}, short D-A distance, and a D-H-A angle approaching 180°, was found to directly enhance solubility, density, and boiling point. Conversely, this robust network facilitated a greater depression of the melting point and, within the studied systems, was associated with lower viscosity, suggesting that a well-connected network may enable more cooperative molecular motion under shear.

The composition and molar ratio of components were critical in modulating this network. In the thymol-menthol systems, compositional asymmetry (Thy-Men 2:1) favoured a higher hydrogen bond count and more favourable properties compared to the equimolar ratio. For the menthol-lauric acid series, the superior hydrogen bond acceptor capacity of lauric acid, with its two acceptor sites, drove the strongest network formation and most significant property effects at the lowest menthol ratio (Men-Lau 0.5:1). As the proportion of menthol increased, the hydrogen bond network weakened, leading to a predictable trend in all properties.

In summary, this work establishes clear, quantitative correlations between molecular-scale hydrogen bonding interactions and the bulk physicochemical properties of Type V NADES. These findings provide a computational framework for the rational design of sustainable solvents, where targeted manipulation of component chemistry and molar ratio can be used to tailor properties like solvation power, volatility, and fluidity for specific industrial applications within green chemistry.

References

1. Prabhune, A., Dey, R.: Green and sustainable solvents of the future: deep eutectic solvents. J. Mol. Liquids **379** (2023)
2. Perna, F.M., Vitale, P., Capriati, V.: Deep eutectic solvents and their applications as green solvents. In: Current Opinion in Green and Sustainable Chemistry, vol. 21, pp. 27–33. Elsevier B.V (2020)
3. Rodríguez, Y.F., Benito, C., Aparicio, S., Trenzado, J.L.: Unveiling the high-pressure behavior of Thymol+Carvone NADES: a combined experimental-computational approach. J. Supercritical Fluids **215** (2025)
4. Bochek, A.M.: Effect of hydrogen bonding on cellulose solubility in aqueous and nonaqueous solvents. Russ. J. Appl. Chem. **76**(11), 1711–1719 (2003)
5. Fan, C., Liu, Y., Sebbah, T., Cao, X.: A theoretical study on terpene-based natural deep eutectic solvent: relationship between viscosity and hydrogen-bonding interactions. Global Chall. **5**(3), 2000103 (2021)
6. Sun, Q., et al.: Unraveling the direct effect of hydrogen bonding on density and thermostability of energetic materials through isomerism. Chem. Eng. J. **444**, 136539 (2022)
7. Chevé-Kools, E., et al.: Natural deep eutectic solvents (NaDES): green solvents for pharmaceutical applications and beyond. Green Chem. **27**(28), 8360–8385 (2025)

A Data-driven Analysis of Backwash Interval on Ultrafiltration Membrane Performance

Mohamad Fikkri Sarif[1], Noorhazirah Sunar[1](✉), Norhaliza Abdul Wahab[1], Anita Ahmad[1], Nurul Adilla Mohd Subha[1], Mashitah Che Razali[1,2], and Alya Afiqah Shakhar[3]

[1] Universiti Teknologi Malaysia (UTM), 81300 Johor Bahru, Johor, Malaysia
noorhazirah@utm.my

[2] Fakulti Teknologi dan Kejuruteraan Elektrik, Universiti Teknikal Malaysia Melaka, Hang Tuah Jaya, 76100 Durian Tunggal, Melaka, Malaysia

[3] GP Batteries Malaysia, 81400 Senai, Johor, Malaysia

Abstract. Membrane-based filtration systems, particularly ultrafiltration (UF), are widely adopted for water treatment because of their ability to remove pathogens from organic matter, and particulates. However, membrane fouling remains a persistent challenge that reduces efficiency, lowers permeating quality, and increases operating costs. This study evaluates a lab-scale UF membrane system under three conditions: continuous filtration without backwash, filtration with a 10-min. Backwash interval, and filtration with a 20-min. Backwash interval. River water was used as feed, and performance was monitored with real-time pressure, flow, and turbidity sensors integrated with LabVIEW. Continuous operation without backwash led to rapid flux decline and rising transmembrane pressure (TMP), with a strong negative Pearson correlation value 0.928 between flux and TMP. Both 10 and 20-min. Intervals improved flux recovery, but the 10-min. Interval provided more stable flux and controlled fouling buildup, while the 20-min. Interval yielded higher short-term recoveries but greater fluctuations. Absolute and relative flux recovery analyses highlight dynamic, non-linear backwash behavior, suggesting that fixed schedules are sub-optimal. Findings emphasize the importance of optimizing backwash frequency for operational stability, with implications for both small and large-scale water treatment systems.

Keywords: Ultrafiltration · Membrane fouling · Backwash interval · Flux recovery

1 Introduction

Ensuring access to clean and safe water is a global priority, with water treatment technologies playing an essential role in meeting municipal and industrial needs [1, 2]. Among available technologies, membrane filtration has gained prominence due to its effectiveness in removing pathogens, natural organic matter, oil emulsions, and fine particulates from raw water resources [3, 4]. Ultrafiltration (UF) is widely adopted because of its low chemical requirements, ability to consistently produce high-quality effluent, and relatively compact system footprint.

N. A. Mohd Subha et al. (Eds.): SMS 2026, CCIS 2956, pp. 126–139, 2026.
https://doi.org/10.1007/978-981-92-0306-2_10

Despite these advantages, UF systems face critical challenges: membrane fouling. Fouling occurs when particles, natural organic matter, and microorganisms accumulate on the membrane surface or within its pores, causing a decline in flux, deterioration of effluent quality, and an increase in transmembrane pressure (TMP). The operational and economic impacts are significant: fouling reduces membrane lifespan, increases energy consumption, and raises cleaning costs [5, 6].

To mitigate fouling, backwashing is a common strategy. In backwashing, permeate water is reversed through the membrane to flush out accumulated particles. Traditionally, fouling was addressed through manual cleaning, requiring operators to remove membrane units from their housing. While effective, such manual approaches were labor-intensive and caused extended downtime. Advances in control systems introduced automated backwashing triggered by preset time schedules or pressure thresholds. Although this automation reduced manual involvement, it did not fully solve the issue: preset schedules cannot dynamically adjust to changing feedwater quality or fouling rates, resulting in either excessive cleaning (and wasted water) or insufficient cleaning (and rapid fouling). The complexity of foulant-membrane interactions makes predicting and mitigating fouling particularly challenging [7–9].

Research has therefore explored optimal backwash intervals and durations across different membrane types, foulant sources, and system scales [10–13]. While some studies recommend intervals as short as 5 min, others suggest much longer gaps depending on the raw water and membrane characteristics. The lack of consensus highlights the highly system-specific nature of backwashing strategies. Moreover, few studies have investigated the cycle-to-cycle dynamics of backwash performance, often assuming uniformity across cycles. In practice, recovery and decline can vary significantly from one cycle to the next.

In membrane filtration, flux represents the volumetric flow rate of permeate per unit membrane area and is one of the key indicators of hydraulic performance. Flux behavior can be described in terms of flux recovery and flux decline. Flux recovery refers to the restoration of flux after it decreases due to fouling, typically following a backwashing process. In contrast, flux decline describes the reduction in flux during filtration as foulants accumulate on the membrane surface. It is commonly evaluated by comparing the flux at the beginning and end of filtration cycle. Flux recovery reflects the effectiveness of backwash process [14, 15], while flux decline indicates the extent of fouling build up during filtration [15, 16].

This study contributes by experimentally comparing two practical backwash intervals-10 and 20 min using a lab-scale UF pilot plant. The objectives are (1) to evaluate the impact of backwash interval on flux recovery and decline, (2) to assess TMP behavior under different cleaning strategies, and (3) to highlight the dynamic behavior of recovery across multiple cycles. The results inform operational strategies for UF systems and provide a foundation for future optimization approaches.

2 Materials and Methods

2.1 Pilot Plant Setup

Figure 1 shows a lab-scale UF pilot plant that was constructed to evaluate membrane performance under controlled conditions. The membrane was a proprietary polyether sulfone (PES) UF module, operated at a filtrate flow rate of 5–10 L/min with a feed flow rate of 10 L/min. The system was designed to operate within a pressure of range 2.1–5 bar. Feedwater was sourced form a river in Skudai, Johor, Malaysia. To protect the UF module from coarse particulates, a cartridge pre-treatment unit was installed upstream to reduce sediment load and delay fouling onset.

A pressurized centrifugal pump (12 L/min capacity) supplied feedwater to the system. Flow and pressure regulation were achieved using Bürket proportional control valves (model 2835), paired with Bürket Type 8605 digital controllers, ensuring stable operation during both filtration and backwash cycles. The materials for the plant is listed out in Table 1.

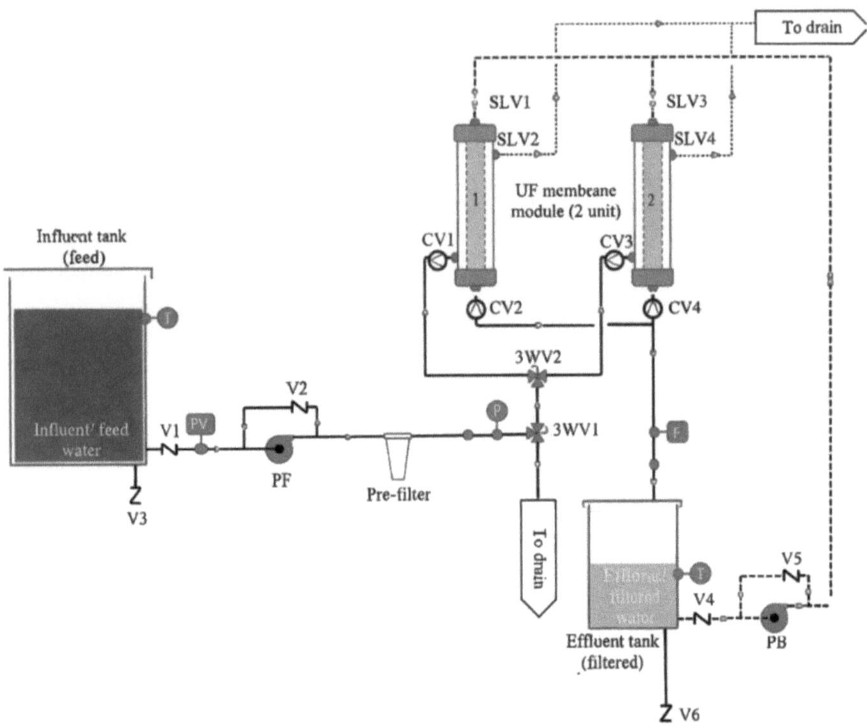

Fig. 1. Schematic UF pilot plant system

Table 1. List of materials for UF pilot plant system

Items	Description
T	Turbidity sensor
V	Valve
PV	Proportional Valve
PF	Pump Feed
P	Pressure Sensor
3WV	3-Way Valve
CV	Check Valve
SLV	Solenoid valve
F	Flow sensor
PB	Pump Backwashing

2.2 Instrumentation and Monitoring

As shown in Fig. 1, systems' performance was monitored in real time using three types of sensors:

- Label P: A WIKA A-10 pressure transmitter measured TMP at the inlet.
- Label F: A Keyence FD-X series clamp-on micro-flow sensor measured flow rate.
- Label T: A DFRobot SEN0189 turbidity sensor tracked feed and permeated turbidity.

Signals from these instruments were processed using LabView, which also managed proportional valve actuation. Data were logged both locally in CSV formatted and, in a MySQL, database hosted using XAMPP, chosen for its ability to function as a local server without requiring internet connectivity. This dual storage approach ensured both immediate analysis and redundancy.

2.3 Experimental Conditions

Three conditions were tested:

- **Condition 1**: Continuous filtration with no backwash (baseline).
- **Condition 2**: Backwash every 10 min, backwash duration = 30 s.
- **Condition 3**: Backwash every 20 min, backwash duration = 30 s.

The total runtime for each experiment ranged from 2,500–4,500 s (40–85 min), ending when TMP approached the manufacturer's upper limit. The 10-min interval and 30-s duration were based on recommendations for similar UF systems [17], and other UF backwash studies [7, 10] where interval 5–5 min are commonly applied to prevent irreversible fouling. The 20-min interval was included as stress test to assess fouling effects under extended accumulation, which reflects less conservative but realistic scenarios observed in some field operation [13]. A backwash duration of 30 s was adopted following [9, 17], where 20–60 s is the standard effective range. This duration balances recovery efficiency with operational downtime and water usage. The rationale underlying the experiment parameters is summarized in Table 2.

Table 2. Experimental parameters and rationale.

Condition Interval	Interval	Duration	Rationale	Reference
Continuous	-	-	Baseline (no cleaning)	-
10-min backwash	10 min	30 s	Effective in similar UF setups	[17]
20-min backwash	20 min	30 s	Longer accumulation period test	[17]

2.4 Performance Metrics

The relationship between flux (J) and transmembrane pressure (TMP) was assessed. Using the Pearson correlation coefficient:

$$r = \frac{\sum_{i=1}^{n}(x_i - \overline{x})(y_i - \overline{y})}{\sqrt{\sum_{i=1}^{n}(x_i - \overline{x})^2}\sqrt{\sum_{i=1}^{n}(y_i - \overline{y})^2}} \tag{1}$$

where x_i and y_i represent the paired data points of flux and TMP, respectively, and x $\overline{x}$ and $\overline{y}$ are their corresponding mean values, and n is the total number of observations.

Absolute FR is calculated using the initial flux of the first filtration cycle as a fixed baseline for all subsequent cycles. This approach reflects the cumulative recovery efficiency relative to the original membrane condition and allows long-term performance degradation to be observed.

$$\%FR_{abs,n} = \frac{J_{w,n}}{J_0} \times 100 \tag{2}$$

Similarly, absolute FD represents the total flux reduction relative to the initial clean reference, indicating overall fouling progression throughout the experiment.

$$\%FD_{abs,n} = \frac{J_0 - J_{f,n}}{J_0} \times 100 \tag{3}$$

In contrast, relative FR compares the recovered flux of each cycle to that of the immediately preceding cycle, capturing short-term recovery behavior and cycle to cycle variations.

$$\%FR_{rel,n} = \frac{J_{w,n}}{J_{i,n}} \times 100 \tag{4}$$

Relative FD quantifies incremental flux loss between consecutive cycles, providing insight into dynamic fouling development over time.

$$\%FD_{rel,n} = \frac{J_{i,n} - J_{f,n}}{J_{i,n}} \times 100 \tag{5}$$

where J_0 is the initial flux of the first filtration cycle (global reference), $J_{w,n}$ is the recovered flux immediately after backwashing in cycle n, $J_{i,n}$ is the initial flux at the beginning of cycle n, and $J_{f,n}$ is the flux measured at the end of filtration in cycle, n. The subscript n denotes the cycle number.

Therefore, absolute values reflect cumulative performance relative to the initial membrane state, while relative values describe transient fluctuations and progressive changes between cycles.

3 Results and Discussion

3.1 Continuous Operation

Figure 2 shows continuous operation without backwash produced rapid flux loss and a steady rise in TMP. Flux declined rapidly and dropped below 10 $Lm^{-2}h^{-1}$ within the first 500 s, while TMP continued rising, indicating progressive membrane resistance growth. While Fig. 3 shows a flux-TMP scatter with a linear fit with strong negative correlation (r = −0.928) between flux and TMP confirmed the link between fouling buildup and hydraulic resistance under no-leaning operation. This trend consistent with earlier studies [17], which noted similar flux-TMP coupling in UF systems without cleaning.

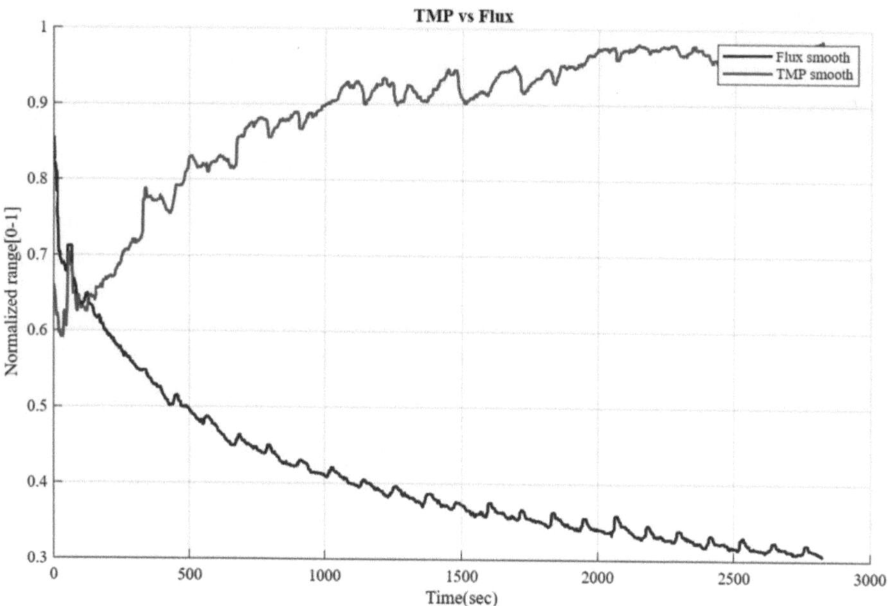

Fig. 2. Condition 1 (no backwash): Filtration process operates until reaching TMP threshold limit.

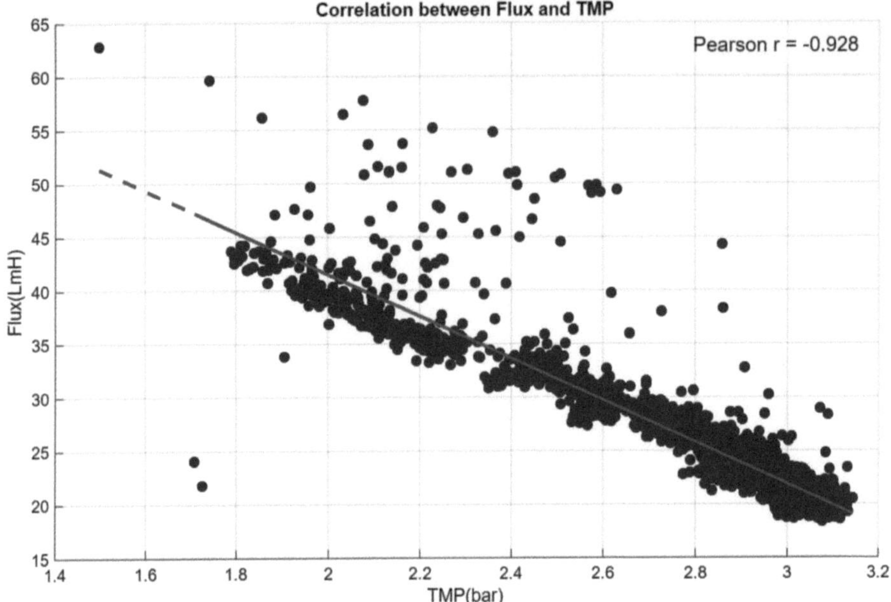

Fig. 3. Condition 1: flux-TMP scatter plotted diagram.

3.2 Backwash at 10-Min Interval

Figure 4 presents the filtration process under condition 2, where backwashing was performed every 10 min of filtration with a duration of 30 s. The pattern exhibited recovery peaks at 600 s intervals, with a relatively stable long-term flux between 12–16 $Lm^{-2}h^{-1}$. A more detailed analysis is provided in Fig. 5. Absolute flux recovery (FR) declined from 94.01% in the first cycle (the highest observed FR) to 74.15% in the fifth cycle, with a partial improvement in the fourth cycle (87.58%) compared to the third cycle (79.71%). In contrast, the fouling degree (FD) increased steadily from 16.96% to 37.05% across successive backwash cycles, indicating progressive fouling. Figure 6 shows that relative FR exceeds 100% in cycle 4 (109.88%), indicating that recovered flux in this cycle exceeded that of the preceding cycle. This does not imply that membrane performance surpassed its original clean condition, rather, it reflects temporary redistribution or partial detachment of foulant layers, resulting in short term enhancement relative to the previous cycle. Therefore, its highlighting non-linear cycle behavior. These results are consistent with literature suggesting that shorter backwashing intervals are effective in managing reversible fouling while mitigating irreversible buildup [10].

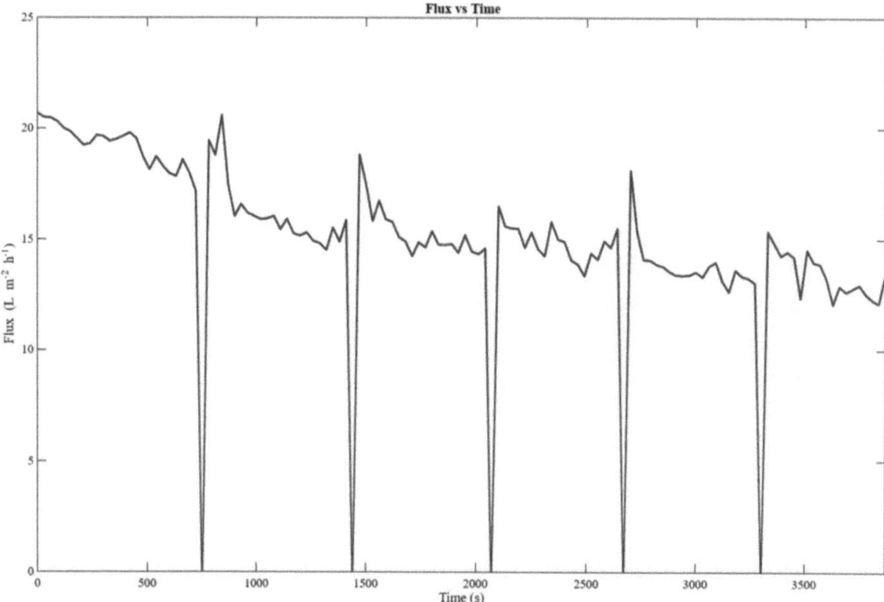

Fig. 4. Condition 2: Flux graph over time with backwash performed every 10 min and lasted for 30 s.

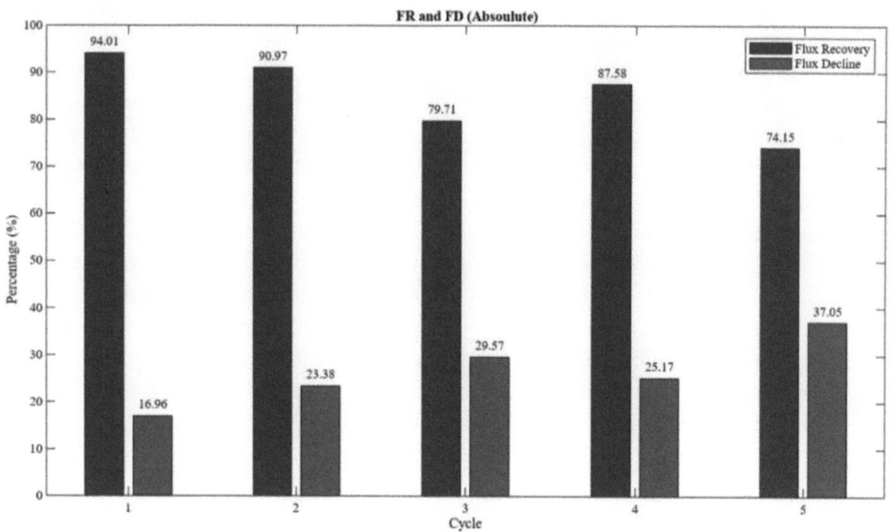

Fig. 5. Condition 2: Absolute FR and FD trends over cycles.

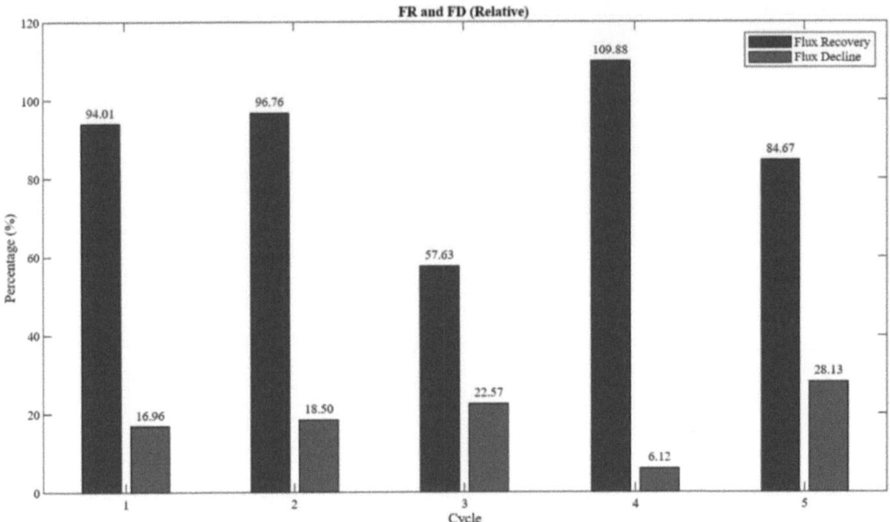

Fig. 6. Condition 2: Relative FR and FD dynamic behavior within each cycle.

3.3 Backwash at 20-Min Interval

Figure 7 presents the filtration process under condition 3, where backwashing was performed every 20 min. of filtration with a duration of 30 s. The recovery peaks were less frequent, and the average flux declined from ~20 to 10–12 $Lm^{-2}h^{-1}$, with larger fluctuations due to longer foulant accumulation. A more detailed analysis is provided in Fig. 8, which shows that the absolute FR in the initial cycle was 71.97% and improved across the cycles, reaching 85.61%. In contrast, absolute FD peaked in the first cycle at 40.19% and showed consistent behavior, dropping to 32.56% in the second cycle but rising again to 38.12% in the third. This indicates an unpredictable trend due to longer intervals, which heightens the risk of persistent fouling buildup. Figure 9 shows a peak relative FR of 118.12% in cycle 2, which a similar phenomenon of relative FR exceeding 100% was observed, consistent with the dynamic behavior discussed previously, followed by a decline. This suggests that longer accumulation allowed more fouling to develop, producing unstable performance compared to the 10-min interval. Literature supports that extended intervals increase the risk of irreversible fouling despite occasional high recovery [13].

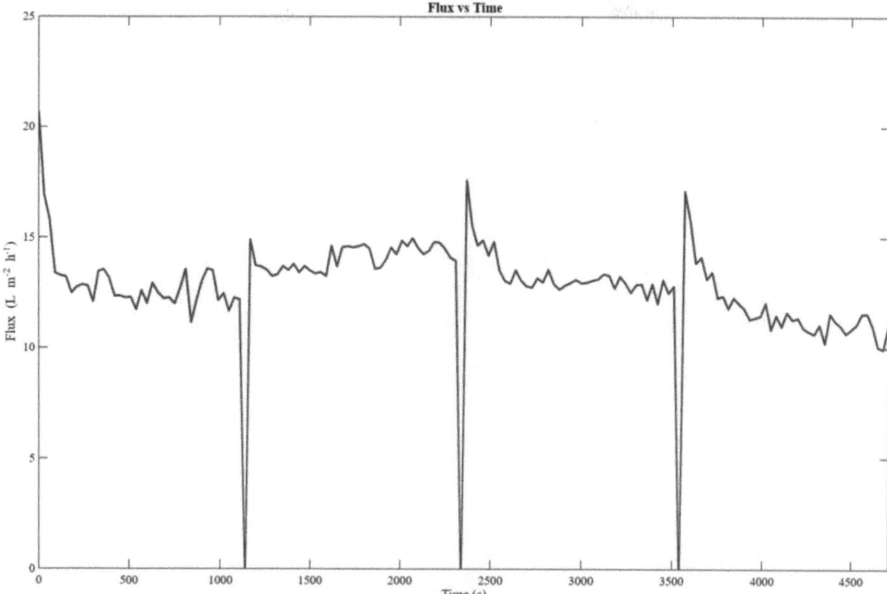

Fig. 7. Condition 3: Flux graph over time with backwash performed every 20 min and lasted for 30 s.

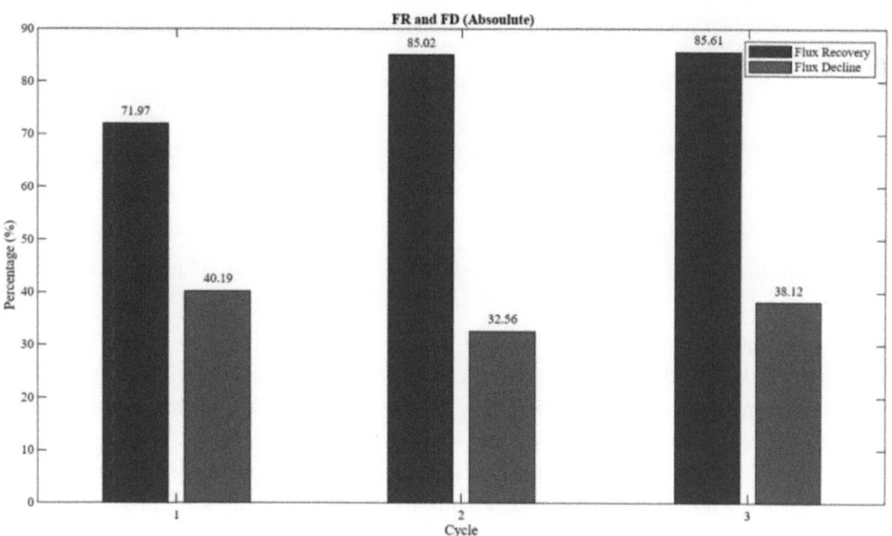

Fig. 8. Condition 3: Absolute FR and FD trends over cycles.

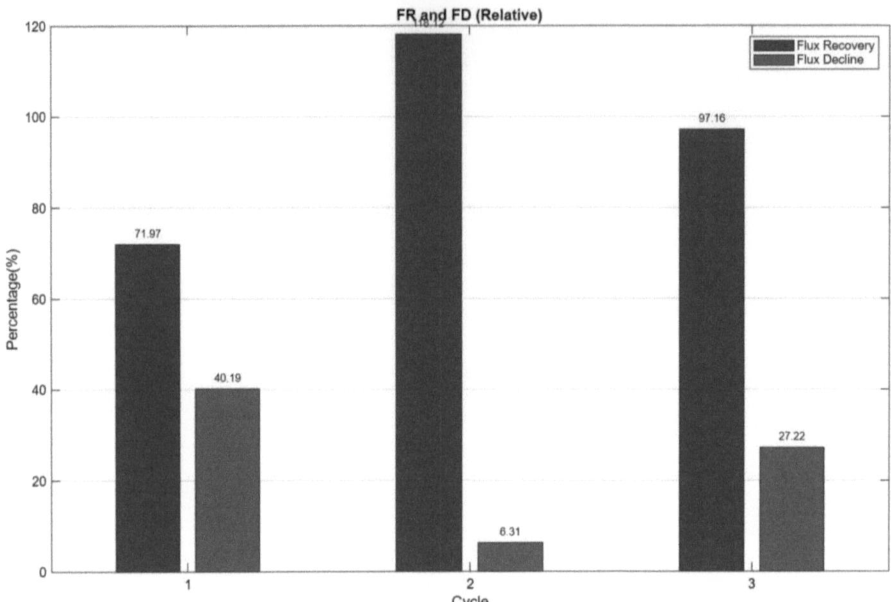

Fig. 9. Condition 3: Relative FR and FD dynamic behavior within each cycle.

3.4 Performance Summary

Table 3. Performance summary across conditions.

Condition Avg, flux ($Lm^{-2}h^{-1}$)	FR (%)	FD (%)	TMP trend
Continuous <10 (rapid decline)	-	>50	Steady rise
10-min BW 12–16 (stable)	74–94	17–37	Controlled rise
20-min BW 20 to 10–12 (decline)	72–85	33–41	Fluctuating rise

The comparison in Table 3 shows that 10-min intervals offer more stable long-term performance, while 20-min intervals risk greater fouling accumulation despite some high short-term recoveries. For operators, this suggests that 10-min backwash interval provides better membrane protection and long-term cost efficiency.

Figures 10 and 11 present comparison of both operating in terms of average flux, FR and FD. As shown in Fig. 10, the average flux for 10-min interval is slightly higher than average flux compared to the 20-min interval. Although the difference in average flux is moderate, the cumulative performance indicators in Fig. 11 provide clearer distinctions. The 10-min interval achieved higher average FR and lower average FD, indicating more effective fouling control and improved operational stability. In contrast, the 20-min interval displays greater flux decline, suggesting increased foulant accumulation between cleaning cycles. These results confirm that 10-min backwash intervals provide better long-term membrane protection despite similar overall runtime durations.

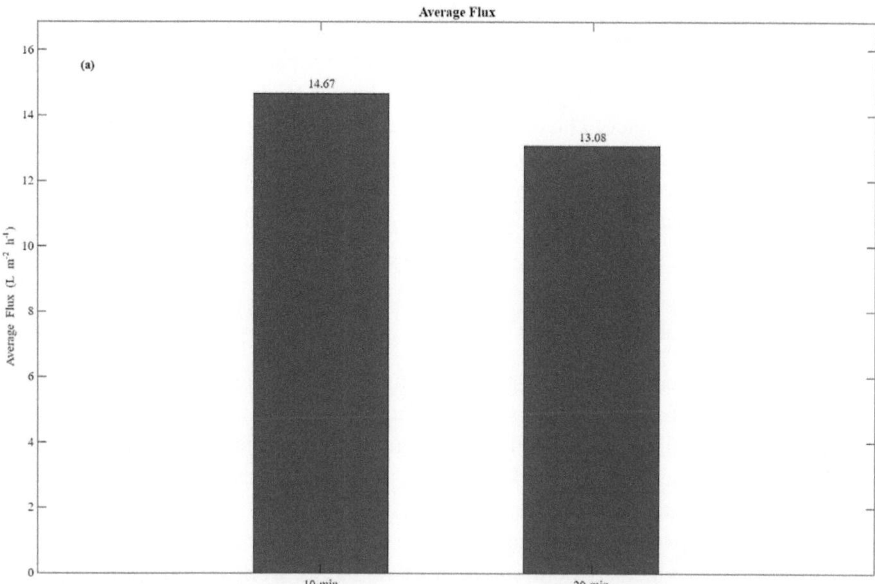

Fig. 10. Comparison of average flux under 10-min and 2-min backwash interval.

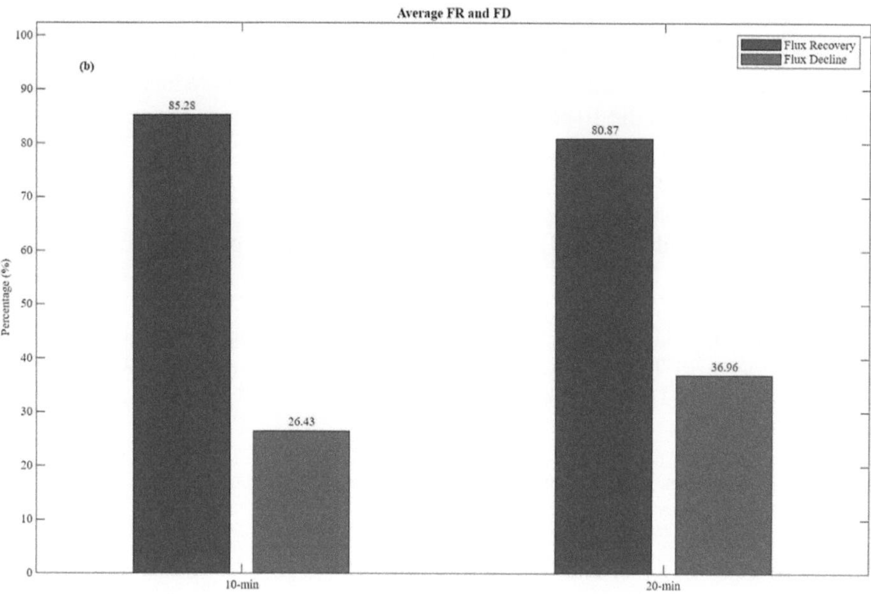

Fig. 11. Comparison of average flux recovery (FR) and flux decline (FD) for both operating conditions.

Based on Fig. 4 and Fig. 7, the graph suggests that 20-min interval experiment exhibit longer total runtime before reaching the TMP limit compared to the 10-min interval. However, the FR and FD analysis indicate that longer runtime does not necessarily imply superior performance. The extended runtime observed under the 20-min interval was associated with larger flux fluctuations and greater fouling accumulation. Therefore, when evaluating the optimal backwash interval, the total runtime alone should not be considered, and it must be assessed together with operational stability and fouling control performance.

3.5 Implications and Limitations

This study emphasizes that backwash scheduling cannot be universally prescribed but must consider system-specific factors such as foulant type and water quality. For small-scale plants, the findings suggest that a 10-min backwash interval balances operational cost with fouling control. For larger-scale plants, optimization could benefit from advanced control strategies that adjust intervals dynamically.

Limitation only two backwash intervals were tested. The study did not include chemical cleaning cycles which are common in practice. Future work should apply design of experiments (DOE) methods [18, 19] to test a wider range of intervals (5–30 min) and durations, mapping response surfaces for flux recovery and TMP stability.

4 Conclusion

Continuous UF without backwash caused rapid flux decline and TMP growth, demonstrating severe fouling. Both 10 and 20-min backwash intervals reduced fouling and improved recovery, but the 10-min intervals maintained stable long-term performance, while the 20-min interval showed unstable fluctuations. Flux recovery and decline analyses revealed dynamic, cycle-dependent behaviors, highlighting the need for adaptive backwash strategies rather than rigid schedules. The findings provided practical guidance for plant operators and motivated future optimization studies to generalize results across systems.

Acknowledgments. This research was funded by Ministry of Higher Education Malaysia (FRGS/1/2024/TH07/UTM/02/22-R. J130000.7823.5F709) and supported by University Teknologi Malaysia (UTMFR Q.J130000.3823.23H91).

Disclosure of Interests. The author declares no competing interests relevant to this article.

References

1. Moran, S.: Clean water characterization and treatment objectives. An applied guide to water and effluent treatment plant design (2018). https://doi.org/10.1016/b978-0-12-811309-7.00006-0
2. Zhao, W., Chen, I.W., Huang, F.: Toward large-scale water treatment using nanomaterials. Nano Today **27**, 11–27 (2019)

3. Ochoa, N.A., Masuelli, M., Marchese, J.: Development of charged ion exchange resin-polymer ultrafiltration membranes to reduce organic fouling. J. Membrane Sci. **278**(1–2), 457–463 (2006)
4. Thamaraiselvan, C., Noel, M.: Membrane processes for dye wastewater treatment: recent progress in fouling control. Crit. Rev. Environ. Sci. Technol. **45**(10), 1007–1040 (2015)
5. Ahmed, M., et al.: "Recent developments in hazardous pollutants removal from wastewater and water reuse within a circular economy. NPJ Clean Water **5**(1), 12 (2022)
6. Latif, E.F., Khaled, E.: Overcoming fouling challenges in ultrafiltration for drinking water treatment. Eng. Res. Express **6**(4), 042101 (2024)
7. Hillis, P., et al.: Effects of backwash conditions on out-to-in membrane microfiltration. Desalination **118**(1–3), 197–204 (1998)
8. Charcosset, C.: Classical and recent developments of membrane processes for desalination and natural water treatment. Membranes **12**(3), 267 (2022)
9. Smith, P.J., et al.: Design of a generic control system for optimising back flush durations in a submerged membrane hybrid reactor. J. Membrane Sci. **255**(1–2), 99–106 (2005)
10. Yang, L., et al.: Identification of performance and cost in a new backwash method to clean the UF membrane: backwashing with low dosage of NaClO. Environ. Sci. Pollut. Res. **30** (58), 121983–121992 (2023)
11. Ellouze, F., et al.: Optimal control of backwash scheduling for pumping energy saving: Application to the treatment of urban wastewater. J. Water Process Eng. **56**, 104378 (2023)
12. Alamoudi, T., et al.: Optimization of osmotic backwashing cleaning protocol for produced water fouled forward osmosis membranes. J. Membrane Sci. **663**, 121013 (2022)
13. Lee, J., et al.: Comparison of different cleaning strategies on fouling mitigation in hollow fiber nanofiltration membranes for river water treatment. J. Cleaner Prod. **380**, 134764 (2022)
14. Abdelrasoul, A., Doan, H., Lohi, A.: Fouling in membrane filtration and remediation methods. Mass transfer-advances in sustainable energy and environment oriented numerical modeling **195** (2013)
15. Xu, B., et al.: A review of temperature effects on membrane filtration. Membranes **14**(1), 5 (2023)
16. AlSawaftah, N., et al.: A comprehensive review on membrane fouling: mathematical modelling, prediction, diagnosis, and mitigation. Water **13**(9), 1327 (2021)
17. Cho, H., et al.: Optimal water backwashing condition in combined water treatment of alumina microfiltration and PP beads. Membranes **12**(1), 92 (2022)
18. Montgomery, D.C.: Design and Analysis of Experiments. Wiley (2017)
19. Wu, C.J., Hamada, M.S.: Experiments: Planning, Analysis, and Optimization. Wiley (2011)

Spare-Buffer Cell Rewiring for Efficient Timing Engineering Change Order

Nik Muhammad Afif Nik Rumzi[1], Mohd Shahrizal Rusli[2(✉)], Ab Al-Hadi Ab Rahman[1], Muhammad Nadzir Marsono[1], Haslina Md. Sarkan[2], and Fakhrul Zaman Rokhani[3]

[1] Department of Electronic and Computer Engineering, Faculty of Electrical Engineering, Universiti Teknologi Malaysia, Johor Bahru, Malaysia

[2] Department of Intelligence Informatics, Faculty of Artificial Intelligence, Universiti Teknologi Malaysia, Johor Bahru, Malaysia

shahrizalr@utm.my

[3] Department of Computer and Communication Systems Engineering, Faculty of Engineering, Universiti Putra Malaysia, Serdang, Malaysia

Abstract. Timing closure during the Engineering Change Order (ECO) stage, remains a major challenge in advanced integrated circuit design as circuit complexity increases. Although pre-placed spare cells support post-layout fixes, many spare-cell ECO methods emphasize allocation policies rather than maximizing timing recovery using accurate net-delay impact. This paper presents an automated timing-driven spare-buffer rewiring ECO framework that tightly couples the static timing analysis stage with the physical design stage to perform localized, delay-optimized timing repair. The framework identifies the worst-slack nets along the data path, constructs path-specific bounding boxes to constrain candidate spare buffers, estimates source-to-spare and spare-to-sink delays using Manhattan-distance-based wire-delay models and applies a spare-locking mechanism to ensure conflict-free reuse across iterations. The approach is evaluated in 32 nm technology on five benchmark instances spanning a 4 by 4 DCT core, an FP64 unit, and a pipelined MAC design. Overall, spare-buffer rewiring reduces ECO-critical path violations substantially more than conventional gate sizing, achieving an average ECO-path resolution of about 73% across all benchmark instances, compared with about 59% for gate sizing. Consistent with these results, the proposed method delivers higher TNS gain defined as pre-ECO TNS minus post ECO TNS and reduces runtime by up to about 26% versus gate sizing, while maintaining low overhead with total cell area increase below 0.15% and power overhead below 4%.

Keywords: Timing ECO · Static Timing Analysis · Spare-Cell Rewiring · Bounding Box · Timing Closure · Physical Design · Total Negative Slack (TNS)

1 Introduction

Timing closure during the ECO stage is a significant challenge in modern IC design, particularly as chip complexity and routing density increase. After routing, timing violations typically occur on the most critical paths, where placement flexibility and tight timing

N. A. Mohd Subha et al. (Eds.): SMS 2026, CCIS 2956, pp. 140–157, 2026.
https://doi.org/10.1007/978-981-92-0306-2_11

margins limit possible fixes. Engineers must resolve these violations without disturbing optimized regions, often involving repeated cycles of gate resizing, buffer insertion, routing adjustments, and multi-corner STA checks, which lengthen turnaround time. To aid in this process, current design flows use pre-placed spare cells such as buffers, inverters, and logic gates that can be activated with minimal timing violations. While spare cells help achieve faster and more reliable timing closure, their overuse can introduce unnecessary area, power overhead, and routing congestion, particularly in dense regions. Furthermore, activating spare cells without proper timing awareness can degrade performance by adding delay, increasing load, or affecting signal quality. Spare cells must be planned early in the design flow, creating a balance between providing enough resources and avoiding wasted silicon. Too few spares may lead to disruptive changes or multiple closure iterations, while too many can consume valuable area and complicate the physical design [1]. As technology nodes scale and PPA requirements tighten, inefficient spare-cell usage can negatively affect final design quality, making careful planning essential to support timing closure effectively without compromising power, performance, or area.

In physical design flows, the deployment of spare cells has become a widely adopted practice to support efficient post-layout ECO. Prior work has shown that strategically distributing spare cells throughout the layout significantly enhances the ability to implement late-stage timing fixes and functional modifications while minimizing disturbance to the routed design [2]. To improve spare-resource utilization, [3] proposed a negotiation-based framework in which multiple timing paths compete for available spare cells, preventing premature assignment and enabling logic restructuring when beneficial. A proximity-driven solution was introduced in [1], where candidate spare cells are restricted to an essential bounding region and dynamic cost updating is used to account for future availability; activated cells are recycled as new spares to maintain balance. To address the interaction between functional and timing ECO, [4] applied a Hungarian-matching strategy that prioritizes spare selection during functional ECO while preserving sufficient capacity for subsequent timing repairs. When conventional buffering or sizing is insufficient, TRECO [5] leverages spare-cell-guided technology remapping to reconstruct critical subcircuits, using bounding-box-based proximity to reduce routing disturbance. A stable-matching formulation was introduced in [6] to prevent early functional ECO steps from consuming nearby spares required for timing optimization, ensuring conflict-aware spare allocation. Meanwhile, [7] proposed a unified ECO flow that transforms timing fixes into equivalent functional operations and assigns spare cells using a modified Hungarian algorithm while penalizing long-wire detours. Building on this direction, [8] propose a metal-only ECO timing optimization method that considers both data paths and clock paths, instead of modifying data-path gates only [8]. builds a candidate set from elements on the violated data paths and the associated clock paths, applies a cost-driven selection to choose the best ECO action, and iterates until no further timing improvement is achieved. The method also supports using spare cells for buffer insertion or gate sizing during the ECO process. Collectively, these studies highlight that effective spare-cell ECO relies on proximity-aware selection, resource-competition management, and coordinated handling of functional and timing modifications to achieve robust late-stage timing closure in advanced technology nodes.

In addition to spare-cell usage, conventional timing ECO often employs gate sizing to improve critical-path timing by upsizing selected gates and, when combined with buffering, provides low-disruption late-stage fixes [9]. Recent approaches enhance this process with IR-drop-aware frameworks and machine learning: RL-LR-Sizer combines reinforcement learning, Lagrangian relaxation, and a relational graph convolutional network to select gates and discrete sizes that minimize layout disturbance while improving delay–power tradeoffs [10]. Similarly, RL-based methods replace rule-based decisions with learned sizing using local neighborhood information, iteratively updating timing to reduce total negative slack efficiently without full-chip evaluations [11].

While prior studies have advanced spare-cell–assisted ECO by improving spare availability, proximity-based selection, and conflict-aware allocation mechanisms [3–7], they primarily focus on managing how spare cells are shared, reassigned, or matched across different timing and functional tasks rather than optimizing the timing benefit of each spare-cell activation. These methods emphasize resource fairness, logic substitution, and global optimization but do not explicitly perform net-level delay evaluation to select spare buffers based on measurable timing improvement. For method such as gate sizing [9] [10, 11], because stronger cells typically introduce larger input capacitance and higher switching and leakage components, it inherently trades improved slack for increased area and power, requiring careful, path-aware application to avoid excessive overhead. Furthermore, [8] present a metal-only post-silicon ECO approach, where timing is repaired primarily through routing-level modifications. As a result, while the method is effective for post-silicon correction, it is outside the scope of this work, which targets post-layout, pre-tapeout timing ECO using spare-cell-based fixes within the standard implementation flow.

In late-stage ECO, timing violations are often dominated by interconnect delays due to tight placement and routing constraints, which limits the possibility of large physical modifications. In this situation, conventional gate sizing provides limited improvement because it mainly reduces intrinsic cell delay while increasing input capacitance, which may slow down upstream drivers and reduce the overall timing benefit along the critical path. Similarly, traditional buffer insertion requires manual identification of insertion points followed by repeated static timing analysis (STA) verification [12]. As the number of violating paths increases, this process becomes time-consuming and difficult to scale, especially when high-accuracy path-based analysis (PBA) is required in sign-off tools such as Synopsys PrimeTime, leading to excessive ECO turnaround time [13, 14]. To address these issues, this work proposes an automated spare-buffer rewiring framework that reduces the number of STA iterations during ECO. The method uses slack severity and source–sink pin coordinates to identify delay-dominating interconnect segments and restricts the search area using a path-specific bounding box. Candidate spare buffers located within this region are evaluated using Manhattan-distance-based delay estimation while considering timing constraints such as driver strength, fanout, and load capacitance. The buffer that provides the best delay reduction and slack improvement is selected, and a priority-based allocation mechanism ensures that the most critical paths are fixed first while avoiding resource conflicts. By limiting ECO modifications to localized rewiring of pre-placed spare buffers, the proposed approach achieves faster convergence and more efficient timing recovery compared with conventional gate sizing and manual buffer insertion techniques.

The remainder of this paper is organized as follows. Section 2 describes the methodology of the proposed solution. Section 3 present the proposed approaches of the spare-buffer rewiring in detail. Section 4 highlighted the experimental result and evaluation of proposed method. Finally, Sect. 5 highlighted the conclusions.

2 Methodology of Proposed Solution

In this section, we define key terms and the problem formulation, incorporating the bounding box concept that focuses on the critical interconnects between the source and sink pins of the ECO paths. A timing path is defined as: (1) a path from a primary input to a primary output, (2) a path from a primary input to a flip-flop input, (3) a path from a flip-flop output to a primary output, or (4) a path between a flip-flop output and a flip-flop input.

Definition 1: An ECO path is a timing path that violates the specified timing constraints. The bounding box for the ECO path is defined by the source and sink pins, with the longest and most timing-critical interconnects highlighted within this region.

In our approach, we aim to fix all ECO paths using spare-cell rewiring. The source of an ECO path is the flip-flop or primary input at the beginning, and the sink is the flip-flop or primary output at the end. As illustrated in Fig. 1, the bounding box is drawn between the source and sink pins, encompassing the critical interconnects along the ECO path. The spare-cell candidates are placed within this bounding box, allowing the method to address timing violations efficiently by focusing only on the most critical interconnects.

3 Proposed Approaches

The proposed ECO flow integrates physical design and static timing analysis (STA) to perform targeted post-layout timing repair using spare-buffer rewiring. After completing placement, CTS, and routing in the IC Compiler environment, STA in PrimeTime identifies violated setup/hold paths and the worst-slack nets along each critical path. The method then focuses on the most critical net and constructs a bounding-box between the source and sink pins to limit the spare-cell search to physically relevant candidates. For each spare buffer inside this region, the delay from source-to-spare and spare-to-sink is calculated, and the buffer that provides the greatest timing improvement is selected. ECO commands are generated and applied in ICC, requiring routing only since the spare buffers are already placed and legalized. STA is run again to verify closure, and the loop repeats if violations remain. This closed-loop, STA-guided, bounding-box-based approach enables localized, delay-driven timing ECO with minimal design disturbance and faster convergence compared to conventional ECO methods (Fig. 2).

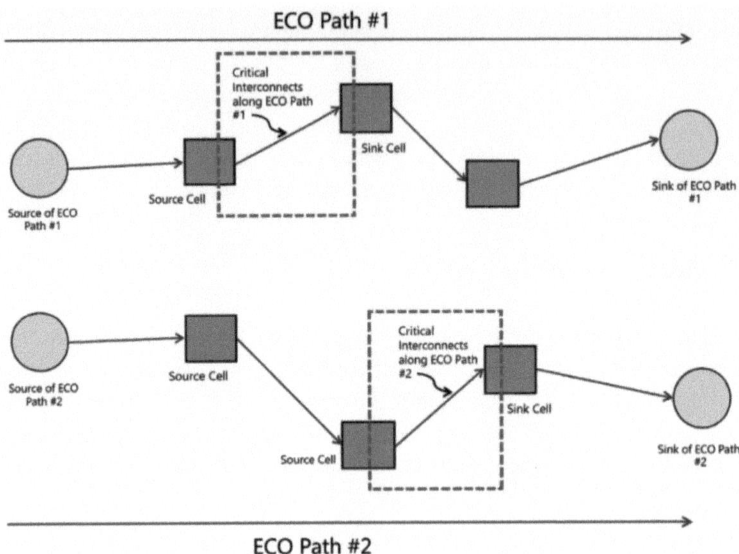

Fig. 1 Path-specific bounding box approach for identifying critical interconnects in timing ECO

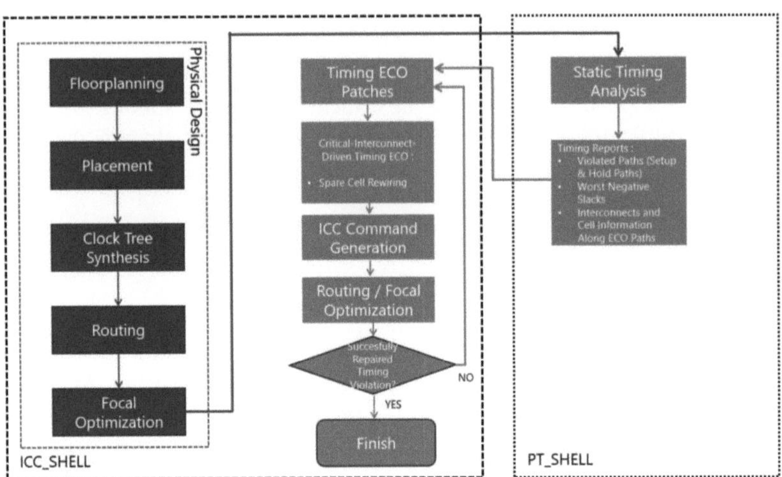

Fig. 2 Overall flow of the proposed spare-buffer rewiring ECO framework

3.1 ECO Buffer-Rewiring Features

The proposed ECO method selects spare buffers using timing and layout features from the design as shown in Table 1. Path slack, Manhattan distance, and the coordinates of the source and sink pins are used to form a bounding-box search region along each critical path. Features such as cell drive strength, fan-out, and input pin capacitance C_{pin} are checked to ensure the added buffer does not overload the driver or worsen signal quality. Wire-load parameters R_{wire} and C_{wire} are also used to estimate delay between the source, candidate spare buffer, and sink. The method then chooses the spare buffer that gives the lowest overall delay while keeping routing changes small by selecting candidates inside or near the bounding box

Table 1. Spare-Buffer Cell Rewiring Decision Features.

Features	Roles in ECO Decision	Impact
Slack	Indicates timing violation severity on the critical path	Larger negative slack provides highest urgency for buffer insertion
Manhattan Distance Between Pins	Measures physical distance between driver and load pins	Longer distances provide higher priority buffer insertion to cut delay
Spare Cell Location (Coordinates)	Physical position of spare buffer cells near the violating path	Closer spares prioritized to minimize routing disturbance
Source/Sink Cell Type	Type & drive strength of original cells	Ensures compatibility and improvement over original driver
Number of Fanouts, N	Number of downstream loads driven by the source	Higher fan-out nets prioritized due to loading impact
Input Pin Capacitance, C_{pin}	Input capacitance of the sink node	Guides buffer strength selection for optimal delay improvement

3.2 Critical-Interconnect-Driven ECO Bounding Box and Spare-Cell Candidate Region

Fig. 3 shows the bounding-box-based spare-cell selection strategy adopted for the proposed timing ECO method. For each violated timing path, the method first identifies the critical interconnect segment contributing to the worst slack, which becomes the focal point for buffer insertion. A bounding box is then constructed using the physical coordinates of the source and sink pins of this critical net, creating a localized search window along the true delay-dominating path. Spare cells located inside this bounding region are considered viable candidates for rewiring, as their proximity helps minimize additional routing parasitics and ensures impactful timing improvement. Spare cells outside the bounding box are excluded since longer detours may worsen delay or increase routing congestion. By targeting the critical interconnect per ECO path and restricting candidate spares to those physically aligned along the source-to-sink corridor, the proposed approach ensures precise, timing-driven, and physically-aware spare-buffer insertion.

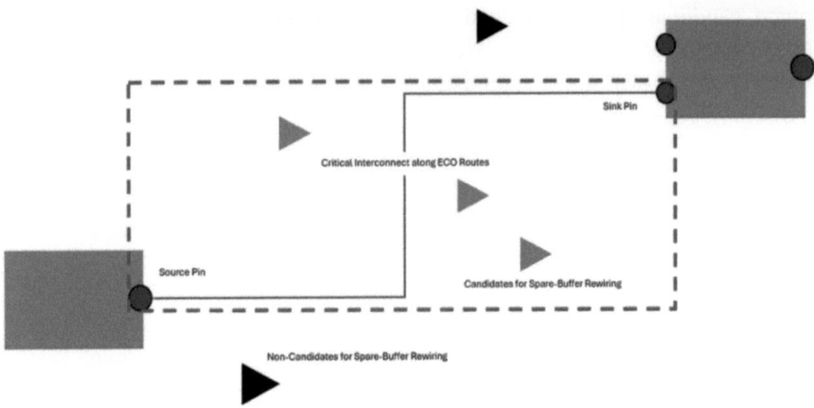

Fig. 3 Bounding box implementation

3.3 Net Delay Calculation Method for Spare Buffer Selection

In the proposed Bounding-Box-Based spare cell rewiring method, accurate net delay estimation is essential for identifying the most effective spare buffer candidate along each critical timing path. To quantify the timing benefit, the interconnect delay from the source cell to the spare cell and from the spare cell to the sink cell is computed as shown in Fig. 4. A routed net can be approximated by a distributed wire resistance R_{wire}, wire capacitance C_{wire}, and a lumped receiver load C_{load}. The net delay of a single segment is modelled as

$$Net\ Delays,\ t_{net} \approx R_{\text{wire}}\left(\frac{C_{\text{wire}}}{2} + C_{\text{load}}\right) \quad (1)$$

Where, in this work, the far-end load is taken as the receiving cell input capacitance, hence

$$C_{pin} = C_{\text{load}} \quad (2)$$

For a multi-fanout net with fanout N, a balanced RC-tree approximation divides the wire parasitics evenly across branches

$$R_{br} = \frac{R_{wire}}{N},\ C_{br} = \frac{C_{wire}}{N} \quad (3)$$

Substituting Eq. 2 and Eq. 3 into Eq. 1 gives the balanced-tree branch net delay that adopts a balanced-tree interconnect approximation, where the wire resistance and capacitance are evenly distributed across N fan-out branches. The net delay for each segment is expressed as:

$$Net\ Delays,\ t_{net} \approx \left(\frac{R_{wire}}{N}\right)\left(\frac{C_{wire}}{2N} + C_{pin}\right) \quad (4)$$

For spare-buffer selection, the total interconnect contribution for a candidate spare cell is computed by summing the two net delays

$$t_{net,total} = t_{source\ to\ spare} + t_{spare\ to\ sink} \tag{5}$$

For reference, a more pessimistic worst-case RC-tree net-delay form can be written as

$$Net\ Delays,\ t_{net,wc} \approx \left(\frac{R_{wire}}{N}\right)\left(\frac{C_{wire}}{2N} + C_{pin}\right) \tag{6}$$

[15]where R_{wire} represents the interconnect resistance, C_{wire} denotes the wire capacitance, C_{pin} is the input capacitance of the receiving cell, and N corresponds to the fan-out count of the source node. This model captures both wiring parasitics and load contribution, ensuring that spare cell selection is driven by true delay impact rather than geometric proximity alone. By comparing the total delay of the "source-to-spare" and "spare-to-sink" connections across all spare candidates within the bounding region, the algorithm selects the spare cell that minimizes cumulative timing delay, thus maximizing improvement in slack while maintaining physical feasibility.

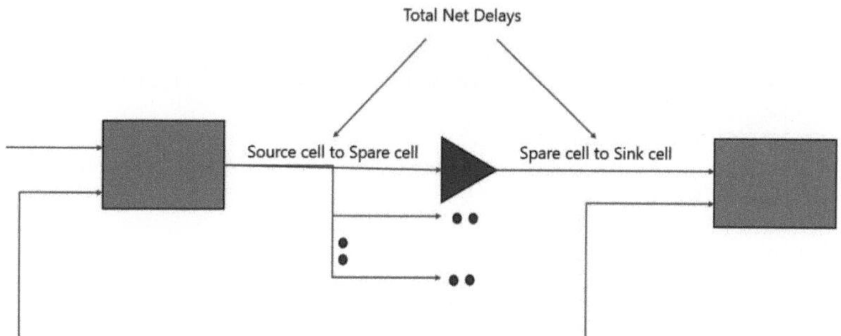

Fig. 4 Net delay calculation for source-to-spare and spare-to-sink connections

3.4 Spare-Buffer Cell Locking Mechanism

The spare-cell locking mechanism can be formally represented using a bipartite matching model, where one partition consists of timing paths and the other consists of available spare cells. Let $P = \{p_1, p_2, \ldots, p_n\}$ denote the ordered set of timing paths, sorted by slack severity so that p_1 is the most critical path, and $S = \{s_1, s_2, \ldots, s_m\}$ denote the set of spare cells. An edge (p_i, s_j) exists if spare cell s_j lies within the feasible physical bounding region of path p_i and satisfies electrical and timing criteria. Unlike classical bipartite matching frameworks such as stable-matching or Hungarian assignment, which globally optimize assignments across all paths, the proposed method performs priority-based greedy matching, assigning the best available spare s_j to the most critical path p_i based on maximum timing improvement. Once assigned, the spare cell is permanently removed from the candidate pool, enforcing a one-to-one constraint where no spare can serve multiple paths. Formally, if $M \subseteq P \times S$ is the matching set, then $(p_i, s_j) \in M \Rightarrow s_j \notin (S \setminus \{s_j\})$ for all subsequent assignments. This locking strategy ensures conflict-free spare usage, prevents overlapping resource allocation, and guarantees ECO feasibility in physical implementation tools by prioritizing path urgency and eliminating spare contention. As a result, the approach combines the structure of bipartite matching with practical timing-driven constraints to enable reliable post-layout ECO execution (Fig. 5).

```
SpareCellLocking(P, S)
1.  Initialize all paths pi ∈ P as unassigned
2.  Initialize all spare cells sj ∈ S as free
3.  for each path pi in order of decreasing criticality do
4.      Generate candidate spare list Ci ⊆ S within bounding box of pi
5.      Evaluate timing improvement Δt(pi, sj) for each sj ∈ Ci
6.      Select sj* = argmax Δt(pi, sj) for each sj ∈ Ci
7.      if sj* is free then
8.          Assign (pi, sj*)
9.          Lock sj* (mark as unavailable for all other paths)
10.     else
11.         // spare sj* already allocated to a more critical path
12.         Skip sj* and choose next best sj from Ci
13.         Repeat steps 6-12 until a free sj is found or Ci is empty
14.     end if
15.     if Ci is empty then
16.         Mark pi as unresolved due to lack of spare resources
17.     end if
18. end for
19. Return final mapping of P → S with locked spares
```

Fig. 5. Pseudocode for spare cell locking algorithm with priority-based assignment

3.5 Resource-Aware Spare-Buffer ECO Strategy

The proposed ECO strategy reduces spare-buffer consumption by focusing on the most timing-critical interconnect segment, rather than inserting buffers for each individual violation. In practical post-layout scenarios, several timing-critical paths often traverse a common long net that dominates overall path delay. As shown in Fig. 6, critical Path #1, #2, and #3 all pass through the same delay-intensive interconnect segment, making it the primary contributor to negative slack. Instead of deploying a separate spare-buffer for each path, the method detects this shared bottleneck and places a single buffer at the most effective position along the net. This single insertion strengthens the signal drive and reduces interconnect delay for all impacted paths, thereby removing the need for additional buffers. Consequently, timing is improved across multiple paths with one buffer instance, enabling efficient utilization of spare resources, minimizing area overhead, and preventing excessive buffer placement or routing congestion. This outcome confirms that the approach achieves timing closure while maintaining a compact ECO footprint and meeting modern physical-design efficiency requirements.

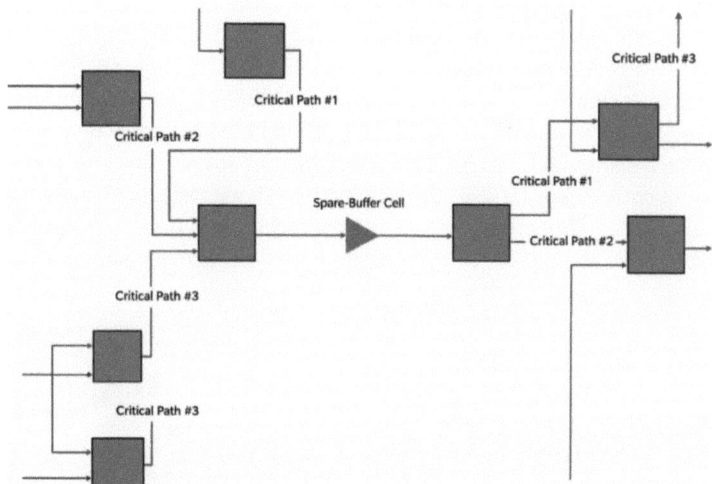

Fig. 6 Multiple critical paths required the same spare-buffer cell

3.6 Tcl IC Compiler Command Generation

Following the adjustment phase, each recommended ECO action is automatically translated into corresponding IC Compiler Tcl commands to be executed in the physical design stage. The generated commands are structured to integrate seamlessly into the existing netlist and routing environment, enabling direct implementation within the sign-off flow. In this work, the ECO strategy focuses on spare-buffer cell rewiring, where a pre-placed spare buffer is inserted along a critical timing path to enhance signal propagation and improve transition characteristics. By connecting the spare buffer between the source and sink pins of the original net, the method effectively partitions a long, high-delay interconnect into two shorter segments, thereby reducing load, improving slew, and recovering timing slack (Fig. 7).

```
disconnect_net n17229 [get_pins U19338/Y]
disconnect_net n17229 [get_pins U19339/A4]
disconnect_net VSS [get_pins spare_0_107/A ]
connect_pin -from [get_pins U19338/Y] -to [get_pins spare_0_107/A ]
connect_pin -from [get_pins spare_0_107/Y] -to [get_pins U19339/A4]
```

Fig. 7 TCL commands for spare-buffer cell rewiring in ICC shell

The spare-buffer rewiring operation shown in Fig. 8, inserts a pre-placed inactive buffer between two sequential logic elements to improve signal propagation and load distribution along a critical path. Initially, cell U1938 directly drives U1939 through the long interconnect net $n17229$, while the spare buffer $spare\_0\_107$ remains unused with its input tied to VSS and its output floating. During ECO execution, the physical design tool issues disconnect-and-reconnect commands to restructure the net. The original net $n17229$ is first detached from both the driver (U1938/Y) and receiver (U1939/A4), and the spare buffer is freed from its default ground connection. New connectivity is then established such that U1938/Y drives the input of $spare\_0\_107$, and the buffer output subsequently drives U1939/A4, forming two shorter nets in place of the original long segment. This restructuring reduces interconnect delay, improves signal transition quality, and restores timing slack by reinforcing drive strength at an intermediate point.

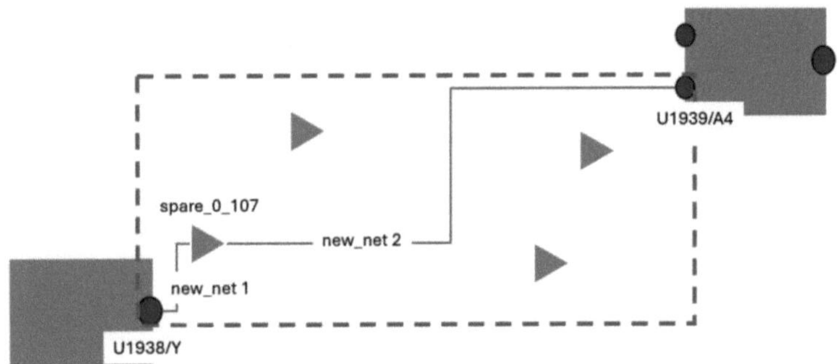

Fig. 8 After spare-buffer rewiring ECO

3.7 Focal Optimization

After spare-buffer rewiring, a focal optimization step refines the post-route design to ensure full timing closure. Using accurate routed parasitic data, it performs localized adjustments to fix remaining setup, hold, transition, or capacitance violations, primarily addressing local skew in critical areas. When necessary, gate sizing is applied to resolve balance violations. The phase involves minimal physical changes, including small cell movements, targeted buffer insertion, selective rerouting, and sizing adjustments, while maintaining full DRC compliance.

4 Experimental Result and Evaluation of Proposed Method

All experiments were conducted on a high-performance Linux workstation with an Intel Xeon Silver 4310 processor running at 2.10 GHz and 124 GB of memory to support synthesis, place-and-route, and timing sign-off. The benchmark designs were implemented using a 32 nm CMOS technology with low-Vt variants and evaluated under a worst-case corner of slow silicon, 0.95 V, and 125 °C to ensure accurate timing analysis. Physical design and ECO implementation, including placement legalization, clock tree synthesis, and routing optimization, were performed using Synopsys IC Compiler [16], while post-layout static timing analysis and final verification were carried out using Synopsys PrimeTime [17]. Three benchmark circuits—a 4 × 4 two-dimensional discrete cosine transform processor, a pipelined multiply–accumulate unit, and a 64-bit double-precision floating-point unit—were used to evaluate the proposed timing ECO framework. The evaluation focuses on validating the spare-buffer rewiring strategy, assessing spare-cell selection for resolving critical timing paths, and verifying the complete ECO flow up to timing sign-off. In addition, power, performance, and area impacts were analyzed to ensure that timing improvements do not introduce significant design overhead.

4.1 Post-Route Design Metrics of Benchmark Circuits and Spare-Buffer Cell Distribution

As shown in Table 2, the study evaluates the proposed ECO prediction methodology using three benchmark circuits, namely a 4 × 4 two-dimensional discrete cosine transform processor DCT2D_4 × 4 [18], a pipelined multiply–accumulate unit PipeMAC, and a 64-bit double-precision floating-point arithmetic unit FP64. To capture intra-design variability, two benchmark instances are considered for DCT2D_4 × 4, denoted as DCT2D_1 and DCT2D_2, and two instances are also used for PipeMAC, denoted as PipeMAC_1 and PipeMAC_2, while FP64 is evaluated using a single instance. Collectively, these benchmarks span a wide range of design scales and logic characteristics, including cell and net counts, buffer and inverter population, and interconnect area, thereby providing a representative and scalable evaluation basis for ECO optimization techniques such as spare-buffer rewiring.

Table 2. Design metrics for the benchmarks and spare-buffer cells distributed.

Metric	DCT2D_1	DCT2D_2	FP64	PipeMAC_1	PipeMAC_2
Total Cells	11,332	11,167	35,267	42,931	45,656
Total Nets	12,878	13,001	36,308	50,701	53,026
Combinational Cells	10,820	10,655	35,136	40,625	43,350
Sequential Cells	512	512	131	2,306	2,306
Buffers/Inverters Cells	3,191	2,903	7,456	4,811	6,451
Spare-Buffer Cell Distributed	250	250	500	500	500

The spare-buffer allocation is selected according to each benchmark's design scale and timing-closure requirements: the larger FP64 and PipeMAC designs are provisioned with 500 spare buffers each to better accommodate deeper logic and denser interconnect, whereas the more compact DCT2D_4 × 4 benchmark uses 250 spare buffers, consistent with its smaller area and reduced routing complexity. The number of spare-buffer cells used in the ECO process is required to be between 10% and 20% of the total cell count in the design [19], in line with industry guidelines, ensuring efficient resource utilization without exceeding typical buffer usage limits.

4.2 Impact of Spare-Buffer Cell Rewiring on Slack Improvement

Table 3 shows the slack for the timing reports for a critical setup-violating path for one of the ECO paths tested, before and after spare-cell rewiring. Initially, the path has a slack of −0.0140 ns, where the data arrival time (2.1815 ns) exceeds the required time (2.1674 ns). This violation occurs between the source cell U19342/Y and the sink cell U27967/A1, with the critical interconnect and accumulated cell delays causing RC-dominant loading, making the path sensitive to timing degradation. To address this, the spare buffer spare_0_161/A was selected from candidates based on the lowest estimated net delay. The buffer was inserted between the source and sink cells within the defined ECO bounding box region, ensuring minimal routing disturbance. After the spare buffer was added, the slack improved to −0.0056 ns, demonstrating a significant reduction in timing violations. While a small violation remains, the improvement confirms that selecting spare cells based on net delay effectively enhances timing closure in complex interconnects (Fig. 9).

Table 3. Timing improvement through spare-buffer cell rewiring

Metric	Before Rewiring	After Rewiring	Improvement
Slack	−0.0140 ns	−0.0056 ns	+0.0084 ns
Data Arrival Time	2.1815 ns	2.1730 ns	−0.0085 ns
Data Required Time	2.1674 ns	2.1674 ns	No Change
Critical Interconnect Path	U19342/Y → U27967/A1	U19342/Y → spare_0_161/A → U27967/A1	Updated connection
Buffer Used	None	spare_0_161/A	Buffer Inserted

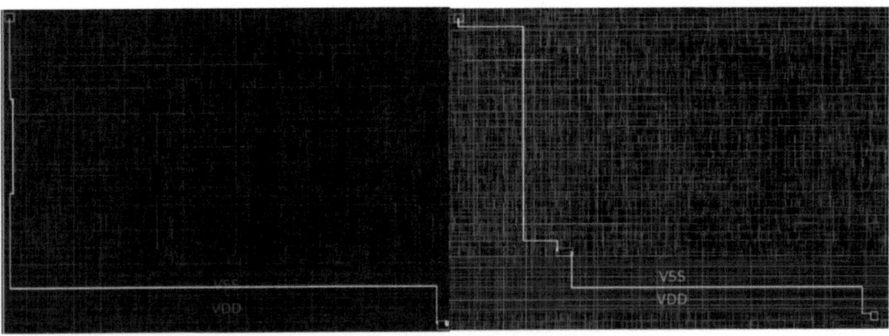

Fig. 9 Critical interconnect segment before and after spare buffer cell rewiring along an ECO-targeted critical path

4.3 Validation and Analysis Based on Timing Performance

The proposed ECO framework is evaluated on three benchmarks such DCT2D_4 × 4, PipeMAC, and FP64 representing different design complexities. Timing improvements are measured using Total Negative Slack (TNS), Worst Negative Slack (WNS), and the number of remaining ECO paths. Compared with gate-sizing-based ECO using spare cells [1], the proposed spare-buffer rewiring consistently achieves better timing recovery, reducing TNS, WNS, and violating paths while often requiring shorter runtime. It attains full closure in simpler DCT2D cases and stronger slack recovery in more complex scenarios, with similar benefits for FP64 and PipeMAC. These results demonstrate that the framework efficiently improves timing closure and scales across diverse designs, making it suitable for practical ASIC ECO optimization (Table 4).

Table 4. Timing closure results across benchmarks using spare-buffer-based and gate sizing ECO techniques

Benchmarks						Spare-Buffer Rewiring					Gate Sizing [1]			
Design	#Cells	#Spare-Buffer Cells	ECO Paths	TNS (ns)	WNS (ns)	ECO Paths	TNS (ns)	WNS (ns)	#Spare-Buffer Cells Usage	Runtime (s)	ECO Paths	TNS (ns)	WNS (ns)	Runtime (s)
DCT2D_1	11,332	250	24	0.6035	0.0736	0	0	0	6	37.1	0	0	0	41.3
DCT2D_2	11,167	250	530	80.6103	0.1937	270	33.3629	0.2558	184	120.7	483	60.3892	0.2167	128.5
FP64	35,267	500	44	2.9825	0.0998	16	0.1390	0.0193	4	94.2	21	0.1716	0.0225	127.3
PipeMaC _1	42,931	500	68	1.4801	0.0967	8	0.0173	0.0080	39	154.8	12	0.0264	0.0093	210
PipeMAC_2	45,656	500	188	4.3232	0.1263	67	2.6053	0.1205	71	332.4	95	2.8553	0.1661	438

Figure 10 compares the TNS recovery across DCT2D_1, DCT2D_2, FP64, PipeMAC_1, and PipeMAC_2, where each bar reports the percentage of TNS reduction relative to the corresponding pre-ECO baseline. The TNS gain is defined as $\text{TNS}_{pre} - \text{TNS}_{post}$, and the plotted metric is $\frac{\text{TNS}_{pre} - \text{TNS}_{post}}{\text{TNS}_{pre}} \times 100\%$. The higher values indicate better recovery, and 100% corresponds to complete elimination of the initial TNS violation. Both spare-buffer rewiring and gate sizing achieve full recovery on DCT2D_1. In DCT2D_2, spare-buffer rewiring recovers 65.8% of TNS versus 25.1% for gate sizing, highlighting its advantage when interconnect delays dominate. For FP64 and PipeMAC_1, both methods perform similarly, while in PipeMAC_2, spare-buffer rewiring again outperforms gate sizing (86.0% vs. 73.3%) due to its ability to locally reroute critical nets through nearby spare buffers. Overall, spare-buffer rewiring provides stronger and more consistent TNS improvement across designs. Spare-buffer rewiring directly addresses violations driven by interconnect delay and parasitics by locally rerouting the critical connection through nearby pre-placed spare buffers, while gate sizing mainly improves intrinsic cell delay and may increase downstream loading, which reduces its effectiveness when interconnect delay dominates.

Figure 11 compares the ECO path gain across DCT2D_1, DCT2D_2, FP64, PipeMAC_1, and PipeMAC_2, where ECO path gain is defined as the reduction in the number of violating ECO paths after applying an ECO technique. Overall, spare-buffer cell rewiring shows a stronger reduction in violating paths than gate sizing, indicating more effective path-level timing cleanup. The two techniques exhibit similar behaviour in the less challenging case, while the advantage of spare-buffer rewiring becomes more evident as the benchmark difficulty increases. For the remaining designs, the trend remains consistent, with spare-buffer rewiring maintaining a modest improvement over gate sizing in reducing the residual set of violating paths.

Figure 12 compares ECO runtime for the same benchmarks. Spare-buffer rewiring consistently requires less runtime than gate sizing, with differences increasing for larger, more complex designs. Gate sizing involves multiple iterations and candidate evaluations across logic stages, which grows more costly when interconnect delay dominates. In contrast, spare-buffer rewiring applies localized reconnections through pre-placed spare buffers, reducing the number of modifications and iterations needed for convergence. As a result, it provides faster, more scalable timing repair as design size and routing complexity increase.

Based on the result, our method achieves about 73% average ECO-path resolution compared with about 59% for gate sizing, delivers higher TNS improvement, and reduces runtime by up to roughly 26% while maintaining low implementation overhead with area increase below 0.15% and power overhead below 4%. Gate sizing primarily reduces cell delay and can increase capacitance and power, so its effectiveness diminishes when the dominant limitation arises from net delay, whereas spare-buffer rewiring improves timing by splitting long RC-heavy interconnects through nearby pre-placed buffers, enabling efficient late-stage closure with minimal physical disruption.

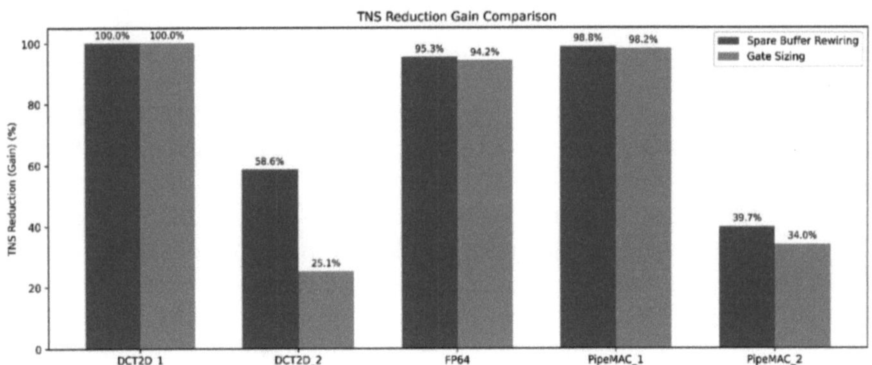

Fig. 10 Total negative slack (TNS) reduction gain comparison for spare-cell rewiring and gate sizing across benchmark circuits

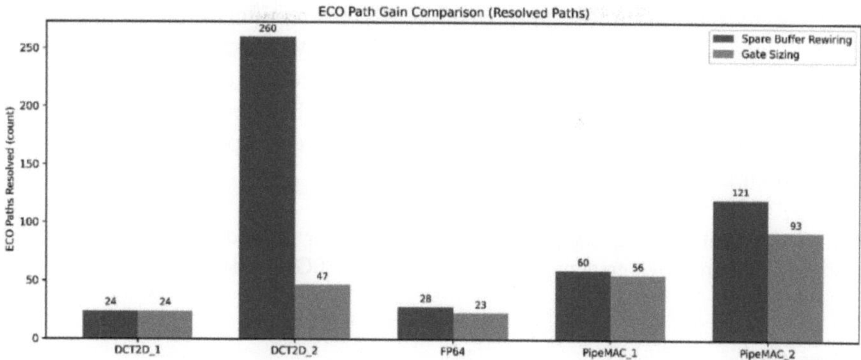

Fig. 11 ECO path gain comparison between spare-buffer cell rewiring and gate sizing.

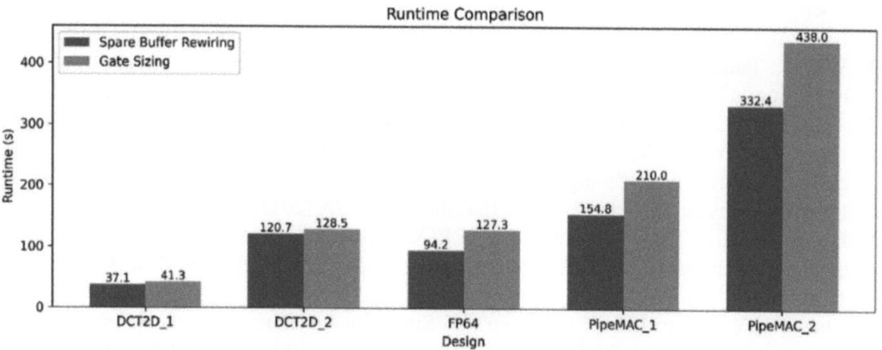

Fig. 12 Runtime comparison between spare-buffer rewiring and gate sizing across benchmark circuits.

5 Conclusion

In conclusion, the proposed automated timing ECO framework demonstrates that spare-buffer cell rewiring is a more effective and efficient post-layout timing repair method than conventional gate sizing across designs of varying complexity. By integrating static timing analysis with physical design, the framework selects candidate spares based on worst-slack impact, path-specific bounding boxes, Manhattan-distance and wire-delay evaluation, and employs a spare-locking mechanism to avoid conflicts. This approach consistently achieves higher TNS reduction, faster runtime, and fewer ECO-critical paths while keeping area increase below 0.15% and power overhead under 4%, confirming that spare-buffer rewiring provides robust timing recovery with minimal implementation cost, making it a practical solution for modern ASIC flows.

Future work will focus on improving the accuracy of pre-ECO delay estimation by incorporating deeper routing awareness into the spare-selection framework. The current

Manhattan-distance-based RC approximation assumes rectilinear routing and uniform parasitics, which can differ from actual post-route delays due to congestion and track availability. Enhancements will include congestion-aware cost evaluation, early global-routing feedback, and incremental parasitic refinement to better predict net delays, reduce ECO iterations, and achieve more consistent timing recovery, especially for interconnect-dominated violations.

Acknowledgement. The authors would like to thank UTM for the funding of this project, with the Flagship CoE/RG research grant number Q.J130000.5023.10G05 and Q.J130000.5023.10G09.

References

1. Ho, K.H., Chen, Y.P., Fang, J.W., Chang, Y.W.: ECO timing optimization using spare cells and technology remapping. IEEE Trans. Comput.-Aided Des. Integr. Circ. Syst. **29**(5), 697–710 (2010)
2. Chen, Y.P., Fang, J.W., Chang, Y.W.: ECO timing optimization using spare cells. In: IEEE/ACM International Conference on Computer-Aided Design, pp. 530–535. IEEE (2007)
3. Wei, X., Tang, W.C., Diao, Y., Wu, Y.L.: ECO timing optimization with negotiation-based re-routing and logic re-structuring using spare cells. In: In: 17th Asia and South Pacific Design Automation Conference 2012, pp. 511–516. IEEE (2012)
4. Hung, J.H., Yeh, Y.K., Tseng, Y.S., Hsieh, T.M.: A new ECO technology for functional changes and removing timing violations. In: In: 12th International Symposium on Quality Electronic Design 2011, pp. 1–5. IEEE, California, USA (2011)
5. Ho, K.H., Jiang, J.H.R., Chang, Y.W.: TRECO: dynamic technology remapping for timing engineering change orders. IEEE Trans. Comput.-Aided Des. Integr. Circ. Syst. **31**(11), 1723–1733 (2012)
6. Chang, H.Y., Jiang, I.H.R., Chang, Y.W.: Simultaneous functional and timing ECO. In: Proceedings of the 48th Design Automation Conference 2011, pp. 140–145. IEEE (2011)
7. Hung, J.H., Lin, Y.C., Cheng, W.K., Hsieh, T.M.: Unified approach for simultaneous functional and timing ECO. IET Circuits Devices Syst. **10**(6), 514–521 (2016)
8. Tang, J.X., Huang, S.H., Hung, J.H.: ECO timing optimization with data paths and clock paths considered. In: International Symposium on Intelligent Signal Processing and Communication Systems (ISPACS) 2021, pp. 1–2. IEEE, Taiwan (2021)
9. Kahng, A.B., Lienig, J., Markov, I.L., Hu, J.: VLSI Physical Design: from Graph Partitioning to Timing Closure, 1st edn. Springer, Netherlands (2011)
10. Chhabria, V.A., Jiang, W., Sapatnekar, S.S.: IR-aware ECO timing optimization using reinforcement learning. In: In: 6th Symposium on Machine Learning for CAD (MLCAD) 2024. IEEE/ACM, Utah, USA (2024)
11. Lu, Y.C., Nath, S., Khandelwal, V., Lim, S.K.: 2021, December. RL-sizer: VLSI gate sizing for timing optimization using deep reinforcement learning. In: 58th ACM/IEEE Design Automation Conference (DAC) 2021, pp. 733–738. IEEE/ACM, California, USA (2021)
12. Wu, H., Huang, Z., Li, X., Zhu, W.: AiTO: simultaneous gate sizing and buffer insertion for timing optimization with GNNs and RL. Integration. **98**, 102211 (2024)
13. Blaauw, D., Chopra, K., Srivastava, A., Scheffer, L.: Statistical timing analysis: from basic principles to state of the art. IEEE Trans. Comput.-Aided Des. Integr. Circ. Syst. **27**(4), 589–607 (2008)
14. Guo, G., Huang, T.W., Lin, Y., Guo, Z., Yellapragada, S., Wong, M.D.: A GPU-accelerated framework for path-based timing analysis. IEEE Trans. Comput.-Aided Des. Integr. Circ. Syst. **42**(11), 4219–4232 (2023)

15. Bhasker, J., Chadha, R.: Static Timing Analysis for Nanometer Designs: a Practical Approach, 1st edn. Springer Science & Business Media, New York, USA (2009)
16. I. Synopsys.: IC Compiler II: Industry Leading Place and Route System, https://www.synopsys.com/content/dam/synopsys/implementation%26signoff/datasheets/ic-compiler-ii-ds.pdf, last accessed 2025/12/21.
17. I. Synopsys.: PrimeTime Datasheet, https://www.synopsys.com/content/dam/synopsys/implementation%26signoff/datasheets/primetime-ds.pdf, last accessed 2025/12/10.
18. Zabidi, N.M., Ab Rahman, A.A.H.: VLSI design of a fast pipelined 8x8 discrete cosine transform. International. J. Electr. Comput. Eng. **7**(3), 1430 (2017)
19. Alpert, C.J., Hrkić, M., Quay, S.T.: A fast algorithm for identifying good buffer insertion candidate locations. In: Proceedings of the International Symposium on Physical Design 2004, pp. 47–52. ACM, New York, USA (2004)

Statistical Meta-process Modeling for Fidelity Selection in Model-Based Optimization

R. Sala(✉)

Department of Mechanical and Process Engineering, RPTU, Kaiserslautern, Germany
ramses.sala@mytum.de

Abstract. Model and simulation-based optimization has emerged as a powerful methodology in science, engineering, and industrial systems. A recurring fundamental challenge is selecting appropriate model fidelity that balances accuracy against computational cost and the resulting constraints on optimization iterations under limited resources. Contrasting and complementary to dynamic multi-fidelity approaches, this work focuses on selecting a single fidelity level per optimization run. This work presents a statistical meta-process modeling framework for analyzing accuracy-cost trade-offs to support offline a priori fidelity or budget selection in model-based optimization. The framework combines three components: a modeling error model that represents accuracy as a function of fidelity, a cost model that describes computational expense per evaluation, and an optimization error model that characterizes optimality-gap convergence across iterations. The corresponding model parameters can be estimated from empirical data obtained from previous experiments on representative problems. An open-source Python implementation is provided for numerical approximations of optimal-fidelity curves and budget-performance plots. Results from case studies demonstrate that the framework can serve as a practical tool for fit-for-purpose model selection in budget-aware optimization settings.

Keywords: Efficient optimization · Fidelity selection · Meta-process modeling · Simulation-based optimization

1 Introduction

Mathematical modeling and simulation combined with optimization are widely used in science, engineering, and technological systems [1,14]. The quest for resource-efficient computing [13] in simulation-based optimization (SBO) can be approached from various perspectives: simulation methods, solvers [27], numerical optimization algorithms [23], and computing hardware. Much progress has been made in these areas [12] to reach the current state of the art, and further progress is anticipated [24,33]. One recurring open challenge that spans these

N. A. Mohd Subha et al. (Eds.): SMS 2026, CCIS 2956, pp. 158–175, 2026.
https://doi.org/10.1007/978-981-92-0306-2_12

perspectives is budget-aware model fidelity selection. How to select an appropriate simulation model resolution or fidelity that balances accuracy with computational cost, while considering the resulting optimization iteration constraints under limited resources? Typically, higher-fidelity models are more accurate but incur higher computational costs. In contrast, low-fidelity approximation models incur lower function evaluation costs and thus enable more optimization iterations, but at the cost of higher modeling error. This inherent accuracy-cost tradeoff occurs in many optimization settings and scales ranging from large-scale High Performance Computing (HPC) problems [8] and Computational Fluid Dynamics [18] to recurring smaller-scale problems in robotics and model predictive control [9]. The fidelity selection problem transcends SBO and is relevant in hyperparameter optimization [28,30] and neural network training [16,17].

The remainder of this paper is organized as follows. Section 2 reviews related work. Section 3 formulates the fidelity selection problem. Section 4 develops the meta-process modeling framework. Section 5 presents analytical results. Section 6 describes calibration and implementation. Section 7 presents case studies, and Sect. 8 concludes.

2 Related Work

Research on fidelity selection has focused on multi-fidelity optimization (MFO), where multiple models with different fidelities are used dynamically throughout the optimization process [15,21], as well as surrogate-based optimization approaches that iteratively construct and exploit data-driven models to guide evaluations [11]. While MFO can achieve superior efficiency, it introduces additional complexity overhead. In practice, most optimization problems are still solved using a single objective function with a fixed fidelity.

In contrast to MFO, this paper addresses the *offline fidelity selection problem*: given a model-based optimization task with a fixed computational budget, which single fidelity model should be selected and used throughout an optimization run? This setting is relevant when resources or infrastructure require committing to a single model fidelity upfront.

Related work on fixed-fidelity selection is scattered across fields, including discrete event simulation [10], stochastic simulations under limited budgets [19], distributed computer simulation [5], and material property prediction [32]. For mesh-based simulations, discretization error typically follows power-law decay with refinement [22]. Optimization algorithm performance can be characterized through convergence theory [20], complexity theory [31], and empirical benchmarking [26]. However, general quantitative frameworks for single-fidelity selection integrating the simulation accuracy-cost tradeoff appear absent.

Optimal Computing Budget Allocation (OCBA) [4] addresses *how many* simulation replications to allocate among alternatives. This framework addresses a complementary question: *at what fidelity* should simulations be conducted? The two perspectives are related through the trade-off between accuracy per evaluation and affordable evaluation count.

The present work develops a trans-disciplinary meta-process modeling approach combining simulation accuracy, optimization performance, and computational cost. Key contributions include: (1) a parametric meta-model decomposing total error into calibrated modeling and optimization components; (2) uncertainty quantification via analytical distributions with empirically estimated coefficient of variation; (3) analytical conditions for optimal fidelity existence and fidelity-budget relationships; (4) validation through structural and aerodynamic optimization case studies; and (5) an open-source Python implementation with interactive GUI.

3 Problem Formulation

Consider minimizing a function $f : \mathcal{X} \to \mathbb{R}$ over compact $\mathcal{X} \subseteq \mathbb{R}^d$, where evaluating f is expensive. In this setting, f represents the ground truth or reference model of the target system. Instead of f, an approximate model $\tilde{f}_\lambda$ with fidelity λ is used during the optimization process.

Let Λ denote a *fidelity index set*, and let $\{\tilde{f}_\lambda : \mathcal{X} \to \mathbb{R}\}_{\lambda \in \Lambda}$ be a family or set of approximation models indexed by $\lambda \in \Lambda$. Each model $\tilde{f}_\lambda$ has an associated evaluation cost $c_\lambda > 0$. These are collected into a set of model-cost pairs $\mathcal{S} = \{(\tilde{f}_\lambda, c_\lambda)\}_{\lambda \in \Lambda}$. The fidelity index λ can represent various characteristics: resolution parameters, mesh refinement levels, sample sizes, or model type labels. Three important cases are distinguished:

Continuous fidelity: $\Lambda = [0, 1]$, with a parameterized model family $\tilde{f}(\cdot, \lambda)$. Practical examples include mesh resolution or simulation time step.

Discrete ordered fidelity: $\Lambda = \{\lambda_1, \ldots, \lambda_K\}$ with $\lambda_1 < \cdots < \lambda_K$. Typically costs increase with fidelity: $c_{\lambda_1} \leq \cdots \leq c_{\lambda_K}$.

Unstructured models: $\Lambda = \{1, \ldots, K\}$ with no inherent ordering. Models represent fundamentally different approximation methods.

3.1 Optimization Process and Total Regret

Let $\mathcal{A}$ denote an optimization algorithm that, given query access to a function $\tilde{f}_\lambda$ and an iteration budget $n \in \mathbb{N}$, produces a candidate solution: $x_{n,\lambda} = \mathcal{A}(\tilde{f}_\lambda, n) \in \mathcal{X}$. Let $f^* = \min_{x \in \mathcal{X}} f(x)$ denote the optimal value of the true objective. The quantity of ultimate interest is the *total error* or total regret measuring suboptimality when the approximate solution $x_{n,\lambda}$ obtained from approximate model optimization is evaluated on the true objective:

$$\varepsilon_{\text{tot}}(n, \lambda) = f(x_{n,\lambda}) - f^* \geq 0. \tag{1}$$

Given a computational budget $B > 0$, the iterations budget n and fidelity λ are coupled through $n \cdot c_\lambda \leq B$, so the maximum affordable iterations is $n_\lambda(B) = \lfloor B / c_\lambda \rfloor$.

3.2 Fidelity Selection Problem

In this work, we focus on the *fidelity selection problem*: given a fixed algorithm $\mathcal{A}$ and budget B, determine the fidelity $\lambda^* \in \Lambda$ that minimizes the total error. Since the algorithm is fixed, it is absorbed it into the functional. Define the *solution map*:

$$\mathbf{x} : (\mathcal{X} \to \mathbb{R}) \times \mathbb{N} \longrightarrow \mathcal{X}, \qquad \mathbf{x}(g, n) = \mathcal{A}(g, n), \tag{2}$$

and the *error functional*:

$$\mathcal{E} : (\mathcal{X} \to \mathbb{R}) \times (\mathcal{X} \to \mathbb{R}) \times \mathbb{N} \longrightarrow \mathbb{R}, \qquad \mathcal{E}(f, g, n) = f\big(\mathbf{x}(g, n)\big) - f^*, \tag{3}$$

which represents the true suboptimality when the optimizer runs on model g for n iterations, and the corresponding result is evaluated on f. The total error under fidelity λ and budget B is:

$$\mathcal{E}_{\text{total}}(\lambda; B) = \mathcal{E}\left(f, \tilde{f}_\lambda, \left\lfloor \frac{B}{c_\lambda} \right\rfloor\right). \tag{4}$$

The fidelity selection problem is then:

$$\min_{\lambda \in \Lambda} \mathcal{E}\left(f, \tilde{f}_\lambda, \left\lfloor \frac{B}{c_\lambda} \right\rfloor\right). \tag{5}$$

This formulation captures the essential trade-off: lower fidelity allows more iterations but introduces larger modeling error, while higher fidelity reduces modeling error but limits affordable iterations. The following sections develop quantitative models for the accuracy-cost trade-off.

4 Meta-process Modeling Framework

The fidelity selection problem (5) can only be practically addressed with additional structure relating error to fidelity, cost, and iteration count. This section develops statistical parametric approximation models whose parameters can be estimated from pilot data on representative problems.

4.1 Total Error Approximation

Let $x_{n,\lambda}$ denote the solution obtained by running an optimizer for n iterations on the fidelity-λ model. The expected total error is modeled as:

$$\mathbb{E}[\varepsilon_{\text{tot}}(\lambda, n)] \approx \mu_{\text{mod}}(\lambda) + \mu_{\text{opt}}(\lambda, n), \tag{6}$$

where $\mu_{\text{mod}}(\lambda)$ captures the expected error penalty for using an approximate model at fidelity λ, and $\mu_{\text{opt}}(\lambda, n)$ represents the expected penalty for incomplete optimization after n iterations.

Throughout this paper, ε (with subscripts) denotes a random variable representing error, while μ denotes its expected value. The additive form (6) is adopted for parsimony: it is the simplest structure that captures the two dominant effects with minimal parameters. The additive structure allows analysis of how fidelity and iteration count each contribute to total error. As demonstrated in Sect. 7, this form provides useful predictive accuracy.

The optimization error component $\mu_{\text{opt}}(\lambda, n)$ is written as a function of both fidelity and iterations to capture the general case where optimization convergence behavior varies with fidelity. In some cases, optimization behavior is approximately independent of fidelity, and $\mu_{\text{opt}}(\lambda, n) \approx \mu_{\text{opt}}(n)$.

Modeling Rationale. The additive form (6) is adopted as a tractable approximation based on the following considerations:

- *Parsimony*: The additive form is the simplest structure that captures the two dominant effects—model fidelity and optimization effort—with minimal parameters.
- *Interpretability*: The additive structure allows analysis of how fidelity and iteration count each contribute to total error, facilitating practical decision-making.
- *Empirical validation*: As demonstrated in Sect. 7, this functional form provides useful predictive accuracy across the tested case studies and problem instances.

In the following sections, each error component is modeled as a nonnegative random variable with a parametric mean and calibrated uncertainty.

4.2 Modeling Error Model

The expected modeling error is represented by a model that decreases with increasing fidelity. For continuous fidelity $\lambda \in [0, 1]$, a power-law form is adopted:

$$\mu_{\text{mod}}(\lambda) = \varepsilon_{m,0} + \alpha_m (1 - \lambda + \rho)^{\beta_m}, \tag{7}$$

where $\alpha_m > 0$ is the error scale at low fidelity, $\beta_m > 0$ is the decay exponent, $\rho > 0$ is a small regularization ensuring finite error at $\lambda = 1$, and $\varepsilon_{m,0} \geq 0$ is an irreducible error floor.

For mesh-based simulations, discretization error scales as h^p where h is mesh size and p is the scheme order [22]. With fidelity $\lambda \propto h^{-d}$ in d dimensions, error follows $(1 - \lambda)^{p/d}$, motivating the power-law form. An exponential alternative $\mu_{\text{mod}}(\lambda) = \alpha_m \exp(-\beta_m \lambda) + \varepsilon_{m,0}$ may provide better fit for some applications.

For continuous fidelity, the parameter $\lambda \in [0, 1]$ is a normalized scale. In the case studies, fidelity is defined as $\lambda = n_{\text{elem}}/n_{\max}$ (ratio of element count to maximum resolution) or similar normalizations appropriate to the problem.

4.3 Cost Model

Computational cost is modeled to increase with fidelity. For continuous fidelity $\lambda \in [0, 1]$:

$$c(\lambda) = c_0 + \alpha_c \lambda^\gamma, \tag{8}$$

where $c_0 \geq 0$ is fixed overhead, $\alpha_c > 0$ is the cost scale, and $\gamma > 0$ is the growth exponent. The exponent γ is determined empirically; common values range from $\gamma \approx 1$ (linear) to $\gamma \approx 3$ (qubic) cost growth. For discrete fidelity levels, the expected costs are simply tabulated values.

In this framework, the cost c_λ and budget B can be interpreted as abstract computational effort, analogous to wall-clock time. For convenience, cost is expressed in *normalized units* relative to the minimal-fidelity evaluation: $c_{\text{norm}}(\lambda) = c_\lambda / c_{\lambda_{\min}}$. Thus, the cheapest option has cost 1, and the budget B represents equivalent minimal-fidelity evaluations.

4.4 Optimization Error Model

The optimization error, or optimality gap, is modeled to capture the expected suboptimality due to incomplete optimization after n iterations. The functional form is modeled to decrease with iterations. The decay depends on the algorithm-problem combination and reflects underlying convergence characteristics. Two canonical forms are used, motivated by convergence theory and computational complexity.

Power-Law Decay. For problems admitting polynomial-time convergence guarantees, or when using algorithms with established polynomial convergence rates, the optimization error follows a power-law decay:

$$\mu_{\text{opt}}(\lambda, n) = \varepsilon_{o,0} + \alpha_o(\lambda)(n + \kappa)^{-\beta_o(\lambda)}, \tag{9}$$

where $\alpha_o(\lambda) > 0$ is the initial error scale, $\beta_o(\lambda) > 0$ is the convergence rate exponent, $\kappa > 0$ is a regularization parameter preventing singularity at $n = 0$, and $\varepsilon_{o,0} \geq 0$ is an optional irreducible error floor. This form can be theoretically grounded in convergence analysis [20].

Logarithmic Decay. For combinatorial optimization or global optimization problems or settings where the search space and difficulty grow exponentially with problem size, the optimization error may follow logarithmic decay:

$$\mu_{\text{opt}}(\lambda, n) = \varepsilon_{o,0} + \frac{\alpha_o(\lambda)}{\log(n + \kappa(\lambda))}, \tag{10}$$

where the parameters have analogous interpretations: $\alpha_o(\lambda) > 0$ is the error scale, $\kappa(\lambda) \geq e$ ensures the logarithm is well-defined and positive, and $\varepsilon_{o,0} \geq 0$ is the irreducible floor. The logarithmic decay form arises from inverting the time-to-accuracy relationship for problems requiring exponential effort [31].

Fidelity-Dependent Parameters. For different fidelity models, the parameters of the optimization models above can be fitted separately. Alternatively for continuous fidelity $\lambda \in [0, 1]$, a practical parameterization for the parameters $\alpha_o(\lambda)$ and $\beta_o(\lambda)$ that captures how optimization behavior varies with fidelity is:

$$\alpha_o(\lambda) = \alpha_{o,0}(1 + s_\alpha(1 - \lambda)), \quad \beta_o(\lambda) = \beta_{o,0}(1 + s_\beta\lambda). \tag{11}$$

In some special cases, it might be justified or practical to use a reduced fidelity-independent form $\mu_{\text{opt}}(n) = \varepsilon_{o,0} + \alpha_o(n + \kappa)^{-\beta_o}$.

The optimization error model typically achieves lower R^2 values than modeling error models; this is expected and reflects inherent run-to-run variance in optimization performance, well documented in the literature [2,7]. The parametric form above models the expectation of the optimization error, while run-to-run variability is captured through the coefficient of variation CV as discussed in the next section.

4.5 Uncertainty Quantification

Each error component is modeled as a nonnegative random variable with variance structure:

$$\text{Var}[\varepsilon] = \sigma_0^2 + \mu^2 \cdot \text{CV}^2, \tag{12}$$

where $\sigma_0 \geq 0$ captures irreducible base uncertainty and CV is the proportional coefficient of variation. Setting $\sigma_0 = 0$ recovers the simpler constant-CV model. A nonnegative distribution (gamma, log-normal, or inverse-gamma) can be fitted via moment matching; the gamma distribution provides a tractable default.

Under independence, the total error $\varepsilon_{\text{tot}} = \epsilon_{\text{mod}} + \epsilon_{\text{opt}}$ has moments:

$$\mu_{\text{tot}} = \mu_{\text{mod}}(\lambda) + \mu_{\text{opt}}(\lambda, n), \quad \sigma_{\text{tot}}^2 = \sigma_{m,0}^2 + \mu_{\text{mod}}^2 \text{CV}_m^2 + \sigma_{o,0}^2 + \mu_{\text{opt}}^2 \text{CV}_o^2. \tag{13}$$

The sum is approximated by moment-matching to a gamma distribution for computing confidence intervals and quantiles.

4.6 Model Applicability

The parametric meta-models developed in Sect. 4 relating fidelity λ to modeling error and cost are directly applicable when a meaningful fidelity parameter exists, either continuous or discrete ordered. For unstructured model sets where no natural ordering exists, two approaches are available: (1) tabulate empirical modeling error and cost values for each model directly, bypassing the parametric relationships; or (2) use normalized cost as a proxy fidelity index $\lambda_k = c_k/c_{\max}$, enabling application of the parametric framework with the understanding that the resulting "fidelity" reflects computational effort rather than inherent model structure.

5 Optimal Approximate Fidelity Analysis

This section establishes the theoretical foundation for fidelity selection, addressing: (1) optimal fidelity for a given budget, (2) optimal fidelity as a function of budget, and (3) budget-performance curves.

The exact fidelity selection problem (5) requires access to the true error functional $\mathcal{E}(f, \tilde{f}_\lambda, n)$, which is generally intractable. Instead, we solve the *approximate fidelity selection problem* using the additive decomposition from Sect. 4. The resulting approximate optimal fidelity is denoted $\hat{\lambda}^*(B)$ to distinguish it from the ideal $\lambda^*(B)$.

5.1 Optimal Fidelity for Fixed Budget

Given budget $B > 0$, the approximate fidelity selection problem seeks the model that minimizes expected total error under the additive decomposition. For a finite set of available models $\Lambda = \{\lambda_1, \ldots, \lambda_K\}$ with associated costs $\{c_1, \ldots, c_K\}$, the approximate optimal fidelity is defined by:

$$\hat{\lambda}^*(B) = \arg\min_{\lambda_k \in \Lambda} \left\{ \mu_{\text{mod}}(\lambda_k) + \mu_{\text{opt}}\left(\left\lfloor \frac{B}{c_{\lambda_k}} \right\rfloor ; \lambda_k \right) \right\}. \tag{14}$$

This formulation applies regardless of whether the models admit a natural ordering. For discrete ordered fidelity, the parametric models from Sect. 4 provide $\mu_{\text{mod}}(\lambda_k)$; for unstructured model sets, the modeling errors are tabulated, and the optimization is obtained by exhaustive evaluation in $O(K)$ operations. When fidelity λ varies continuously, the formulation (14) extends by replacing the finite set Λ with the interval $[\lambda_{\min}, \lambda_{\max}]$, and corresponding associated $c(\lambda)$, $\mu_{\text{mod}}(\lambda)$ and $\mu_{\text{opt}}(\lambda)$ models.

The floor function in the expression for the iteration budget n introduces discontinuities; a *continuous relaxation* treating iteration count $n = B/c(\lambda)$ as real-valued restores continuity. Under this relaxation, and conditions satisfied by the parametric models introduced in the previous sections, the existence of a minimizer follows from the extreme value theorem on a compact domain. Uniqueness holds under convexity; for general parameters, multiple local minima may exist. Despite potential non-convexity, the problem is computationally tractable due to its one-dimensional structure: for example, grid search achieves ϵ-accuracy in $O(1/\epsilon)$ evaluations. The relaxation error from treating n as continuous is bounded by one iteration's worth of optimization progress.

5.2 Optimal Fidelity Curves

By solving (14) for a range of budgets, we obtain the (approximate) *optimal fidelity curve* $\hat{\lambda}^*(B)$. This curve may exhibit non-monotonic behavior at intermediate budgets: lower fidelity can outperform higher fidelity when the iteration benefit outweighs the accuracy loss.

The monotonicity of $\hat{\lambda}^*(B)$ is governed by an elasticity condition. Define the *elasticity of marginal optimization error*:

$$\eta(n) = \frac{n \cdot \mu''_{\text{opt}}(n)}{-\mu'_{\text{opt}}(n)}. \tag{15}$$

For the power-law model $\mu_{\text{opt}}(n) = \alpha_o(n+\kappa)^{-\beta_o}$, this simplifies to $\eta(n) = (\beta_o + 1)n/(n+\kappa)$. The sign of $d\hat{\lambda}^*/dB$ is determined by $\text{sgn}(1-\eta(n^*))$: the approximate optimal fidelity is locally increasing when $\eta < 1$ and decreasing when $\eta > 1$. Despite potential intermediate non-monotonicity, the limiting behavior is well-characterized:

- Small budgets ($B \to 0$): $\hat{\lambda}^*(B) \to \lambda_{\min}$
- High budgets ($B \to \infty$): $\hat{\lambda}^*(B) \to \lambda_{\max}$

For practical implementation, the approximate optimal fidelity curve can be discretized into a lookup table: for each budget B_j in a grid, solve the optimization to obtain $\hat{\lambda}^*(B_j)$ and store the mapping.

5.3 Budget-Performance Curves

The *budget-performance curve* $\varepsilon^*(B)$ describes minimum achievable error as a function of budget:

$$\varepsilon^*(B) = \min_{\lambda \in [\lambda_{\min}, \lambda_{\max}]} \varepsilon(\lambda, B) = \varepsilon(\lambda^*(B), B). \tag{16}$$

Using the described modeling framework, this can be approximated by $\hat{\varepsilon}^*(B)$ as:

$$\hat{\varepsilon}^*(B) = \min_{\lambda \in \Lambda} \left\{ \mu_{\text{mod}}(\lambda) + \mu_{\text{opt}}\left(\frac{B}{c_\lambda}\right) \right\} = \mu_{\text{mod}}(\hat{\lambda}^*(B)) + \mu_{\text{opt}}\left(\frac{B}{c_{\hat{\lambda}^*(B)}}\right). \tag{17}$$

Theorem 1. *(Budget-Performance Monotonicity) Suppose $\mu_{\text{opt}}(n, \lambda)$ is non-increasing in n for each fixed λ. Then $\hat{\varepsilon}^*(B)$ is non-increasing in B.*

Proof. Let $B_1 \leq B_2$ and let $\hat{\lambda}_1^* = \hat{\lambda}^*(B_1)$ be optimal at budget B_1. At budget B_2, using fidelity $\hat{\lambda}_1^*$ yields iteration count $n_2 = \lfloor B_2/c_{\hat{\lambda}_1^*} \rfloor \geq n_1$. Since μ_{opt} is non-increasing in n: $\hat{\varepsilon}(\hat{\lambda}_1^*, B_2) \leq \hat{\varepsilon}(\hat{\lambda}_1^*, B_1) = \hat{\varepsilon}^*(B_1)$. Since $\hat{\lambda}_2^* = \hat{\lambda}^*(B_2)$ minimizes error at B_2: $\hat{\varepsilon}^*(B_2) \leq \hat{\varepsilon}(\hat{\lambda}_1^*, B_2) \leq \hat{\varepsilon}^*(B_1)$.

The proof relies on observing that any fidelity choice available at budget B_1 remains available at $B_2 \geq B_1$ with at least as many iterations. As $B \to \infty$, the predicted error converges to $\varepsilon_{m,0} + \varepsilon_{o,0}$ (irreducible floors). The rate of convergence for large budgets follows $\hat{\varepsilon}^*(B) - \hat{\varepsilon}^*(\infty) = O(B^{-\beta_o})$.

Budget-performance curves support several decision-making scenarios: budget selection for target accuracy by inverting the curve to find $\hat{B}_{\min} = \inf\{B : \hat{\varepsilon}^*(B) \leq \varepsilon_{\text{tol}}\}$; efficiency assessment by comparing achieved error against $\hat{\varepsilon}^*(B)$; resource justification by quantifying expected performance gain from additional resources; and algorithm comparison by comparing curves across different optimizers.

5.4 Summary of Framework Outputs

The analysis yields three complementary tools for fidelity selection based on the meta-model approximation: (1) Approximate optimal fidelity $\hat{\lambda}^*(B)$: the recommended model choice for a given budget, with existence guaranteed under mild continuity assumptions; (2) Fidelity selection curves: the mapping from budget to approximate optimal fidelity, which may exhibit non-monotonic behavior at intermediate budgets; (3) Budget-performance curves $\hat{\varepsilon}^*(B)$: the predicted achievable error as a function of budget, always non-increasing with diminishing returns at high budgets.

6 Model Calibration and Implementation

6.1 Software

A basic open-source Python implementation of the framework is available at `tinyurl.com/metafi-test`. It features an interactive GUI for visualization of the presented models, approximate optimal-fidelity and budget-performance plots for fitted parameters, as well as support for continuous and discrete fidelity modes. The example plots in the section with the case study results are generated with the software.

6.2 Model Calibration

The optimization meta-process model framework requires data or quantitative experience to calibrate the model parameters. This can be historical data from representative optimization problems, or data obtained from specific calibration runs on the available fidelity models, or representative surrogate or synthetic problems.

The model for the expectation of the modeling error μ_{mod} has $p \in [2, 4]$ parameters (dependent on if $\varepsilon_{m,0}$ and ρ are set to zero or free parameters), and can thus be calibrated with p or more fidelity levels. To approximate the expected error, or its distribution for each fidelity level s, uniform random samples can be used (with s typically larger than the problem dimensionality).

Optimization error curves can be estimated from a few optimization run histories per fidelity level, though statistical characterization benefits from $n_{\text{rep}} \approx 5$–15 repeated runs, consistent with standard benchmarking practice [7]. In the case that this is cost-prohibitive, as a last resort, a single iteration vs optimization gap model can be used, fitted on optimization run history data of a low-fidelity model.

Cost model parameters require only a few function evaluation timing measurements per fidelity that can be obtained from the data to calibrate the model and optimization error models.

Typically, more data will improve the quality of the models for the expectation and uncertainty distributions. In industrial and engineering optimization contexts, it is often realistic to have data from optimization runs on similar or representative problem types. When calibration data is severely limited, the usual statistical caveats regarding low-sample inference naturally apply. Nevertheless, the parsimonious nature of the model, combined with interactive curve visualization, offers a structured approach to making the most of scarce data, bridging the gap between purely qualitative expert judgment and comprehensive empirical calibration.

For each fidelity level λ_j $(j = 1, \ldots, J)$:

1. Run K optimizations with different random seeds or problem variants
2. Record convergence histories $\{C_{jk}(n)\}_{n=1}^{N}$
3. Compute final objectives C_{jk}^{final} and timing t_{jk}

To estimate modeling error, designs optimized at lower fidelity are re-evaluated at the highest fidelity:

$$\hat{\epsilon_{\text{mod}\,j}} = \frac{1}{K}\sum_{k=1}^{K}\left[C_{\text{high}}(x_{jk}^{*}) - C_j(x_{jk}^{*})\right] \tag{18}$$

where x_{jk}^{*} is the design from run k at fidelity j, and $C_{\text{high}}(\cdot)$ evaluates at highest fidelity.

Parameters are estimated via nonlinear least squares. The modeling error model has 3–4 parameters and can be calibrated with as few as $k \geq 4$ fidelity levels. Optimization error curves can be estimated from a few optimization run histories per fidelity level, though statistical characterization benefits from approximately 5–15 repeated runs, consistent with standard benchmarking practice [7]. The typically moderate R^2 values of the optimization error model [6,29] reflect intrinsic run-to-run variance in optimization performance, not model misspecification. This variability is explicitly captured CV and propagated in the modeling framework. The CV is estimated as the average relative standard deviation across fidelity levels.

7 Case Studies

The use of the described optimization process model framework is demonstrated and validated through two case studies: structural topology optimization and aerodynamic airfoil optimization, spanning different application domains, solvers, and fidelity sources.

7.1 Case Study 1: Topology Optimization

Study Design. A rectangular cantilever beam (size ratio 2×1) with left edge fixed is optimized for minimum compliance using the SIMP method [3] for topology optimization. Eight mesh resolutions define discrete fidelities, partitioned into fitting and validation sets:

Calibration set: 20×10, 40×20, 60×30 , 80×40.

Validation set: 32×16 $(\lambda = 0.10)$, 48×24 $(\lambda = 0.23)$, 72×36 $(\lambda = 0.52)$.

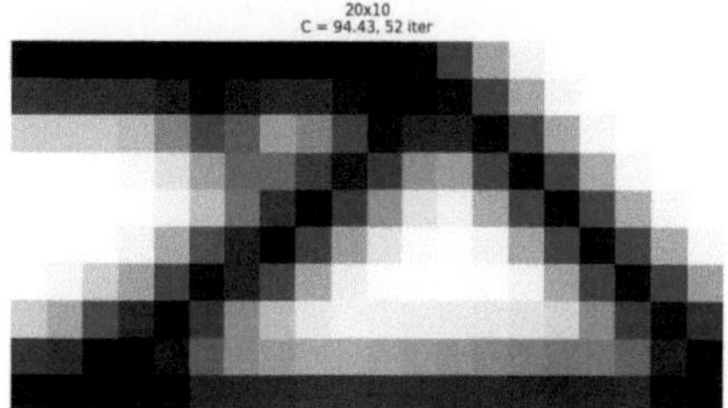

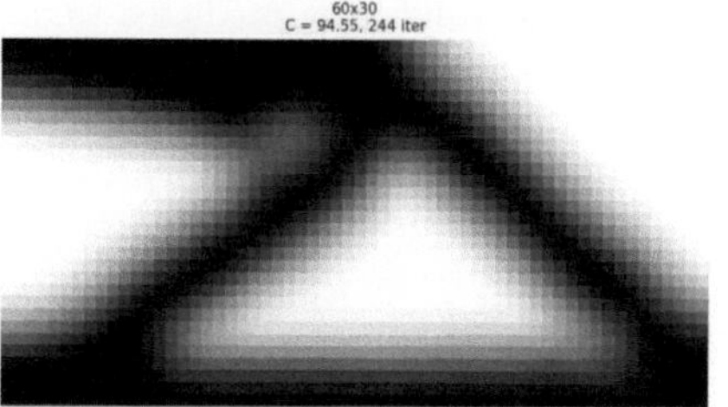

Fig. 1. Optimized topologies at different fidelity levels for cantilever beam. **Left**: Low fidelity (20×10). **Right**: High fidelity (60×30). White = solid, black = void.

Table 1. Empirical modeling error statistics at discrete fidelity levels.

Resolution	λ	Cost (norm.)	Mean Error	Set
20×10	0.04	1.00	21.17 ± 21.6	FIT
32×16	0.10	2.16	8.45 ± 8.6	VAL
40×20	0.16	3.47	5.62 ± 5.8	FIT
48×24	0.23	5.19	3.76 ± 3.9	VAL
60×30	0.36	9.89	2.36 ± 2.4	FIT
72×36	0.52	16.27	1.47 ± 1.5	VAL
80×40	0.64	25.18	1.03 ± 1.0	FIT
100×50	1.00	50.66	0.00	REF

Reference: 100×50 ($\lambda = 1.0$).

To capture cross-problem uncertainty: 3 load cases (point load at bottom-right, center-right, distributed bottom) and 3 volume fractions $v_f \in \{0.4, 0.5, 0.6\}$ were used in combination with different random seeds.

As an illustrative example, Fig. 1 compares optimized topologies at low and high fidelity for a representative problem. The coarse design captures the essential load-bearing structure but with discretization artifacts; the fine mesh shows resolved features with sharper boundaries. Despite similar reported compliance, re-evaluating the coarse design on the fine mesh reveals a modeling error.

Model Selection and Results. Two functional forms were compared for modeling error. The exponential form $\mu_{\text{mod}} = \alpha \exp(-\beta\lambda) + \varepsilon_0$ achieved fitting $R^2 = 0.984$ and was selected based on AIC.

The optimization error model achieves $R^2 = 0.56$ averaged across fidelities, reflecting heterogeneous convergence across the optimization runs. The scaling factor $s_\alpha = 0.11$ indicates that α_o increases approximately 11% at the lowest fidelity relative to the highest. Table 1 shows empirical modeling errors at all fidelity levels.

Table 2. Validation: predicted vs. actual modeling error at held-out fidelities.

Resolution	λ	Actual	Predicted	Rel. Error
32×16	0.10	8.45	10.35	22.6%
48×24	0.23	3.76	3.21	14.7%
72×36	0.52	1.47	1.52	3.7%
Validation R^2			0.84	
Validation MAPE			13.6%	

Table 3. Fitted meta-model parameters from airfoil optimization.

Component	Functional Form	Parameters	CV	R^2
Modeling Error	Power law	$\alpha = 6.24$, $\beta = 1.62$	0.54	0.971
Cost	Power law	$c_0 = 0.82$, $\alpha_c = 0.97$, $\gamma = 1.62$	0.11	0.981
Optimization	Power law	$\alpha_o = 10.0$, $\beta_o = 1.09$	0.46	0.34

Model Validation. Table 2 shows validation results on held-out fidelities, including Mean Absolute Percentage Error (MAPE). The validation $R^2 = 0.84$ demonstrates good generalization—the model accurately predicts modeling error at fidelity levels not used during fitting.

7.2 Case Study 2: Airfoil Shape Optimization

Study Design. Airfoil shapes are optimized using the panel method to maximize lift-to-drag ratio (L/D). Four model sizes define fidelities: `small` ($\lambda = 0.33$), `medium` ($\lambda = 0.50$), `large` ($\lambda = 0.67$), `xlarge` ($\lambda = 0.83$, reference). To capture cross-problem variation, the following variants were combined: 3 base airfoils (NACA 0012, 2412, 4412) $\times$ 3 Reynolds numbers (5×10^5, 10^6, 2×10^6). The airfoil shapes are parameterized using 8 CST coefficients and optimized with evolutionary optimization.

Results. Table 3 summarizes the fitted parameters. The modeling error model achieves $R^2 = 0.971$ using a power-law form. The lower optimization $R^2 = 0.34$ reflects the stochastic nature of evolutionary optimization across diverse problem variants. Table 4 shows empirical statistics at each fidelity level. The modeling error represents the absolute difference in L/D when a design optimized at lower fidelity is re-evaluated at the reference fidelity. An extended validation study using 6 model sizes with held-out validation achieves validation $R^2 = 0.811$ and MAPE $= 19.4\%$.

7.3 Comparative Analysis

Table 5 compares the two case studies. Both case studies demonstrate that the meta-model framework captures fidelity-accuracy trade-offs across different application domains.

Table 4. Empirical modeling error and cost at discrete fidelity levels (airfoil).

Model Size	λ	Cost (norm.)	Mean Error (ΔL/D)	Std
`small`	0.33	1.00	3.18	$\pm$1.76
`medium`	0.50	1.10	2.13	$\pm$1.70
`large`	0.67	1.36	1.21	$\pm$0.99
`xlarge`	0.83	1.53	0.00	0.00

Table 5. Comparison of case studies.

Characteristic	Topology Opt.	Airfoil Opt.
Objective	Minimize compliance	Maximize L/D
Fidelity parameter	Mesh resolution	Panel Fidelity
Cost range	1×–51×	1×–1.5×
Modeling error R^2	0.984 (exponential)	0.971 (power law)
Validation R^2	0.84	0.811
Optimization R^2	0.56 (avg)	0.34
Problem variants	9 (3 loads $\times$ 3 v_f)	9 (3 airfoils $\times$ 3 Re)

7.4 Fidelity Selection Results

Figure 2 presents the calibrated meta-model and fidelity selection analysis for topology optimization. Key observations: (1) At $B = 250$, optimal fidelity $\lambda^* \approx 0.33$ balances modeling and optimization error, corresponding approximately to the 60×30 mesh. (2) The optimal fidelity curve increases with budget, transitioning from low to high fidelity across $B \in [10, 10000]$. (3) The budget-performance curve shows diminishing returns at high budgets. (4) Wide 95% CI reflects substantial problem-to-problem variance, properly captured by the gamma distribution model. For airfoil optimization, the narrower cost range of the panel method makes the highest fidelity affordable even at modest budgets (B¿100).

7.5 Result Discussion

Summary. Both case studies provide complementary validation: topology optimization achieves model fit $R^2 = 0.984$ and validation $R^2 = 0.84$ (MAPE = 13.6%); airfoil optimization achieves fitting $R^2 = 0.988$ and validation $R^2 = 0.811$ (MAPE = 19.4%). Despite simplifying assumptions, the parsimonious model is effective.

Limitations. The additive decomposition assumes separable errors; in practice, modeling bias affects the optimization landscape. The moderate optimization R^2 values [6,29] reflect inherent run-to-run variance rather than model misspecification.

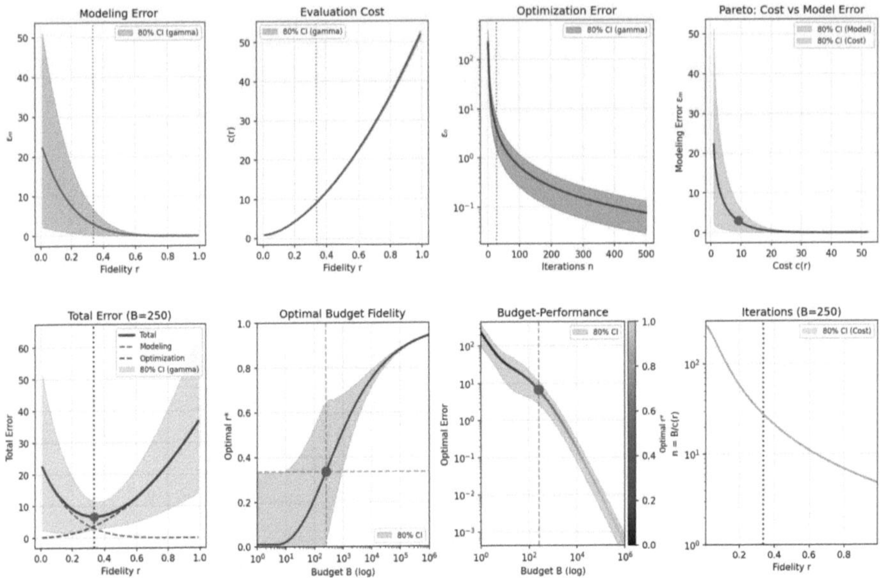

Fig. 2. Meta-model visualization for topology optimization. **Top row**: (a) modeling error vs. fidelity, (b) cost vs. fidelity, (c) optimization error vs. iterations, (d) Pareto front. **Bottom row**: (e) error decomposition at $B = 250$, (f) optimal fidelity vs. budget, (g) budget-performance curve (color indicates λ^*), (h) iteration count vs. fidelity. Shaded regions: 80% CI. Red markers: $B = 250$ operating point. (Color figure online)

8 Discussion and Outlook

The results presented in the previous section demonstrate that the proposed statistical meta-process modeling framework can effectively approximate and capture the complex interplay between modeling fidelity, computational cost, and optimization convergence. By decomposing the approximate total error into parsimonious components that can be calibrated, the framework transforms the often intuitive or ad-hoc process of fidelity selection into a quantitative decision-making problem.

A key insight revealed by the analysis is the non-trivial nature of the optimal fidelity curve $\hat{\lambda}^*(B)$. Contrary to the common heuristic that one should simply "use the highest fidelity afforded by the budget," the model and corresponding analysis enable the identification of budget regimes where intermediate fidelities yield superior total error performance.

The presented modeling framework, like other "a priori" approaches, is limited in that it requires data or estimates to calibrate its parameters. In many industrial settings, prior data on different but similar representative problem instances is available from previous modeling comparisons or optimization runs. In other cases where such data is not available, this "cold start" problem requires spending computational resources to save resources. However, this cost can be justifiable in two common scenarios: (1) *Recurring optimization tasks*, where instances of a problem class are solved repeatedly. Here, the calibration cost is amortized over many future runs. (2) *High-stakes optimization*, where the computational budget and iteration cost are extremely large (e.g., weeks of HPC time), making a smaller pilot study negligible in comparison to the potential waste of a failed full-scale run.

Future work includes investigations to create improved and extended models that capture additional aspects and complexity from model-based optimization, such as more realistic computational resource usage considering parallel distributed computing, memory, and license restrictions. Besides statistical modeling, as in the present work, optimization meta-process simulation can also be explored for testing and assessing online multi-fidelity optimization approaches. Furthermore, investigations into efficient Design of Experiments, data collection, and sampling methods to guide and improve parameter fitting would be useful. Finally, for more affordable calibration of fidelity-dependent optimization progress models and meta-process simulations, investigations on interactions with representative surrogate-model-based [25] approaches are relevant.

9 Conclusions

This work introduced a statistical meta-model framework for offline fidelity selection in model-based optimization. The parametric decomposition into modeling error, cost, and optimization error enabled useful analysis of the fidelity-budget trade-off. The framework is applicable to settings where previous pilot data or experience can inform parameter estimation for the models. An accompanying open-source software implementation of the framework with a graphical user interface is provided and publicly available. Two case studies using simulation-based optimization were presented and used to demonstrate and validate the methodology.

Key contributions include a practical framework and software implementation tool that helps to select the "fit-for-purpose" model resolution, moving beyond ad hoc qualitative heuristics. Additionally, useful general methodological tools such as budget vs. optimal-fidelity curves and budget-performance plots have been introduced. These enable the visualization of expected model-fidelity trade-offs and diminishing returns for increasing computational budgets. Furthermore, the integration of uncertainty quantification models enables risk-aware decision-making.

As model-based optimization problems grow in complexity and cost, optimization meta-process modeling approaches that leverage available data to inform quantitative decisions for related future tasks can enable and improve rational resource-efficient computational workflows and design systems [24]. Future work directions include extending the models to capture parallel computing constraints and investigating efficient calibration methods.

Acknowledgments. This work was funded by the Deutsche Forschungsgemeinschaft (DFG, German Research Foundation) Project number 465355776. To support writing and code development, AI-assisted tools (Grammarly, Claude, Gemmeni, and GPT5) were used; all input and outputs were reviewed and edited by the authors.

References

1. Amaran, S., Sahinidis, N.V., Sharda, B., Bury, S.J.: Simulation optimization: a review of algorithms and applications. Ann. Oper. Res. **240**(1), 351–380 (2016). https://doi.org/10.1007/s10479-015-2019-x
2. Bartz-Beielstein, T., et al.: Benchmarking in optimization: best practice and open issues. arXiv preprint arXiv:2007.03488 (2020)
3. Bendsøe, M.P., Sigmund, O.: Topology Optimization: Theory, Methods, and Applications. Springer (2003)
4. Chen, C.H., Lin, J., Yücesan, E., Chick, S.E.: Simulation budget allocation for further enhancing the efficiency of ordinal optimization. Discrete Event Dyn. Syst. **10**(3), 251–270 (2000)
5. Colaso, A., Prieto, P., Herrero, J.A., Abad, P., Puente, V., Gregorio, J.A.: Accuracy vs. computational cost tradeoff in distributed computer system simulation. arXiv preprint arXiv:1902.02837 (2019)
6. Ghasempour, A.: Support vector regression to predict power consumption. Technical report, University of New Mexico (2024), eCE Publications, UNM Digital Repository
7. Hansen, N., Auger, A., Ros, R., Mersmann, O., Tušar, T., Brockhoff, D.: COCO: a platform for comparing continuous optimizers in a black-box setting. Optim. Methods Softw. **36**(1), 114–144 (2021)
8. Kavetski, D., Qin, Y., Kuczera, G.: The fast and the robust: trade-offs between optimization robustness and cost in the calibration of environmental models. Water Resour. Res. **54**(11), 9432–9455 (2018)
9. Khimani, D., Mate, S., Bhartiya, S.: Model predictive control: trade-offs between performance and computation time. IFAC-PapersOnLine **57**, 77–82 (2024)
10. Kim, H., McGinnis, L.F., Zhou, C.: On fidelity and model selection for discrete event simulation. SIMULATION **88**(1), 97–109 (2012)
11. Koziel, S., Ciaurri, D.E., Leifsson, L.: Surrogate-based methods. In: Computational optimization, methods and algorithms, pp. 33–59. Springer, Cham (2011)
12. Krzywanski, J., Sosnowski, M., Grabowska, K., Zylka, A., Lasek, L., Kijo-Kleczkowska, A.: Advanced computational methods for modeling, prediction and optimization-a review. Materials **17**(14), 3521 (2024)
13. Lannelongue, L., et al.: GREENER principles for environmentally sustainable computational science. Nat. Comput. Sci. **3**(6), 514–521 (2023)
14. Lavin, A., et al.: Simulation intelligence: towards a new generation of scientific methods. arXiv arXiv:2112.03235 (2021)
15. Li, K., Li, F.: Multi-fidelity methods for optimization: a survey. arXiv preprint arXiv:2402.09638 (2024)
16. Li, Z., et al.: Train big, then compress: Rethinking model size for efficient training and inference of transformers. In: International Conference on machine learning, pp. 5958–5968. PMLR (2020)
17. Liu, L., Deng, J.: Dynamic deep neural networks: Optimizing accuracy-efficiency trade-offs by selective execution. In: Proceedings of the AAAI Conference on Artificial Intelligence, vol. 32 (2018)

18. Moradinia, Z., Vandierendonck, H., Murphy, A.: Navigating speed-accuracy trade-offs for multi-physics simulations. Meccanica **60**(6), 1583–1599 (2025)
19. Mullins, J., Mahadevan, S.: Variable-fidelity model selection for stochastic simulation. Reliab. Eng. Syst. Safety **131**, 40–52 (2014)
20. Nesterov, Y.: Lectures on Convex Optimization. Springer, 2 edn. (2018)
21. Peherstorfer, B., Willcox, K., Gunzburger, M.: Survey of multifidelity methods in uncertainty propagation, inference, and optimization. SIAM Rev. **60**(3), 550–591 (2018)
22. Roache, P.J.: Verification and Validation in Computational Science and Engineering. Hermosa Publishers (1998)
23. Sala, R.: Primitive agentic first-order optimization. In: 2024 10th International Conference on Control, Decision and Information Technologies (CoDIT), pp. 2756–2762. IEEE (2024)
24. Sala, R.: Computational rational engineering and development: synergies and opportunities. In: In: Arai, K. (eds.) Proceedings of SAI Intelligent Systems Conference, pp. 744–763. Springer, Cham (2021). https://doi.org/10.1007/978-3-030-82193-7_50
25. Sala, R., Baldanzini, N., Pierini, M.: Representative surrogate problems as test functions for expensive simulators in multidisciplinary design optimization of vehicle structures. Struct. Multidiscip. Optim. **54**(3), 449–468 (2016)
26. Sala, R., Müller, R.: Benchmarking for metaheuristic black-box optimization: perspectives and open challenges. In: 2020 IEEE Congress on Evolutionary Computation (CEC), pp. 1–8. IEEE (2020)
27. Sala, R., Schlüter, A., Sator, C., Müller, R.: Unifying relations between iterative linear equation solvers and explicit Euler approximations for associated parabolic regularized equations. Results Appl. Math. **13**, 100227 (2022)
28. Takenaga, S., Ozaki, Y., Onishi, M.: Dynamic fidelity selection for hyperparameter optimization. In: Proceedings of the Companion Conference on Genetic and Evolutionary Computation, pp. 2304–2307 (2023)
29. Tharwat, A.: Classification assessment methods. Appl. Comput. Inf. **17**(1), 168–192 (2021). https://doi.org/10.1016/j.aci.2018.08.003
30. Wistuba, M., Kadra, A., Grabocka, J.: Supervising the multi-fidelity race of hyperparameter configurations. Adv. Neural. Inf. Process. Syst. **35**, 13470–13484 (2022)
31. Yudin, D.B.: Problem complexity and method efficiency in optimization. Wiley (1983)
32. Zhu, D., Xin, Z., Zheng, S., Wang, Y., Yang, X.: Addressing the accuracy-cost trade-off in material property prediction using a teacher-student strategy. J. Chem. Theory Comput. **20**(13), 5743–5750 (2024)
33. Xhu, S., et al.: Intelligent computing: the latest advances, challenges, and future. Intell. Comput. **2**, 0006 (2023)

Simplified Rcone-Programming Scheme for Restraining Delay-Error Peaks in Variable-Delay Systems

Tian-Bo Deng(✉)

Dept. Information Science, Toho University, Miyama 2-2-1, Funabashi 274-8510, Chiba, Japan
deng@is.sci.toho-u.ac.jp

Abstract. The paper formulates a rotated-quadratic cone (Rcone) technique for producing a variable allpass-delay filter (AP-DF) while significantly containing the delay-error overshoots. After showing that both frequency-response (FR) error and delay-error have denominators and numerators being linear functions of the AP-DF parameters, the paper reveals a minimization scheme for designing a variable AP-DF. The design technique uses the Rcone-programming to minimize the FR-error peak while effectively containing the delay-error overshoots. Since the Rcone-programming design reduces to a convex minimization, performing the convex minimization yields a convergent AP-DF solution. The paper provides an AP-DF design example to exemplify the utility of the Rcone-programming formulation. As compared to other design techniques, this new scheme can reduce the design to a convex-programming problem, rendering it easy to solve. The design simulations also verify that this Rcone-programming formulation can further suppress the delay overshoots as compared with other design techniques. More precisely, the Rcone-programming design causes the smallest delay-error peak (0.002960) among all the designs.

Keywords: Rcone-programming · Rcone constraint · Delay-error peak · Allpass-delay filter (AP-DF) · Variable AP-DF

1 Introduction

The filters with tunable delay are exploited in a broad range of fields, including signal interpolation as well as timing-offset compensation in communications. A typical application is to use such a filter to compensate timing offsets in digital receivers, leading to high-fidelity communications. Because timing offsets are usually not integers, timing-offset compensation requires a filter with tunable non-integer delay.

Either non-recursive or recursive structures can be utilized to design a filter with fractional-delay, but the design using a non-recursive structure must

N. A. Mohd Subha et al. (Eds.): SMS 2026, CCIS 2956, pp. 176–188, 2026.
https://doi.org/10.1007/978-981-92-0306-2_13

approximate both magnitude and delay, rendering the design significantly challenging [1–8]. To the contrary, the design using a recursive allpass structure is less challenging because the allpass structure holds a unity-amplitude response [9–18]. Hence, it suffices to approximate the delay only.

This paper considers designing a variable allpass-delay filter (AP-DF) using such an allpass structure. Most existing techniques design AP-DFs, aiming to minimize either the peak of frequency-response (FR) error or the peak of phase-response error [9–15], while those designs are unable to effectively restrain the delay-error peak. Such large delay-error overshoots may cause severe deterioration of filter performance. The paper aims to investigate a new technique for restraining the delay-error overshoots.

This paper begins with reviewing the FR-error and delay-error of the variable AP-DF. The numerators and denominators of both errors are related to the filter parameters linearly, which enables the design to be formulated using the rotated-cone (Rcone) programming. The design minimizes the FR-error peak while effectively suppressing the delay-error peak. Despite the fact that the Rcone-programming constraint is an approximate one, given that the constraint on the delay errors is feasible, performing the Rcone-programming yields a convergent design. This paper shows an AP-DF design for validating the utility of the novel technique. Comparisons with other designs are also carried out to verify the superiority of this novel technique. As compared with other design techniques, computer simulations confirm that the novel Rcone-programming formulation can further suppress the delay overshoot, achieving the smallest delay-error peak among all of the design solutions.

2 Two Error Functions

This section states both variable FR-error and variable delay-error. Both errors are the rational functions whose denominators and numerators are linear functions of the AP-DF parameters.

The ideal variable AP-DF should have the delay

$$\rho_{\mathrm{d}}(\omega, q) = J + q \tag{1}$$

with J being an integer, and q being a fraction,

$$\begin{aligned} q &\in Q = [q_{\min}, q_{\max}] \\ q_{\max} &= 1 + q_{\min}. \end{aligned} \tag{2}$$

Typically, $q_{\min}$ is set to $q_{\min} = -0.5$. Here, $\rho_{\mathrm{d}}(\omega, q)$ is approximated by using the Jth-order AP-DF

$$H(z, q) = \frac{D(z^{-1}, q)}{D(z, q)} \cdot z^{-J} \tag{3}$$

with

$$D(z,q)=\sum_{i=0}^{J}a_i(q)z^{-i}$$
$$a_i(q)=\sum_{m=1}^{M}b_{im}q^m,\quad a_0(q)=1. \tag{4}$$

The ideal delay $\rho_{\rm d}(\omega,q)$ is related to ideal phase

$$\Theta_{\rm d}(\omega,q)=-\omega(J+q). \tag{5}$$

It also corresponds to the ideal FR

$$F(\omega,q)=e^{j\Theta_{\rm d}(\omega,q)} \tag{6}$$

while the true FR is

$$H(\omega,q)=\frac{\overline{D}(\omega,q)}{D(\omega,q)}\cdot e^{-jJ\omega} \tag{7}$$

with

$$D(\omega,q)=\sum_{i=0}^{J}a_i(q)e^{-ji\omega}=1+\sum_{i=1}^{J}\sum_{m=1}^{M}b_{im}e^{-ji\omega}q^m$$
$$\overline{D}(\omega,q)=\sum_{i=0}^{J}a_i(q)e^{ji\omega}=1+\sum_{i=1}^{J}\sum_{m=1}^{M}b_{im}e^{ji\omega}q^m. \tag{8}$$

The derivation in [16] shows the FR-error

$$\begin{aligned}e_H(\omega,q)&=H(\omega,q)-F(\omega,q)\\&=je^{-j(J+\frac{q}{2})\omega}\cdot\frac{\sum_{i=0}^{J}2a_i(q)\sin\left(i+\frac{q}{2}\right)\omega}{D(\omega,q)}\end{aligned} \tag{9}$$

and thus

$$\begin{aligned}|e_H(\omega,q)|&=\left|\frac{\sum_{i=0}^{J}2a_i(q)\sin\left(i+\frac{q}{2}\right)\omega}{D(\omega,q)}\right|\\&=\frac{\left|\sum_{i=0}^{J}2a_i(q)\sin\left(i+\frac{q}{2}\right)\omega\right|}{\left|\overline{D}(\omega,q)\right|}.\end{aligned} \tag{10}$$

The AP-DF phase is

$$\Theta(\omega,q)=2\alpha(\omega,q)-J\omega \tag{11}$$

while

$$\alpha(\omega,q)=\angle\overline{D}(\omega,q)$$

is the angle of $\overline{D}(\omega, q)$. The derivative of $\alpha(\omega, q)$ is

$$\frac{\partial \alpha(\omega, q)}{\partial \omega} = \frac{\sum_{i=1}^{J} i a_i(q) \cos[i\omega - \alpha(\omega, q)]}{|\overline{D}(\omega, q)|} \tag{12}$$

leading to the entire delay

$$\rho(\omega, q) = -\frac{\partial \Theta(\omega, q)}{\partial \omega} = J + g(\omega, q) \tag{13}$$

with

$$g(\omega, q) = \frac{\sum_{i=1}^{J} (-2i) a_i(q) \cos[i\omega - \alpha(\omega, q)]}{|\overline{D}(\omega, q)|}.$$

Obviously, $g(\omega, q)$ relates to the AP-DF coefficients nonlinearly because it involves $\overline{D}(\omega, q)$ and $\alpha(\omega, q)$ in (13). Since $2\alpha(\omega, q)$ must be close to $(-\omega q)$, i.e.,

$$2\alpha(\omega, q) \approx -\omega q \tag{14}$$

this results in the approximation

$$\alpha(\omega, q) \approx -\frac{\omega q}{2} \tag{15}$$

as well as the approximation of $g(\omega, q)$

$$g(\omega, q) \approx \frac{\sum_{i=1}^{J} (-2i) a_i(q) \cos\left(i + \frac{q}{2}\right) \omega}{|\overline{D}(\omega, q)|}. \tag{16}$$

Here, the nonlinear term $|\overline{D}(\omega, q)|$ can be linearized as the functions of b_{im} as follows.

Let

$$\begin{aligned} \overline{D}(\omega, q) &= x(\omega, q) + jy(\omega, q) \\ x(\omega, q) &= \sum_{i=0}^{J} a_i(q) \cos(i\omega) \\ y(\omega, q) &= \sum_{i=0}^{J} a_i(q) \sin(i\omega). \end{aligned} \tag{17}$$

Figure 1 shows how $|D(\omega, q)|$ is related to $\alpha(\omega, q)$. Expressing

$$\begin{aligned} |D(\omega, q)| &= x(\omega, q) \cdot \cos \alpha(\omega, q) + \sin \alpha(\omega, q) \cdot y(\omega, q) \\ &= \sum_{i=0}^{J} a_i(q)[\cos \alpha(\omega, q) \cdot \cos(i\omega) + \sin \alpha(\omega, q) \cdot \sin(i\omega)] \\ &= \sum_{i=0}^{J} a_i(q) \cos[i\omega - \alpha(\omega, q)] \end{aligned} \tag{18}$$

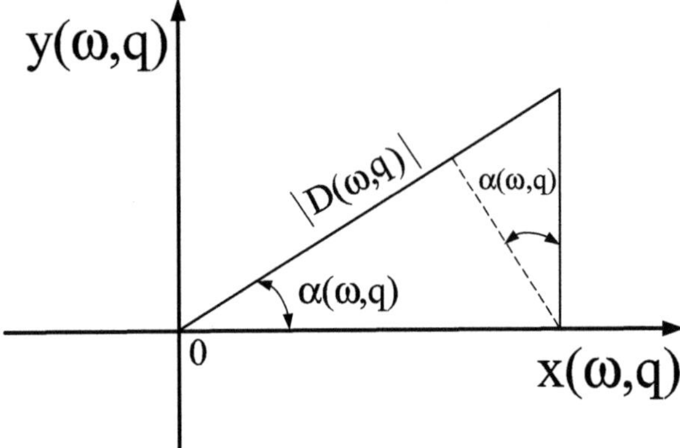

Fig. 1. Deriving $|D(\omega, q)|$.

the relation in (15) leads to

$$|D(\omega, q)| \approx \sum_{i=0}^{J} a_i(q) \cos\left(i + \frac{q}{2}\right)\omega. \tag{19}$$

Thus, $|e_H(\omega, q)|$ and $g(\omega, q)$ are approximated as

$$\begin{aligned} |e_H(\omega, q)| &\approx \frac{\left|\sum_{i=0}^{J} 2a_i(q) \sin\left(i + \frac{q}{2}\right)\omega\right|}{\sum_{i=0}^{J} a_i(q) \cos\left(i + \frac{q}{2}\right)\omega} \\ g(\omega, q) &\approx \frac{\sum_{i=1}^{J} (-2i) a_i(q) \cos\left(i + \frac{q}{2}\right)\omega}{\sum_{i=0}^{J} a_i(q) \cos\left(i + \frac{q}{2}\right)\omega}. \end{aligned} \tag{20}$$

Instead of minimizing $|e_H(\omega, q)|$, the design simply needs to minimize

$$\begin{aligned} \epsilon_H(\omega, q) &= \frac{\sum_{i=0}^{J} a_i(q) \sin\left(i + \frac{q}{2}\right)\omega}{\sum_{i=0}^{J} a_i(q) \cos\left(i + \frac{q}{2}\right)\omega} \\ &= \frac{\sin\left(\frac{\omega q}{2}\right) + \sum_{i=1}^{J} \sum_{m=1}^{M} b_{im} \sin\left(i + \frac{q}{2}\right)\omega \cdot q^m}{\cos\left(\frac{\omega q}{2}\right) + \sum_{i=1}^{J} \sum_{m=1}^{M} b_{im} \cos\left(i + \frac{q}{2}\right)\omega \cdot q^m}. \end{aligned} \tag{21}$$

Here, $\epsilon_H(\omega, q)$ is similar to the approximated phase-error function in [15], where the design minimizes the integral squared error while suppressing the FR-error peak. Moreover, since $g(\omega, q)$ is approximately equal to

$$\begin{aligned}\hat{g}(\omega, q) &= \frac{\sum_{i=1}^{J}(-2i)a_i(q)\cos\left(i+\frac{q}{2}\right)\omega}{\sum_{i=0}^{J}a_i(q)\cos\left(i+\frac{q}{2}\right)\omega} \\ &= \frac{\sum_{i=1}^{J}(-2i)a_i(q)\cos\left(i+\frac{q}{2}\right)\omega}{\cos\left(\frac{\omega q}{2}\right)+\sum_{i=1}^{J}a_i(q)\cos\left(i+\frac{q}{2}\right)\omega}\end{aligned} \tag{22}$$

the approximated delay error becomes

$$\begin{aligned}\epsilon_\rho(\omega, q) &= q - \hat{g}(\omega, q) \\ &= q + \frac{\sum_{i=0}^{J}(2i)a_i(q)\cos\left(i+\frac{q}{2}\right)\omega}{\sum_{i=0}^{J}a_i(q)\cos\left(i+\frac{q}{2}\right)\omega} \\ &= \frac{\sum_{i=0}^{J}(q+2i)a_i(q)\cos\left(i+\frac{q}{2}\right)\omega}{\sum_{i=0}^{J}a_i(q)\cos\left(i+\frac{q}{2}\right)\omega} \\ &= \frac{q\cos\left(\frac{\omega q}{2}\right)+\sum_{i=1}^{J}(q+2i)a_i(q)\cos\left(i+\frac{q}{2}\right)\omega}{\cos\left(\frac{\omega q}{2}\right)+\sum_{i=1}^{J}a_i(q)\cos\left(i+\frac{q}{2}\right)\omega} \\ &= \frac{q\cos\left(\frac{\omega q}{2}\right)+\sum_{i=1}^{J}\sum_{m=1}^{M}b_{im}\cos\left(i+\frac{q}{2}\right)\omega\cdot q^m(q+2i)}{\cos\left(\frac{\omega q}{2}\right)+\sum_{i=1}^{J}\sum_{m=1}^{M}b_{im}\cos\left(i+\frac{q}{2}\right)\omega\cdot q^m}.\end{aligned} \tag{23}$$

However, the real error should be

$$\begin{aligned}e_\rho(\omega, q) &= \rho(\omega, q) - \rho_{\mathrm{d}}(\omega, q) \\ &= q - g(\omega, q)\end{aligned} \tag{24}$$

where $g(\omega, q)$ denotes the delay in (13). This paper employs both (21) and (23) to formulate the design.

3 Rcone-Programming Design

To effectively restrain the delay-error peak while minimizing the FR-error peak, the design formulation is stated as

$$\begin{aligned} &\text{Minimize} \quad \epsilon \\ &\text{Subject to} \quad \begin{cases} |\epsilon_H(\omega, q)| \le \epsilon \\ |\epsilon_\rho(\omega, q)| \le \gamma \end{cases} \end{aligned} \tag{25}$$

with γ introduced for containing the delay-error peak.

In (19), since

$$\sum_{i=0}^{J} a_i(q) \cos\left(i + \frac{q}{2}\right)\omega > 0 \tag{26}$$

the constraint

$$|\epsilon_H(\omega, q)| \le \epsilon$$

can be manipulated to

$$\left|\sum_{i=0}^{J} a_i(q) \cdot \sin\left(i + \frac{q}{2}\right)\omega\right| \le \epsilon \cdot \sum_{i=0}^{J} a_i(q) \cdot \cos\left(i + \frac{q}{2}\right)\omega. \tag{27}$$

Approximating the above constraint as

$$\left|\sum_{i=0}^{J} a_i(q) \cdot \sin\left(i + \frac{q}{2}\right)\omega\right|^2 \le \epsilon \cdot \sum_{i=0}^{J} a_i(q) \cdot \cos\left(i + \frac{q}{2}\right)\omega \tag{28}$$

yields

$$\begin{aligned} &\left|\sin\left(\frac{\omega q}{2}\right) + \sum_{i=1}^{J}\sum_{m=1}^{M} b_{im} \sin\left(i + \frac{q}{2}\right)\omega \cdot q^m\right|^2 \\ &\le \epsilon \left[\cos\left(\frac{\omega q}{2}\right) + \sum_{i=1}^{J}\sum_{m=1}^{M} b_{im} \cos\left(i + \frac{q}{2}\right)\omega \cdot q^m\right]. \end{aligned} \tag{29}$$

To simplify the double-indexed b_{im}, we map

$$\begin{aligned} &(i, m) \longrightarrow k \\ &b_{im} \longrightarrow x_k \\ &\cos\left(i + \frac{q}{2}\right)\omega \cdot q^m \longrightarrow d_k(\omega, q) \\ &\cos\left(\frac{\omega q}{2}\right) \longrightarrow d_0(\omega, q) \\ &\sin\left(i + \frac{q}{2}\right)\omega \cdot q^m \longrightarrow s_k(\omega, q) \\ &\sin\left(\frac{\omega q}{2}\right) \longrightarrow s_0(\omega, q). \end{aligned} \tag{30}$$

This simplifies (29) as

$$\left| s_0(\omega, q) + \sum_{k=1}^{K} s_k(\omega, q) x_k \right|^2 \leq \epsilon \left[d_0(\omega, q) + \sum_{k=1}^{K} d_k(\omega, q) x_k \right] \tag{31}$$

with $K = JM$. This is the Rcone constraint

$$\boldsymbol{\Upsilon} = \begin{bmatrix} \epsilon \\ d_0(\omega, q) + \sum_{k=1}^{K} d_k(\omega, q) x_k \\ \sqrt{2} \left(s_0(\omega, q) + \sum_{k=1}^{K} s_k(\omega, q) x_k \right) \end{bmatrix} \in \mathcal{K}_r \tag{32}$$

where $\mathcal{K}_r$ is Rcone. Let

$$\begin{aligned} \boldsymbol{c}_1 &= \begin{bmatrix} 0 \\ d_0 \\ \sqrt{2} s_0 \end{bmatrix}, \quad \boldsymbol{y} = \begin{bmatrix} \epsilon \\ x_1 \\ x_2 \\ \vdots \\ x_K \end{bmatrix} \\ \boldsymbol{A}_1^T &= - \begin{bmatrix} 1 & 0 & 0 & \cdots & 0 \\ 0 & d_1 & d_2 & \cdots & d_K \\ 0 & \sqrt{2} s_1 & \sqrt{2} s_2 & \cdots & \sqrt{2} s_K \end{bmatrix} \end{aligned} \tag{33}$$

with $d_k(\omega, q)$ and $s_k(\omega, q)$ abbreviated as d_k, s_k respectively. Thus, the constraint in (32) reduces to

$$\boldsymbol{\Upsilon} = \boldsymbol{c}_1 - \boldsymbol{A}_1^T \boldsymbol{y} \in \mathcal{K}_r. \tag{34}$$

Next, let us consider the second constraint

$$|\epsilon_\rho(\omega, q)| \leq \gamma$$

in (25), i.e.,

$$\begin{aligned} &\left| q \cos\left(\frac{\omega q}{2}\right) + \sum_{i=1}^{J} \sum_{m=1}^{M} b_{im} \cos\left(i + \frac{q}{2}\right) \omega \cdot q^m (q + 2i) \right| \\ &\leq \gamma \left[\cos\left(\frac{\omega q}{2}\right) + \sum_{i=1}^{J} \sum_{m=1}^{M} b_{im} \cos\left(i + \frac{q}{2}\right) \omega \cdot q^m \right]. \end{aligned} \tag{35}$$

Since γ is a constant, the constraint becomes linear. Exploiting the same one-to-one mappings in (30) gets

$$\begin{aligned} &\cos\left(i + \frac{q}{2}\right) \omega \cdot q^m (q + 2i) \longrightarrow f_k(\omega, q) \\ &q \cos\left(\frac{\omega q}{2}\right) \longrightarrow f_0(\omega, q). \end{aligned} \tag{36}$$

Thus, the constraint in (35) can be reshaped to

$$\begin{cases} \sum_{k=1}^{K} (f_k - \gamma d_k)\, x_k \leq \gamma d_0 - f_0 \\ \sum_{k=1}^{K} (-f_k - \gamma d_k)\, x_k \leq \gamma d_0 + f_0 \end{cases} \tag{37}$$

where f_k and d_k stand for $f_k(\omega, q)$ and $d_k(\omega, q)$, respectively. Let

$$\begin{aligned} g_k &= f_k - \gamma d_k \\ \hat{g}_k &= -f_k - \gamma d_k. \end{aligned} \tag{38}$$

The matrix form of (37) becomes

$$\boldsymbol{c}_2 - \boldsymbol{A}_2^T \boldsymbol{y} \in \mathcal{K}_l \tag{39}$$

where $\mathcal{K}_l$ stands for the non-negative orthant, and

$$\boldsymbol{c}_2 = \begin{bmatrix} -g_0 \\ -\hat{g}_0 \end{bmatrix}, \quad \boldsymbol{A}_2^T = \begin{bmatrix} 0 \; g_1 \; g_2 \cdots g_K \\ 0 \; \hat{g}_1 \; \hat{g}_2 \cdots \hat{g}_K \end{bmatrix}. \tag{40}$$

Letting

$$\boldsymbol{b} = \begin{bmatrix} -1 \; 0 \; 0 \cdots 0 \end{bmatrix}^T$$

the problem in (25) reduces to

$$\begin{aligned} &\text{Maximize} \quad \boldsymbol{b}^T \boldsymbol{y} \\ &\text{Subject to} \quad \begin{cases} \boldsymbol{c}_1 - \boldsymbol{A}_1^T \boldsymbol{y} \in \mathcal{K}_r \\ \boldsymbol{c}_2 - \boldsymbol{A}_2^T \boldsymbol{y} \in \mathcal{K}_l. \end{cases} \end{aligned} \tag{41}$$

4 Design Validation

The ideal delay in (1) is approximated by setting

$$\begin{aligned} &\omega \in W = [0, 0.9\pi] \\ &q \in Q = [q_{\min}, q_{\max}] = [-0.5, 0.5] \\ &(M, J) = (5, 35) \\ &\gamma = 0.0035. \end{aligned} \tag{42}$$

It should be noted that the small γ is selected by trial and error. To suppress the delay-error overshoots, a small value should be chosen. However, it cannot be too small beause a too small value of γ may render the Rcone-programming infeasible.

The design is evaluated by using two error metrics

$$
\begin{aligned}
e_{H\max} &= \max\{20\log_{10}|e_H(\omega,q)|\}\ \text{(dB)} \\
e_{\rho\max} &= \max\{|e_\rho(\omega,q)|\}
\end{aligned}
\tag{43}
$$

where (ω, q) represent the samples of ω and q. Note that $e_H(\omega, q)$ represents the FR-error in (10), and the $e_\rho(\omega, q)$ represents the delay-error in (24). To perform the design in (41), $L_1 \times L_2$ uniform samples $(\omega, q) \in W \cup Q$ are used, where $(L_1, L_2) = (201, 31)$. Furthermore, more samples are employed in checking the accuracy, i.e., $(\hat{L}_1, \hat{L}_2) = (201, 61)$. Table 1 provides the delay-error peaks and the numbers of AP-DF coefficients. As compared to the designs in [10]-[13], the novel Rcone-programming technique can effectively restrain the delay-error peak, resulting in the smallest delay-error overshoot .

Table 1. Simulation Results.

Technique	$e_{\rho\max}$	Coeff. Number
WLS (Iterative Scheme) [10]	0.029087	210
WLS (Noniterative Scheme) [11]	0.006717	175
WLS (Iterative Scheme) [12]	0.003328	175
WLS (Iterative Scheme) [13]	0.014292	175
Rcone-Programming	0.002960	175

Figure 2 gives the FR-errors $|e_H(\omega, q)|$ (dB). As indicated in Fig. 2, the peak error is $e_{H\max} = 82.44$ (dB), which is located at $(\omega, q) = (0.0135\pi, 0.5)$. Furthermore, Fig. 3 and Fig. 4 detail the actual delay and delay-errors, respectively.

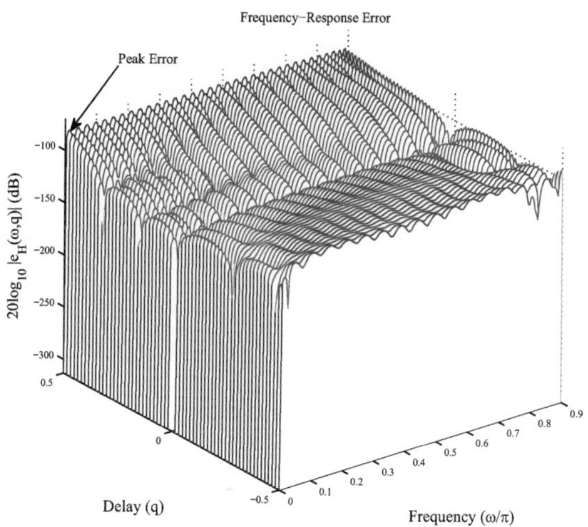

Fig. 2. AP-DF's FR-errors.

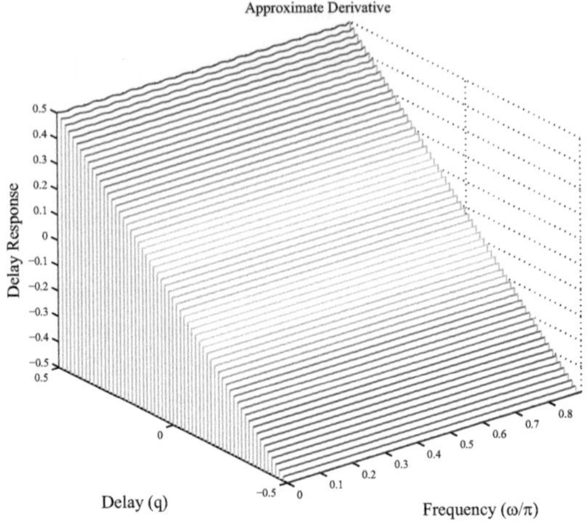

Fig. 3. AP-DF's fractional delay.

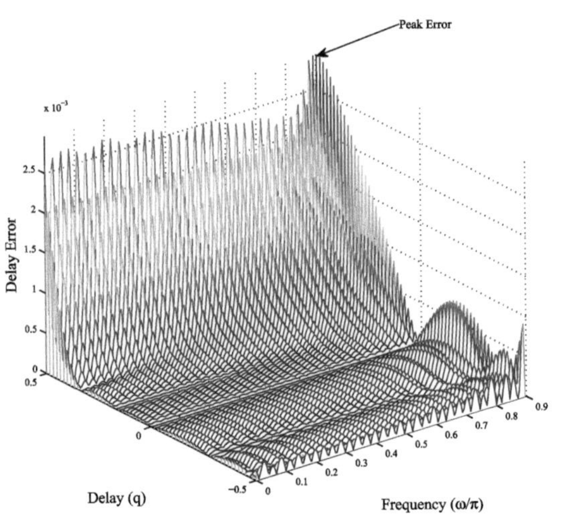

Fig. 4. AP-DF's delay-errors.

As pointed in Fig. 4, the peak of the delay-errors is $e_{\rho\,\max} = 0.002960$, which is located at $(\omega, q) = (0.8910\pi, 0.4833)$. In addition, Fig. 5 shows the largest radius versus q as $q \in Q$ varies. Figure 5 shows that the maximum radius (peak value located at $q = -0.5$) is 0.9422, which is less than unity. Therefore, the variable AP-DF $H(z, q)$ is stable.

Finally, it should be pointed out that the existing design techniques in [10]-[13] do not aim to suppress the overshoots of delay-errors, while the linear-programming scheme in [16] minimizes a delay-response expression being a non-linear function with respect to the AP-DF parameters, resorting to an itera-

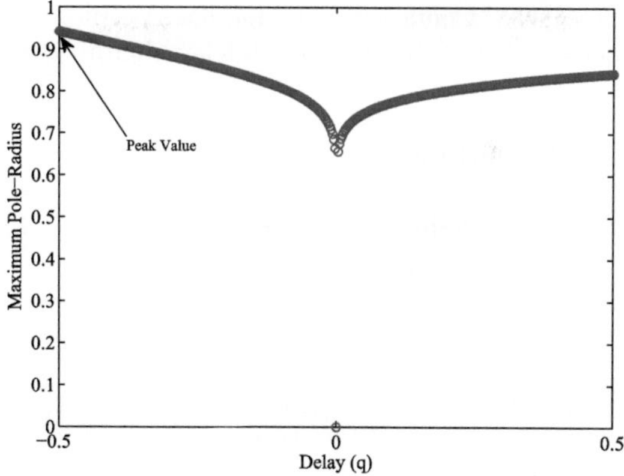

Fig. 5. AP-DF's maximum poles.

tive minimization. In contrast to those design techniques, the proposed one can get a closed-form solution through simplifying the AP-DF design as an Rcone-programming problem.

5 Conclusions

This paper has formulated an Rcone-programming technique for effectively containing the delay-error peak of the variable AP-DF. Through restraining the delay-error peak in formulating the Rcone-programming design, the design technique can further suppress the delay-error overshoot. The AP-DF design has exemplified the utility of the novel technique, showing that it can achieve the smallest delay-error peak among all the designs. The computer simulations have also confirmed the stability of the resulting variable AP-DF.

References

1. Laakso, T.I., Valimaki, V., Karjalainen, M., Laine, U.K.: Splitting the unit delay: tools for fractional delay filter design. IEEE Signal Process. Mag. **13**(1), 30–60 (1996)
2. Deng, T.-B.: Discretization-free design of variable fractional-delay FIR digital filters. IEEE Trans. Circuits Syst. II, Analog Digit. Signal Process. **48**(6), 637–644 (2001)
3. Deng, T.-B.: Closed-form design and efficient implementation of variable digital filters with simultaneously tunable magnitude and fractional-delay. IEEE Trans. Signal Process. **52**(6), 1668–1681 (2004)
4. Deng, T.-B.: Symmetry-based low-complexity variable fractional-delay FIR filters. Proc. IEEE ISCIT **2004**, 194–199 (2004)
5. Deng, T.-B., Lian, Y.: Weighted-least-squares design of variable fractional-delay FIR filters using coefficient symmetry. IEEE Trans. Signal Process. **54**(8), 3023–3038 (2006)

6. Deng, T.-B., Qin, W.: Coefficient relation-based minimax design and low-complexity structure of variable fractional-delay digital filters. Signal Process. **93**(4), 923–932 (2013)
7. Deng, T.-B., Qin, W.: Improved bi-equiripple variable fractional-delay filters. Signal Process. **94**(1), 300–307 (2014)
8. Soontornwong, P., Deng, T.-B., Chivapreecha, S.: Low-complexity and high-modularity structure for implementing transient-free Pascal-delay filter. IEEE Trans. Signal Process. **65**(23), 6233–6243 (2017)
9. Deng, T.-B.: Generalized WLS method for designing allpass variable fractional-delay digital filters. IEEE Trans. Circuits Syst. I, Reg. Papers **56**(10), 2207–2220 (2009)
10. Tseng, C.-C.: Design of 1-D and 2-D variable fractional delay allpass filters using weighted least-squares method. IEEE Trans. Circuits Syst. I, Fundam. Theory Appl. **49**(10), 1413–1422 (2002)
11. Deng, T.-B.: Noniterative WLS design of allpass variable fractional-delay digital filters. IEEE Trans. Circuits Syst. I, Reg. Papers **53**(2), 358–371 (2006)
12. Lee, W.-R., Caccetta, L., Rehbock, V.: Optimal design of all-pass variable fractional-delay digital filters. IEEE Trans. Circuits Syst. I, Reg. Papers **55**(5), 1248–1256 (2008)
13. Shyu, J.-J., Pei, S.-C., Chan, C.-H.: Minimax phase error design of allpass variable fractional-delay digital filters by iterative weighted least-squares method. Signal Process. **89**(9), 1774–1781 (2009)
14. Deng, T.-B.: Minimax design of low-complexity allpass variable fractional-delay digital filters. IEEE Trans. Circuits Syst. I, Reg. Papers **57**(8), 2075–2086 (2010)
15. Dam, H.H.: Design of allpass variable fractional-delay filter. IEEE Trans. Signal Process. **59**(12), 6240–6244 (2011)
16. Deng, T.-B., Soontornwong, P.: Delay-error-constrained minimax design of all-pass variable-fractional-delay digital filters. Signal Process. **120**(3), 438–447 (2016)
17. Deng, T.-B.: Quintic fractional error function for designing cascade-type phase compensators. Signal Image Video Process. **18**, 7247–7254 (2024)
18. Deng, T.-B.: High-order allpass cascade-connected phase-system design utilizing fractional polynomial errors. J. Inf. Telecommun. **8**(4), 570–586 (2024)

KPI-Adaptive DQN Framework for Handover Optimization in 5G Ultra-Dense Networks

Muhammad Umair Munir[1], Ismail Ahmedy[2(✉)], Rafidah Md Noor[2], Saaidal Razalli Azzuhri[2], Sharmin Sharmin[1], and Uzair Iqbal[1]

[1] Faculty of Computer Science and Information Technology, Universiti Malaya, Kuala Lumpur, Malaysia

[2] Center for Mobile Cloud Computing Research, Faculty of Computer Science and Information Technology, Universiti Malaya, Kuala Lumpur, Malaysia

ismailahmedy@um.edu.my

Abstract. Ultra-dense network (UDN) deployment in 5G or B5G systems creates significant handover management challenges due to frequent cell transitions and highly dynamic signal conditions. While existing deep reinforcement learning approaches optimize handover decisions based on signal quality, they treat all scenarios uniformly and focus on single objectives. This paper proposes a KPI-Adaptive Deep Q-Network (DQN) framework that addresses these limitations through three key contributions. First, a multi-objective reward function with adaptive penalty weights is incorporated that follows multi-objective reinforcement learning principles and enable explicit balance between Radio Link Failure (RLF), handover frequency, and signal quality. Second, inside DQN, a phase-based curriculum learning mechanism with three distinct training stages is introduced that progressively shifts priorities from connection stability to balanced optimization. Third, a dynamic weight adaptation mechanism is implemented that automatically adjusts objective weights based on real-time KPI performance deviations from target values. Comprehensive simulation results in a 21-cell ultra-dense environment demonstrate that the proposed method achieves 13.2% RLF reduction compared to baseline 3GPP A3 approach and 5.4% improvement over plain DQN, while simultaneously reducing handover frequency by 1.8% and maintaining stable signal quality above 70 dB throughout 120 training episodes.

Keywords: Handover Management · Machine Learning · Reinforcement Learning · HCP Parameters Tuning

1 Introduction

The deployment of ultra-dense networks (UDNs) in 5G systems has changed the aspect of human activities to communicate with each other. Nowadays, most of the activities shifted to online like business operations, remote meetings, remote monitoring and data collection through IoT sensors, and each of these are dependent on the usage of internet and adoption of fifth generation (5G) networks. In UDNs, multiple base stations (BSs) are deployed within small regions, and these BSs consist of small cells which are

N. A. Mohd Subha et al. (Eds.): SMS 2026, CCIS 2956, pp. 189–203, 2026.
https://doi.org/10.1007/978-981-92-0306-2_14

supported by macro cells. However, this densification introduces substantial challenges for mobility management, specifically in handover decision-making [1]. The increased number of cell boundaries and frequent handover opportunities in UDNs create a complex optimization problem where multiple key performance indicators (KPIs) must be balanced simultaneously: maintaining adequate signal quality (RSRP), minimizing radio link failures (RLF), reducing unnecessary handovers (HO), and preventing ping-pong effects (HOPP) [2].

Traditional handover algorithms are based on 3GPP specifications which apply fixed threshold-based mechanisms such as the A3 event, which triggers handover when the serving cell's Reference Signal Received Power (RSRP) falls below a predefined threshold for a specific Time-to-Trigger (TTT) duration [3]. These rule-based approaches are straightforward to implement and provide deterministic behavior, however, they suffer from several limitations in dynamic network environments which can also lead to several challenges including radio link failure (RLF), handover ping-pong (HOPP), and unnecessary handovers, all of which degrade network performance and user experience. To address these limitations, 5G incorporates Self-Organizing Networks (SON) functionalities which include Mobility Robustness Optimization (MRO), that dynamically adjusts handover control parameters (HCPs) such as Time-to-Trigger (TTT) and Handover Margin (HOM) [4–6]. Recently researchers also started applying machine learning (ML) for intelligent handover optimization, these ML techniques include fuzzy logic controllers [7], neural networks [8], and decision trees that are applied to handover decision-making and parameter optimization. Among these approaches, reinforcement learning (RL) enables agents to learn optimal policies through trial-and-error interaction with the environment [9]. This RL agent get rewards or penalties to maximize based on it's decision which improve it's decision making capabilities. Several studies have explored RL-based approaches for handover optimization in 5G networks. Deep Q-Networks (DQN), which combine RL with deep neural networks for function approximation, have shown success in optimizing handover decisions in complex network scenarios [10, 11]. However, existing DQN-based methods employ single-objective optimization which maximize signal quality or minimize handover frequency independently [2]. These approaches fail to address the fundamental multi-objective nature of the handover problem where competing KPIs must be explicitly balanced. To address these limitations, this paper proposes a novel KPI-Adaptive Phase-Based DQN framework that extends conventional deep reinforcement learning with three key contributions. First, we introduce a multi-objective reward function with proactive safety shaping that explicitly balances signal quality, handover frequency, ping-pong prevention, and RLF mitigation. Second, we implement a phase-based curriculum learning strategy with three distinct training stages that progressively guide the agent through safety establishment, balance optimization, and policy refinement. Third, we develop a real-time KPI error-based weight adaptation mechanism that dynamically adjusts optimization priorities based on observed performance, creating a closed-loop self-optimizing system. The remainder of this paper is organized as follows: Sect. 2 reviews related work on handover optimization and reinforcement learning approaches. Section 3 presents the proposed KPI-Adaptive Phase-Based DQN framework including system model, algorithm design, and implementation details. Section 4 evaluates experimental results comparing our method against baseline approaches. Section 5 concludes the paper and discusses future research directions.

2 Literature Review

In this section, different handover optimization algorithms has been discussed for 5G ultra dense networks. Handover management is a process when UEs change its base station to the newer base station. In literature HO mechanism use several techniques from rule-based to machine learning and deep reinforcement learning approaches. The purpose of all of these approaches is to tune the HCP parameters that include HOM and TTT. Conventional handover approaches apply threshold-based mechanisms as standardized by 3GPP specifications. These approaches apply A3 event which is widely adopted in LTE and 5G systems and trigger handover when the target cell's signal strength exceeds the serving cell's signal by a specific Handover Margin (HOM) for a defined Time-to-Trigger (TTT) duration [3]. These rule-based approaches are simple and light weight but cannot adapt to dynamic environments. For instance, Zhang et al. [12] demonstrated that fixed TTT values lead to either excessive handover delays which causes radio link failures or premature handovers that also leads towards ping-pong effects. Liu et al. [13] showed that static HOM thresholds cannot accommodate dynamic environmental conditions in ultra-dense deployments where cell boundaries overlap significantly. Following these limitations, several adaptive rule-based methods have been proposed. Goyal et al. [14] introduced velocity-aware handover parameter optimization technique where HOM and TTT are dynamically selected based on user equipment speed categories. Chen et al. [15] developed SINR-based categorization where handover parameters adapt to signal quality zones. More recently, Ahmed et al. [16] proposed RSRP-categorized thresholds with emergency mode triggering for extremely poor signal conditions. While these adaptive approaches demonstrate improvements over fixed baselines, they remain constrained by predetermined category boundaries and cannot optimize for multiple competing KPIs simultaneously. Furthermore, some researchers applied fuzzy-logic for Handover optimization like, [17] developed a weighted fuzzy self-optimization (WFSO) approach for 4G/5G HetNets that integrate multiple KPIs including RSRP, UE speed, and cell load to optimize HOM and TTT. [7] proposed a robust handover optimization technique with fuzzy logic controller (RHOT-FLC) for beyond 5G mobile networks and showed significant improvements in HO probability, HO failure, and ping-pong rates across different mobility scenarios. The fundamental limitation is that these methods rely on static rules rather than learning from observed network performance [18].

Based on the above limitations, different ML based algorithms has also been proposed. Such as, Ladosz et al. [19] proposed support vector machines (SVM) to classify handover decisions based on signal measurements and mobility patterns that achieve improved performance over fixed threshold methods. Pagin et al. [20] developed fuzzy logic controllers that adapt handover parameters based on network conditions and user mobility characteristics. However, these methods require extensive labeled datasets for training and cannot easily adapt to new scenarios not represented in the training data. Following these, Reinforcement learning proposes handover optimization in which agents learn through interaction with the environment rather than requiring labeled datasets.

For example, Ulvan et al. [21] applied Q-learning to optimize handover parameters in LTE networks and demonstrate the ability to adapt policies without requiring pre-labeled data. However, tabular Q-learning methods do not scale effectively to high-dimensional state spaces in realistic network scenarios. As the number of state variables increases, the state space also grows exponentially, which makes it impractical to maintain and update Q-tables for all state-action pairs. Deep reinforcement learning addresses the scalability limitations of tabular methods by using deep neural networks that enable effective learning in high-dimensional state spaces. Several researchers have applied DQN and its variants to handover optimization with promising results. For example, Wang et al. [22] demonstrated that DQN outperforms rule-based methods in reducing handover failures for high-mobility scenarios in LTE-A networks that uses a state space consisting of RSRP measurements and handover history. Zhao et al. [23] utilized Double DQN (DDQN) to mitigate Q-value overestimation in handover decision-making and demonstrated more stable learning compared to standard DQN. Na et al. [24] optimize for effective handover ratio but do not explicitly balance multiple competing KPIs. This single-objective approach inherently cannot maintain explicit control over the trade-offs between signal quality, handover frequency, ping-pong prevention, and radio link failure mitigation. More recently, Riaz et al. [8] proposed a data-driven MLP-based method that optimizes HOM and TTT through offline dataset generation and achieved significant improvements by considering UE speed, RSRP, SINR, and cell load. However, this approach relies on offline training with pre-optimized datasets and lacks the adaptive learning capability required for dynamic network conditions. Furthermore, these methods treat all network scenarios uniformly throughout training which lacks the mechanisms to prioritize critical objectives such as connection reliability during dynamic environments. To address these limitations, this paper proposes KPI-Adaptive Phase-Based DQN framework that extends conventional deep reinforcement learning and explained in Sect. 3.

3 Methodology

In this study, four different handover algorithms has been evaluated to optimize handover management in 5G ultra-dense network. These algorithms include conventional rule-based baseline method based on the 3GPP A3 event mechanism as discussed in [3, 25], Adaptive Rule-Based Handover (RSRP-Categorized Thresholds) as discussed in [8], Plain Deep Q-Network (DQN) single objective RL approach following by [12] and [24]. And finally, the proposed KPI-adaptive Phase-Based DQN (Multi-Objective RL with Curriculum Learning) has been discussed. Initially the simulation environment, and then each algorithm will be explained in this section.

3.1 Simulation Environment and System Model

The simulation environment was developed in MATLAB by using ultra-dense 5G network (UDN) deployment with multiple base stations. The simulated environment was operated under realistic mobility conditions in which UEs experience time-varying channel conditions due to path loss, shadowing, and fast fading effects. All of the four algorithms are evaluated in the same UDN environment which ensure to have the fair comparison. Table 1 illustrates the simulation parameters that are used in this research.

Table 1. Simulation Parameters.

Parameter	Value
Number of Base Stations	21 (macro cells with small cells)
Episode Steps (T)	50
Training Episodes	120
User Velocity	50 km/h
Timestep Duration	50 ms
RLF Threshold	65 dB
Discount Factor (γ)	0.95
Learning Rate (α)	0.001

The network topology consists of 21 LTE-A macro cells overlaid with small cells that create high interference and frequent handover scenarios most related to ultra-dense deployments. In the simulation UEs moves at constant velocity of 50 km/h which is common in urban environments. The handover decision problem is formulated as a binary choice at each timestep: remain connected to the current cell (Stay) or execute handover to the target cell (Handover). Each simulation episode consists of 50 timesteps with 50ms duration per timestep which result in 2.5 s total episode duration which is an enough to observe multiple handover events and their consequences.

3.2 Conventional Rule-Base Baseline

The traditional rule-based SON methods rely in threshold-based handover decision mechanism [3]. These approaches does not require any learning mechanism and serve as benchmark to evaluate the performance achieved by other methods. These methods follows the 3GPP A3 event mechanism which uses fixed thresholds and a Time-to-Trigger (*TTT*) mechanism to avoid ping-pong handovers caused by temporary signal fluctuations. These methods follow two main conditions: firstly, the serving cell's RSRP should fall below a predefined threshold ($signal_{Low} = 68\ dB$), and secondly, a specific time has passed since the last handover to satisfy the *TTT* constraint ($TTT_{minSteps} = 3$ which is equivalent to 150ms). The A3 handover triggering condition is mathematically expressed as shown in Eq. 1:

$$RSRP_{serving} < Threshold, for\ duration \geq TTT \tag{1}$$

where $RSRP_{serving}$ represents the Reference Signal Received Power from the serving cell, Threshold is set to 68 dB which is commonly used to balance handover timing, and TTT is the time to trigger the handover event. This traditional algorithm continuously monitors *RSRP* from the serving cell at each timestep. When *RSRP* falls below the *threshold* value, a Time-to-Trigger counter starts incrementing. If the poor signal condition persists for the minimum *TTT* duration, a handover command is executed. If signal quality improves before *TTT* expires, the counter resets to zero and the UE maintains connection with the current serving cell. These approaches have fixed thresholds that cannot adapt to dynamic network conditions, however, these are straightforward to implement and cannot optimize the trade-off between RLF prevention (which leads towards early handovers) and ping-pong reduction (which favors delayed handovers).

3.3 Adaptive Rule-Based Handover

The Adaptive Rule-Based algorithm represents an improvement over the fixed baseline approach by categorizing signal conditions into multiple regions and adjust the handover parameters accordingly. Inspired by recent data-driven approaches in wireless network optimization [8], this method uses RSRP-based categorization to make more context-aware handover decisions rather than relying on a single fixed threshold. Instead of using a single fixed threshold, this algorithm categorizes the measured signal strength into three distinct regions: poor signal zone ($RSRP < 65\ dB$) which indicate the requirement of immediate action, moderate signal zone ($65 \leq RSRP < 75\ dB$) which represent caution zone with adaptive behavior, and good signal zone ($RSRP \geq 75\ dB$) which indicate safe conditions in which maintaining the current connection is preferred. At each timestep the algorithm determines the current *RSRP* category and adjusts the handover triggering behavior based on this categorization. Figure 1 shows the flowchart for this approach.

Figure 1 shows the flowchart for adaptive rule-based approach. In which the algorithm first check for emergency conditions (SINR < -6.5 dB). If detected, it set HOM = 0 dB and TTT = 80 ms for rapid handover. Otherwise, RSRP is categorized into three categories that are, poor (<65 dB), moderate (65–75 dB), or good (≥75 dB) zones, and appropriate HOM and TTT values are selected from predefined discrete sets. This categorization approach provides more granular control compared to the binary threshold approach of the baseline method. However, it still rely on predetermined category boundaries which may not be optimal for all network scenarios and mobility patterns. The method does not explicitly optimize for multiple KPIs simultaneously nor can it adapt its decision strategy based on observed long-term outcomes.

3.4 Plain Deep Q-Network (DQN) Approach

The Deep Q-Network (DQN) approach formulates the handover optimization problem as a Markov Decision Process (MDP), as defined by (S, A, P, R, γ) and use deep reinforcement learning to learn optimal handover policies [9, 26]. Unlike rule-based methods that rely on predetermined logic, DQN learns optimal policies through trial-and-error interaction with the environment using only signal strength as the reward signal. The state vector captures network status: $s_t = [RSRP_{serving}, Load_{norm}, T_{lastHO}, \rho_{RLF}, \rho_{HOPP}]^T$ where *RSRP* represents current signal strength in dB, *Load* indicates normalized cell load in range [0,1], T_{lastHO} denotes time since last handover providing temporal context, ρ_{RLF} captures recent radio link failure rate through moving average, and ρ_{HOPP} represents recent ping-pong rate through moving average. The action space is binary: $A = \{0\text{: Stay, } 1\text{: Handover}\}$. The Q-network consists of input layer (5 neurons), two hidden layers (128 and 64 neurons with ReLU), and output layer (2 neurons). Training uses experience replay (buffer size 5000, mini-batch 64) and target network with soft updates ($\tau = 0.005$). The ε-greedy exploration strategy uses $\varepsilon = 1.0$ for the first 60% of episodes, then $\varepsilon = 0.1$. Figure 2 shows the DQN flowchart.

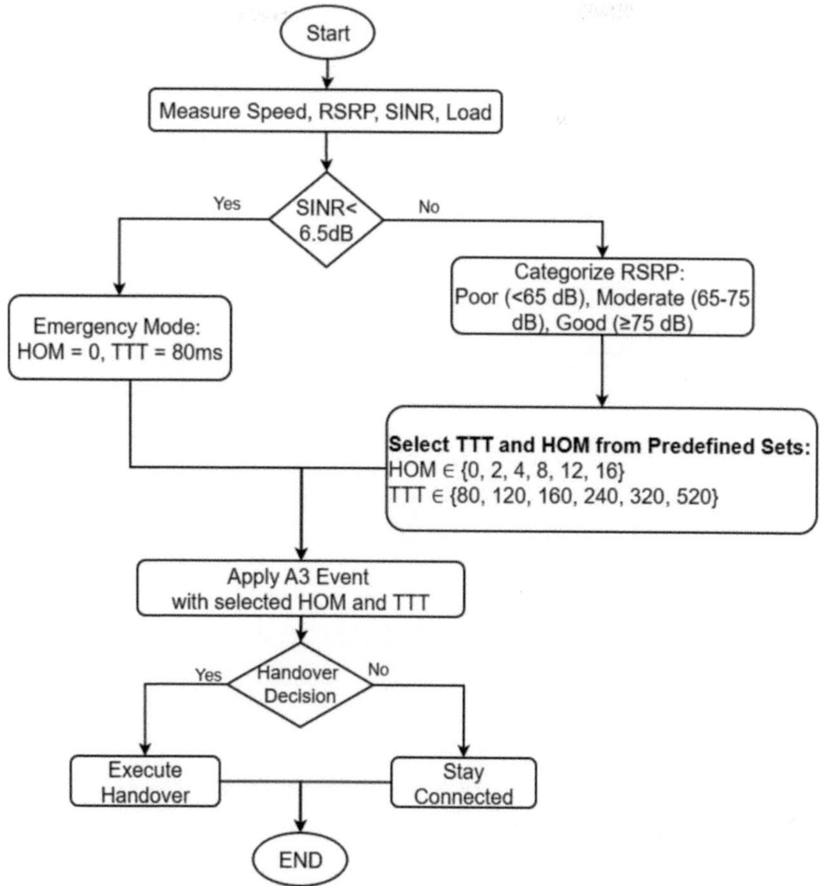

Fig. 1. Adaptive Rule-Based Handover Flowchart

As illustrated in Fig. 2, the DQN algorithm begins by initializing the Q-Network, Target Network, and Experience Replay Buffer. At each time step, the agent observe the current state $s_{t,}$. The agent then follows the ε-greedy policy for action selection and explore by selecting a random action when a randomly generated value is less than ε, or exploits its learned knowledge by selecting the greedy action $a = argmax\ Q(s,a)$ that maximizes the predicted Q-value. The selected action can Stay (means remain connected) or Handover (change the base station), is then executed in the network environment. Following action execution, the agent receives a reward r_t calculated as the difference between the *RSRP* value and the handover penalty, which encourages maintaining good signal quality while penalizing unnecessary handovers. The transition tuple (s_t, a_t, r_t, $s_{(t+1)}$) containing the current state, action, reward, and resulting next state is stored in the replay buffer for future learning. The training phase involves sampling a minibatch of 64 transitions from the buffer and using them to update the Q-Network weights, followed by a soft update of the Target Network parameters. This process repeats each step within

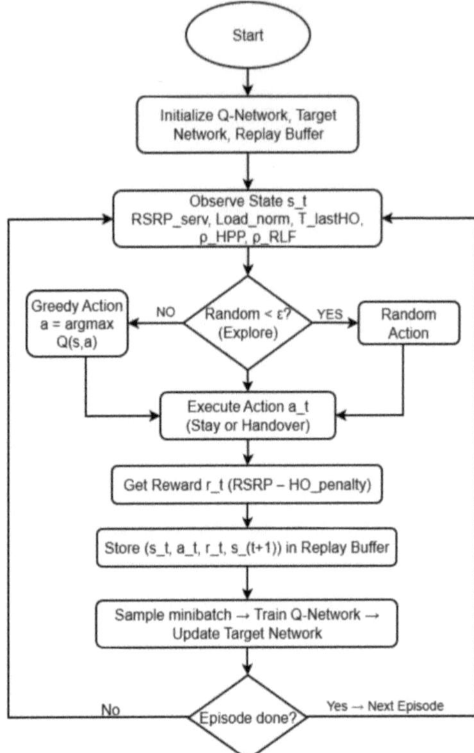

Fig. 2. Deep Q-Network (DQN) Handover Optimization Flowchart

an episode, and upon episode completion, the algorithm proceeds to the next episode, continuing until the training converges to an optimal handover policy.

3.5 KPI-Adaptive Phase-Based DQN Framework

The standard DQN approach that we discussed above optimizes the handover decisions based on signal quality and treat all handover scenarios in the same way. However, in real network dynamic environmental situations occur that require different priorities for example, when connection failures are frequent over that time reducing *RLF* becomes essential. Similarly, minimizing unnecessary handovers matters more when the network is stable. To handle these, we propose a KPI-adaptive phase-based RL framework that extends conventional DQN with three contributions. Firstly, we incorporated multi-objective reward function with adaptive penalty weights following the scalarized MORL approach [27]. Secondly, we introduce phase-based learning with three distinct training stages. And thirdly, we implemented dynamic weight adaptation that automatically adjusts each objective based on how well the system is performing compared to target KPI values as proposed in recent multi-objective deep RL work [28].

Figure 3 illustrates the overall architecture of our proposed KPI-adaptive reinforcement learning framework for handover optimization in ultra-dense 5G networks. The framework integrates five main components: firstly, MATLAB provide state information which include *RSRP*, cell load, time since last handover, and KPI history through moving averages, secondly, Feature Extraction and DQN Agent processes the five-dimensional state through neural network layers ($5 \rightarrow 128 \rightarrow 64 \rightarrow 2$) to output Q-values for Stay and Handover actions. After this, thirdly, Decision selects the action with highest Q-value using ε-greedy policy and determines appropriate *HOM* and *TTT* parameters which are sent as control commands to the Small Cell Network, fourthly, Small Cell Network executes handover decisions, the network reports KPIs (*HO, HOPP, RLF occurrences*) to the KPI-Adaptive Reward Module (fifth step). This module computes the multi-objective reward $r = r_{signal}$ - $w_2 \cdot HO$ - $w_3 \cdot HOPP$ - $(\lambda \cdot w_4) \cdot RLF + r_{safety}$ where the proactive safety *term* r_{safety} = -0.08 × (1 + *proximity*) penalizes risky 'Stay' decisions when *RSRP* approaches the RLF threshold (*65 dB + 4 dB margin*). The adaptive reward signal flows back to the DQN agent for Q-network weight updates through experience replay and target network mechanisms. Training is organized using three distinct phases as shown in the Phase-based Curriculum component of Fig. 3.

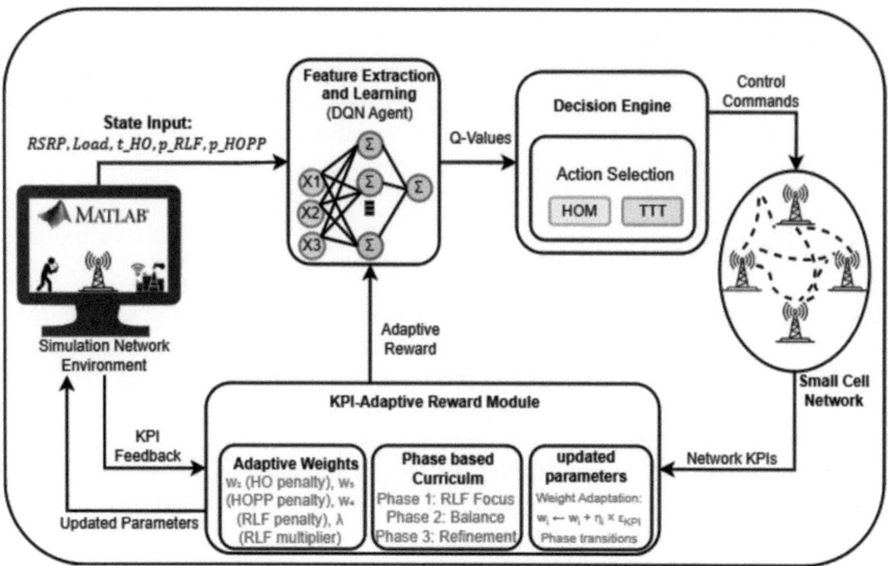

Fig. 3. Proposed KPI-adaptive Reinforcement Learning Framework

After a 20-episode warmup, the module continuously tracks KPI performance using exponential moving averages MA_{RLF} and MA_{HO} with smoothing factor 0.3, computes performance errors $\varepsilon_{RLF} = MA_{RLF}$ - *0.04 and* $\varepsilon_{HO} = MA_{HO}$ - *1.0* against target values, and updates penalty weights based on phase-specific rules. In phase 1 (episodes 1–60) aggressively eliminates *RLF* with high learning rate $\eta_{RLF} = 2.5$ and lambda boost $\lambda \leftarrow min(\lambda \times 1.15, 4.0)$ when *RLF* exceeds target. Phase 2 (episodes 61–96) balances all KPIs with moderate rates $\eta_{RLF} = 1.5$ and $\eta_{HO} = 0.4$. Phase 3 (episodes 97–120) fine-tunes the policy with low rates $\eta_{RLF} = 0.8$ and $\eta_{HO} = 0.3$ while applying lambda decay $\lambda \leftarrow \lambda \times 0.997$. The updated weights flow back to the Simulation Environment as shown in Fig. 3, creating a self-optimizing closed-loop system where the agent learns to balance signal quality, minimize handovers, eliminate ping-pong effects, and prevent radio link failures through adaptive multi-objective optimization guided by real-time KPI feedback.

Table 2. Algorithm Comparison Summary

Algorithm	Type	Adaptation	Optimization
Baseline A3	Rule-based	None	Fixed thresholds
Adaptive Rules	Rule-based	RSRP categories	Predefined rules
Plain DQN	Learning-based	Experience replay	Single-objective
Proposed	Learning-based	Real-time KPI	Multi-objective

Table 2 provides a comprehensive comparison summarizing key characteristics of all four evaluated algorithms. The progression from Baseline through Adaptive Rules and Plain DQN to our proposed method represents increasing sophistication in handling the multi-objective handover optimization problem. Our KPI-Adaptive Phase-Based DQN combines the learning advantages of reinforcement learning with explicit multi-objective optimization and dynamic priority adjustment achieving superior overall network performance as demonstrated in the experimental results section.

4 Results

This section presents the performance evaluation of all four handover algorithms: Baseline A3, Adaptive Rule-Based, Plain DQN, and the proposed KPI-Adaptive Phase-Based DQN. All methods were trained and evaluated over 120 episodes in the same UDN environment to ensure fair comparison. Performance is measured through three key indicators: Handover rate (HO/ep) which represent the average number of handover events per episode, Radio Link Failure rate (RLF/ep) which counts connection failures per episode when RSRP falls below 65 dB threshold, and Average Serving Signal (RSRP in dB) which measure mean signal quality maintained throughout each episode.

4.1 Signal Quality Maintenance

Figure 4 shows the average serving cell RSRP maintained by each algorithm throughout the 120 episodes. All four methods maintain RSRP values between approximately 70–79 dB throughout the evaluation period, well above the RLF threshold of 65 dB. The similarity in signal quality across all methods confirms that performance differences in RLF and handover rates are driven entirely by the optimization strategy, not by differences in signal conditions. Baseline, Adaptive Rules, Plain DQN, and KPI-Adaptive all exhibit similar RSRP behavior with peaks around 78–79 dB and occasional dips to 71–73 dB during challenging mobility scenarios. The KPI-Adaptive framework achieves superior RLF reduction without any sacrifice in signal quality, which demonstrates that adaptive multi-objective optimization improves connection stability independently of raw signal strength.

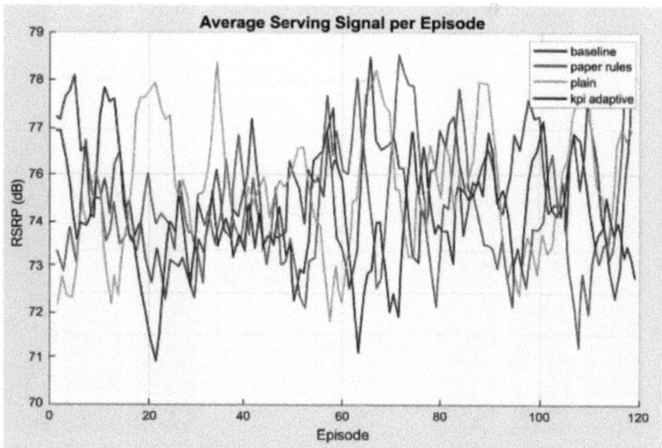

Fig. 4. Average Serving Cell Signal Quality per Episode

4.2 Overall Performance Comparison

Figure 5 presents the cumulative performance of all four algorithms across 120 training episodes. For handover events, the proposed KPI-Adaptive method records the lowest count at 764 events, compared to 778 for Baseline A3, 790 for Plain DQN, and 799 for Adaptive Rules. The fact that Adaptive Rules and Plain DQN both exceed the baseline count reveals that without proper multi-objective control, optimization efforts can actually increase unnecessary switching. The 1.8% reduction achieved by the proposed method reflects the direct effect of the handover penalty term in the reward function.

For RLF events, the differences are more significant. Baseline A3 records 121 failures because its fixed thresholds cannot react to sudden channel drops. Adaptive Rules performs worst at 124 failures, its aggressive triggering in poor signal zones causes premature handovers to weaker cells. Plain DQN improves to 111 failures (8.3% better than baseline) by learning from signal patterns, but it still lacks the ability to balance multiple objectives at the same time. The proposed method achieves the best result at 105 failures, which is a 13.2% improvement over Baseline A3 and 5.4% over Plain DQN. This result confirms the three components of the framework. The multi-objective reward, the phase-based curriculum, and the KPI error-based weight adaptation, each address a specific weakness that other methods fail to handle.

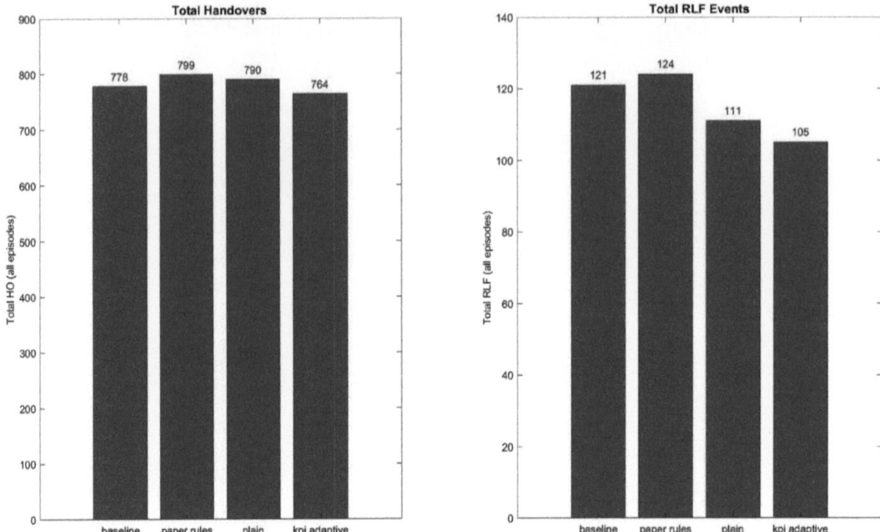

Fig. 5. Cumulative Handovers and RLF Events Across 120 Episodes

4.3 Average KPI Performance with Statistical Analysis

Figure 6 presents mean KPI values with standard deviations across all 120 episodes. For handover rates, all algorithms achieve statistically similar performance: Baseline (6.48 $\pm$ 2.31), Adaptive Rules (6.66 $\pm$ 2.43), Plain DQN (6.58 $\pm$ 2.53), and KPI-Adaptive (6.37 $\pm$ 2.25), with comparable standard deviations (2.3–2.5) which confirm consistent handover patterns across all methods. The slight reduction in the proposed method's mean (6.37 vs. 6.48 for baseline) is a direct result of the handover penalty term, which suppresses unnecessary switching events. For RLF rates, significant performance differences emerge: Baseline exhibits 1.01 $\pm$ 0.90 failures per episode, Adaptive Rules shows 1.03 $\pm$ 0.91, while RL-based methods achieve substantial reduction with Plain DQN at 0.93 $\pm$ 0.77 and the proposed method delivering the lowest at 0.88 $\pm$ 0.79. Notably, the proposed method achieve both the lowest mean RLF and the lowest standard deviation, which indicate more stable and predictable performance across all episodes. The lower standard deviation (0.79 vs. 0.90 for baseline) confirms that the adaptive weight mechanism prevents failure spikes by boosting the RLF penalty whenever performance deviates from the target rate, a capability absent in all other evaluated methods. This confirms that the proposed framework not only achieves the best average RLF performance but also the most consistent results across all episodes, which is essential for real-world deployment where stable service quality is a strict requirement. Regarding computational overhead, the KPI weight adaptation introduces negligible additional computation per timestep, as the moving average and error calculations require O(1) operations and the weight update involves simple scalar arithmetic. The primary computational cost remains the standard DQN forward and backward passes, which confirms that the proposed framework is suitable for real-time 5G SON deployment.

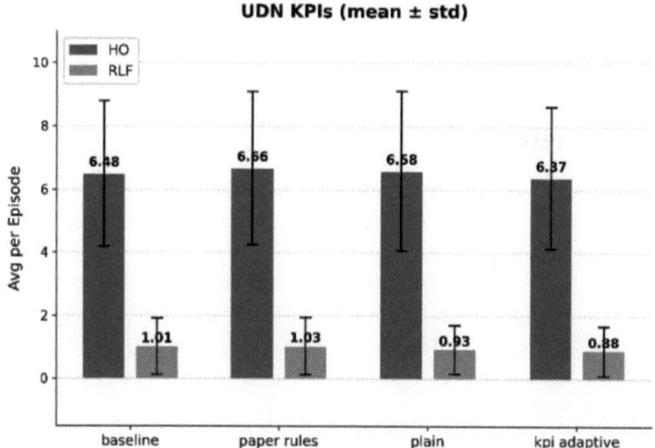

Fig. 6. Average KPI Performance (Mean ± Std) Per Episode

5 Conclusion

This paper presented a KPI-Adaptive Deep Reinforcement Learning framework for handover optimization in 5G ultra-dense networks that addresses the fundamental limitation of existing approaches, the inability to adapt optimization priorities based on dynamic network conditions. Through comprehensive evaluation against traditional rule-based methods and standard deep reinforcement learning approaches, we demonstrated our experimental results that validate our framework's effectiveness across multiple dimensions. The proposed multi-objective reward function reduced Radio Link Failures by 13.2% compared to baseline A3 (105 vs. 121 events) and 5.4% compared to plain DQN (105 vs. 111 events). The statistical analysis revealed that while all algorithms maintain similar handover rates (6.37–6.66 per episode) and signal quality (70–79 dB RSRP), the proposed method achieves more stability with standard deviations comparable to or better than baseline methods. The 1.8% reduction in total handovers (764 vs. 778 events) demonstrates that the proposed framework successfully minimizes unnecessary handovers without compromising connection reliability, addressing the core trade-off in handover optimization. Future work will extend this framework to diverse environments including indoor, urban, and highway scenarios at varying speeds. Additional KPIs such as ping-pong rate and end-to-end latency will be incorporated as explicit evaluation metrics with dedicated performance targets, and performance will be evaluated under varying mobility patterns and user densities. Transfer learning techniques will also be investigated to enable faster adaptation when deploying the trained model to new network configurations with different topologies and user distributions.

Acknowledgement. This research is supported by the Universiti Malaya Research Excellence Grant (Project No. UMREG011-2024) and the Universiti Malaya Center of Research Grant (Project No. CORG008-2024).

References

1. Sinclair, N., Harle, D., Glover, I.A., Irvine, J., Atkinson, R.C.: An advanced SOM algorithm applied to handover management within LTE. IEEE Trans. Veh. Technol. **62**, 1883–1894 (2013)
2. Ullah, Y., Roslee, M.B., Mitani, S.M., Khan, S.A., Jusoh, M.H.: A survey on handover and mobility management in 5G HetNets: current state, challenges, and future directions. Sensors (Basel). **23**, 5081 (2023)
3. Specification # 36.331. https://portal.3gpp.org/desktopmodules/Specifications/SpecificationDetails.aspx?specificationId=2440. Accessed 21 Dec 2025
4. Lee, D.-W., Gil, G.-T., Kim, D.-H.: A cost-based adaptive handover hysteresis scheme to minimize the handover failure rate in 3GPP LTE system. EURASIP J. Wirel. Commun. Netw. **2010**, 750173 (2010)
5. Alraih, S., Nordin, R., Abu-Samah, A., Shayea, I., Abdullah, N.F.: A survey on handover optimization in beyond 5G mobile networks: challenges and solutions. IEEE Access. **11**, 59317–59345 (2023)
6. Sharmin, S., Ahmedy, I., Jabaraj, B.R.P., Munir, M.U., Noor, R.M., Ahmad, M.: A comprehensive review of machine learning-driven self-organizing networks for 5G coverage and capacity. Telecommun. Syst. **89**, 44 (2026)
7. Alraih, S., Nordin, R., Abu-Samah, A., Shayea, I., Abdullah, N.F., Alhammadi, A.: Robust handover optimization technique with fuzzy logic controller for beyond 5G mobile networks. Sensors (Basel). **22**, 6199 (2022)
8. Riaz, H., Öztürk, S., Aldirmaz-Colak, S., Çalhan, A.: A Handover decision optimization method based on data-driven MLP in 5G ultra-dense small cell HetNets. J. Netw. Syst. Manag. **33**, (2025). https://doi.org/10.1007/s10922-025-09903-6
9. François-Lavet, V., Henderson, P., Islam, R., Bellemare, M.G., Pineau, J.: An introduction to deep reinforcement learning. Found. Trends® Mach. Learn. **11**, 219–354 (2018)
10. Chinchali, S., et al.: Cellular network traffic scheduling with deep reinforcement learning. In: Proceedings of the Conference AAAI Artificial Intelligence, vol. 32 (2018). https://doi.org/10.1609/aaai.v32i1.11339
11. Sun, Y., Peng, M., Zhou, Y., Huang, Y., Mao, S.: Application of machine learning in wireless networks: Key techniques and open issues. IEEE Commun. Surv. Tutor. **21**, 3072–3108 (2019)
12. Zhang, H., Chu, X., Guo, W., Wang, S.: Coexistence of Wi-Fi and heterogeneous small cell networks sharing unlicensed spectrum. IEEE Commun. Mag. **53**, 158–164 (2015)
13. Liu, Y., Elkashlan, M., Ding, Z., Karagiannidis, G.K.: Fairness of user clustering in MIMO non-orthogonal multiple access systems. IEEE Commun. Lett. **20**, 1 (2016)
14. Goyal, R.K., Kaushal, S., Sangaiah, A.K.: The utility based non-linear fuzzy AHP optimization model for network selection in heterogeneous wireless networks. Appl. Soft Comput. **67**, 800–811 (2018)
15. Chen, T., Yang, Y., Zhang, H., Kim, H., Horneman, K.: Network energy saving technologies for green wireless access networks. IEEE Wirel. Commun. **18**, 30–38 (2011)
16. Ahmed, A., Boulahia, L.M., Gaiti, D.: Enabling vertical handover decisions in heterogeneous wireless networks: a state-of-the-art and a classification. IEEE Commun. Surv. Tutor. **16**, 776–811 (2014)
17. Alhammadi, A., Roslee, M., Alias, M.Y., Shayea, I., Alriah, S., Abas, A.B.: Advanced handover self-optimization approach for 4G/5G HetNets using weighted fuzzy logic control. In: 2019 15th International Conference on Telecommunications (ConTEL). IEEE (2019). https://doi.org/10.1109/contel.2019.8848507
18. Ye, Q., Rong, B., Chen, Y., Al-Shalash, M., Caramanis, C., Andrews, J.G.: User association for load balancing in heterogeneous cellular networks (2012). http://arxiv.org/abs/1205.2833

19. Ladosz, P., Weng, L., Kim, M., Oh, H.: Exploration in deep reinforcement learning: a survey. Inf. Fusion. **85**, 1–22 (2022)
20. Prados-Garzon, J., Adamuz-Hinojosa, O., Ameigeiras, P., Ramos-Munoz, J.J., Andres-Maldonado, P., Lopez-Soler, J.M.: Handover implementation in a 5G SDN-based mobile network architecture. In: 2016 IEEE 27th Annual International Symposium on Personal, Indoor, and Mobile Radio Communications (PIMRC). IEEE (2016). https://doi.org/10.1109/pimrc.2016.7794936
21. Ulvan, A., Bestak, R., Ulvan, M.: Handover procedure and decision strategy in LTE-based femtocell network. Telecommun. Syst. **52**, 2733–2748 (2013)
22. Bennis, M., Perlaza, S.M., Debbah, M.: Learning coarse correlated equilibria in two-tier wireless networks. In: 2012 IEEE International Conference on Communications (ICC). IEEE (2012). https://doi.org/10.1109/icc.2012.6364308
23. Zhao, N., Liang, Y.-C., Niyato, D., Pei, Y., Wu, M., Jiang, Y.: Deep reinforcement learning for user association and resource allocation in heterogeneous cellular networks. IEEE Trans. Wirel. Commun. **18**, 5141–5152 (2019)
24. He, J., Xiang, T., Wang, Y., Ruan, H., Zhang, X.: A reinforcement learning handover parameter adaptation method based on LSTM-aided digital twin for UDN. Sensors (Basel). **23**, 2191 (2023)
25. Liu, Q., Kwong, C.F., Wei, S., Li, L., Zhang, S.: Intelligent handover triggering mechanism in 5G ultra-dense networks via clustering-based reinforcement learning. Mob. Netw. Appl. **26**, 27–39 (2021)
26. Mnih, V., et al.: Human-level control through deep reinforcement learning. Nature **518**, 529–533 (2015)
27. Vamplew, P., Dazeley, R., Berry, A., Issabekov, R., Dekker, E.: Empirical evaluation methods for multiobjective reinforcement learning algorithms. Mach. Learn. **84**, 51–80 (2011)
28. Abels, A., Roijers, D.M., Lenaerts, T., Nowé, A., Steckelmacher, D.: Dynamic Weights in multi-objective deep Reinforcement Learning (2018). http://arxiv.org/abs/1809.07803

Optimal Trajectory Tracking Based Improved SCA for Gantry Crane System

Mohamed O. Elhabib[1], Herman Wahid[1(✉)], Zaharuddin Mohamed[1], and O. A. Elhabib[2]

[1] Control and Mechatronics Engineering Division, Faculty of Electrical Engineering, Universiti Teknologi Malaysia, 81310 Johor Bahru, Malaysia
herman@utm.my

[2] Department of Mechanical Engineering, College of Engineering, Taif University, Taif 21944, Saudi Arabia

Abstract. The precise manipulation of gantry cranes is inherently complicated by underactuated nonlinear dynamics, where a single input must simultaneously regulate trolley positioning and suppress multi-mode oscillations. Traditional Linear Quadratic Regulator (LQR) approaches often yield steady-state errors due to the absence of integral action, while parameter tuning via standard Sine Cosine Algorithms (SCA) is prone to premature convergence in high-dimensional search spaces. This work addresses these limitations by developing an Integral LQR (ILQR) controller tuned via an Improved Sine Cosine Algorithm (ISCA). Distinguished by a linear search path and empirical best-history retention, the ISCA mechanism balances exploration and exploitation to avoid local optima stagnation, ensuring precise gain optimization. Numerical validations confirm the superiority of the proposed ISCA-ILQR scheme, demonstrating a 12.5% reduction in payload sway compared to standard SCA-tuned counterparts in nominal operations. Moreover, the controller exhibits robust disturbance rejection; under combined parametric uncertainties involving significant cable extension (0.40 m) and payload mass variations (0.50 kg), the system effectively confines residual sway within ±1○ limits, ensuring operational safety in dynamic industrial environments.

Keywords: Improved Sine Cosine Algorithm (ISCA) · Double-Pendulum Gantry Crane · Integral LQR · Optimal control · Robust Trajectory Tracking

1 Introduction

Gantry cranes constitute the operational backbone of contemporary heavy industry, functioning as the primary logistical nodes in maritime ports, construction sites, and manufacturing facilities. These systems typically comprise a hoisting

N. A. Mohd Subha et al. (Eds.): SMS 2026, CCIS 2956, pp. 204–217, 2026.
https://doi.org/10.1007/978-981-92-0306-2_15

mechanism mounted on a trolley-girder or boom structure, tasked with maneuvering heavy and often hazardous payloads with high precision [1]. By coordinating the trolley's translational motion with the cable's varying length, the system navigates payloads through complex workspaces [2]. However, the dynamic behavior of gantry cranes presents a rigorous control challenge due to their inherent underactuated and nonlinear nature. A single control input—the force applied to the trolley—must simultaneously regulate multiple output variables, specifically the trolley displacement and the sway angles of both the hook and the payload. As the trolley accelerates, the suspended load inevitably oscillates; failing to suppress these swings compromises positioning accuracy and increases the risk of catastrophic workplace accidents [3]. Consequently, the development of robust control strategies that ensure precise trajectory tracking while actively damping load sway is a paramount safety requirement.

Over the last decade, diverse control methodologies have been investigated to address these coupled dynamics. Singhose [4] demonstrated that input shaping could mitigate residual oscillations in planar cranes, effectively reducing transient vibrations. Huang et al. [5] extended this with command smoothing techniques for double-pendulum systems, while Chen et al. [6] proposed a hybrid robust framework combining sliding mode control with input shaping to minimize chattering. Furthermore, adaptive control techniques have been applied to handle parametric uncertainties in underactuated setups, achieving accurate positioning despite model mismatches [7].

Among optimal control strategies, the Linear Quadratic Regulator (LQR) remains a dominant approach due to its ability to handle multi-variable systems. Corriga [8] utilized gain-scheduling LQR to accommodate varying cable lengths, though the method exhibited limitations in response speed. Adeli [9] integrated Takagi-Sugeno fuzzy models with LQR to enhance swing reduction, and Zawawi [10] validated LQR's superiority over Proportional-Derivative (PD) schemes in limiting overshoot. However, the standard LQR formulation faces two critical limitations in high-performance trajectory tracking. First, the lack of inherent integral action leads to steady-state errors when tracking time-varying references. Second, the determination of optimal weighting matrices Q and R is non-trivial; relying on trial-and-error often results in suboptimal performance. While meta-heuristic algorithms have been introduced to automate parameter tuning, standard approaches like the Sine Cosine Algorithm (SCA) often suffer from premature convergence and stagnation in local optima when exploring high-dimensional search spaces. To overcome these deficiencies, this paper proposes an Integral LQR (ILQR) controller tuned by an Improved Sine Cosine Algorithm (ISCA). Unlike standard evolutionary methods, the proposed ISCA leverages a linear search path and empirical best-history retention to balance exploration and exploitation effectively. This hybrid ISCA-ILQR approach is designed to eliminate steady-state tracking errors and robustly suppress double-pendulum sway under varying payload and cable conditions.

2 System Modeling and Linearization

This section delineates the mathematical framework governing the dynamics of the double-pendulum overhead crane. The modeling process begins with a precise problem formulation, followed by the derivation of the nonlinear equations of motion in a compact matrix form using the Euler-Lagrange methodology, and concludes with the linearization of the system for optimal controller synthesis.

2.1 Problem Formulation

Consider a two-dimensional double-pendulum overhead crane system operating in a vertical plane as in Fig. 1. The configuration space $\mathcal{Q} \subset \mathbb{R}^3$ is defined by the generalized coordinate vector $\mathbf{q}(t)$:

$$\mathbf{q}(t) = \left[x(t)\ \theta_1(t)\ \theta_2(t)\right]^T \tag{1}$$

where $x(t)$ denotes the translational displacement of the trolley, $\theta_1(t)$ is the sway angle of the hook (mass m_1), and $\theta_2(t)$ is the sway angle of the payload (mass m_2). The system is mechanically constrained by the hoisting rope length l_1 and the rigging cable length l_2, which are assumed to be rigid and constant during the operation. The trolley motion is subjected to non-conservative resistive forces arising from the drive mechanism. In this formulation, the total resistive force is modeled as a constant parameter F_f, representing the aggregate effect of Coulomb and viscous friction components.

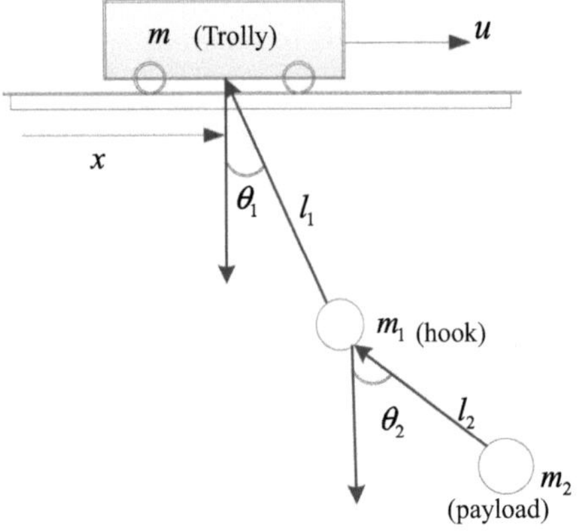

Fig. 1. Gantry Crane System with Double Pendulum.

The primary control problem is to synthesize a control law $u(t) = F(t)$ that drives the system from an initial state $\mathbf{q}(0)$ to a desired target state $\mathbf{q}_d = [x_d, 0, 0]^T$, where x_d is the reference trajectory. Mathematically, the control objectives are twofold:

1. **Trajectory Tracking:** The trolley displacement must asymptotically track a time-varying reference trajectory $x_{ref}(t)$ (e.g., a cycloid curve) such that the tracking error $e_x(t)$ converges to zero:

$$\lim_{t\to\infty} \|x(t) - x_{ref}(t)\| = 0 \tag{2}$$

2. **Sway Suppression:** The residual oscillations of both the hook and the payload must be suppressed to the vertical equilibrium, satisfying the stabilization conditions:

$$\lim_{t\to\infty} \|\theta_1(t)\| = 0, \quad \lim_{t\to\infty} \|\theta_2(t)\| = 0 \tag{3}$$

Consequently, the problem is formulated as the minimization of the generalized error vector $\mathbf{e}(t) = [x(t)-x_{ref}(t), \theta_1(t), \theta_2(t)]^T$ subject to the underactuated nonlinear dynamics and the actuator constraints.

2.2 Nonlinear Dynamic Model

To derive the equations of motion, the energy properties of the system are analyzed. The total kinetic energy K is the summation of the translational energy of the trolley and the kinetic energies of the suspended masses [11]. The velocity vectors for the hook (v_1) and payload (v_2) are derived kinematically, yielding the total kinetic energy expression:

$$K = \frac{1}{2}M_0\dot{x}^2 + \frac{1}{2}m_1 v_1^2 + \frac{1}{2}m_2 v_2^2 \tag{4}$$

The potential energy P, defined relative to the horizontal plane of the trolley, is given by the gravitational displacement of the suspended masses:

$$P = m_1 g l_1(1 - \cos\theta_1) + m_2 g[l_1(1 - \cos\theta_1) + l_2(1 - \cos\theta_2)] \tag{5}$$

Applying the Euler-Lagrange equation $\frac{d}{dt}(\frac{\partial\mathcal{L}}{\partial\dot{q}}) - \frac{\partial\mathcal{L}}{\partial q} = \tau$, where $\mathcal{L} = K - P$ and $q = [x, \theta_1, \theta_2]^T$, yields the highly coupled nonlinear dynamics. These dynamics can be expressed in the standard robotic matrix form:

$$M(q)\ddot{q} + C(q,\dot{q})\dot{q} + G(q) = \tau \tag{6}$$

where $M(q) \in \mathbb{R}^{3\times3}$ is the symmetric positive-definite inertia matrix, $C(q,\dot{q}) \in \mathbb{R}^{3\times3}$ represents the Centrifugal and Coriolis forces, $G(q) \in \mathbb{R}^{3\times1}$ is the gravitational vector, and $\tau \in \mathbb{R}^{3\times1}$ is the generalized force vector.

The components of the inertia matrix $M(q)$ are defined as:

$$M(q) = \begin{bmatrix} M_0 + m_1 + m_2 & (m_1 + m_2)l_1 c_1 & m_2 l_2 c_2 \\ (m_1 + m_2)l_1 c_1 & (m_1 + m_2)l_1^2 & m_2 l_1 l_2 c_{12} \\ m_2 l_2 c_2 & m_2 l_1 l_2 c_{12} & m_2 l_2^2 \end{bmatrix} \tag{7}$$

where $c_i = \cos(\theta_i)$ and $c_{ij} = \cos(\theta_i - \theta_j)$.

The Coriolis and Centrifugal matrix $C(q, \dot{q})$ is given by:

$$C(q,\dot{q}) = \begin{bmatrix} 0 & -(m_1+m_2)l_1 s_1 \dot{\theta}_1 & -m_2 l_2 s_2 \dot{\theta}_2 \\ 0 & 0 & m_2 l_1 l_2 s_{12} \dot{\theta}_2 \\ 0 & -m_2 l_1 l_2 s_{12} \dot{\theta}_1 & 0 \end{bmatrix} \tag{8}$$

where $s_i = \sin(\theta_i)$ and $s_{ij} = \sin(\theta_i - \theta_j)$.

The gravitational vector $G(q)$ and the input vector τ are expressed as:

$$G(q) = \begin{bmatrix} 0 \\ (m_1+m_2) g l_1 s_1 \\ m_2 g l_2 s_2 \end{bmatrix}, \quad \tau = \begin{bmatrix} F - F_f \\ 0 \\ 0 \end{bmatrix} \tag{9}$$

This matrix representation compactly captures the nonlinear coupling between the trolley translation and the double-pendulum swing dynamics.

2.3 Linearization and State-Space Representation

For the design of the Linear Quadratic Regulator (LQR), a linear state-space model is required. The nonlinear system described by (3) is linearized around the stable equilibrium point corresponding to the vertical downward position ($x = x_0, \dot{x} = 0, \theta_1 = 0, \theta_2 = 0, \dot{\theta}_1 = 0, \dot{\theta}_2 = 0$). We invoke small-angle assumptions ($\sin\theta \approx \theta$, $\cos\theta \approx 1$, and $\dot{\theta}^2 \approx 0$).

The linearized equations can be arranged into the form $\mathbf{M}_s \ddot{q} + \mathbf{K}_s q = \mathbf{H} u$, where $u = F$. The linearized mass matrix $\mathbf{M}_s$ and stiffness matrix $\mathbf{K}_s$ are:

$$\mathbf{M}_s = \begin{bmatrix} M_0 + m_1 + m_2 & (m_1+m_2) l_1 & m_2 l_2 \\ (m_1+m_2) l_1 & (m_1+m_2) l_1^2 & m_2 l_1 l_2 \\ m_2 l_2 & m_2 l_1 l_2 & m_2 l_2^2 \end{bmatrix}, \quad \mathbf{K}_s = \begin{bmatrix} 0 & 0 & 0 \\ 0 & (m_1+m_2) g l_1 & 0 \\ 0 & 0 & m_2 g l_2 \end{bmatrix} \tag{10}$$

Defining the state vector $\mathbf{x} = [x, \theta_1, \theta_2, \dot{x}, \dot{\theta}_1, \dot{\theta}_2]^T$, the system is represented in the continuous state-space form $\dot{\mathbf{x}} = A\mathbf{x} + B\mathbf{u}$:

$$A = \begin{bmatrix} \mathbf{0}_{3\times 3} & \mathbf{I}_{3\times 3} \\ -\mathbf{M}_s^{-1}\mathbf{K}_s & \mathbf{0}_{3\times 3} \end{bmatrix}, \quad B = \begin{bmatrix} \mathbf{0}_{3\times 1} \\ \mathbf{M}_s^{-1}\mathbf{E}_1 \end{bmatrix} \tag{11}$$

where $\mathbf{0}$ and $\mathbf{I}$ denote zero and identity matrices of appropriate dimensions, and $\mathbf{E}_1 = [1, 0, 0]^T$ is the input mapping vector. This linear model serves as the foundation for the subsequent Integral LQR design and ISCA optimization.

3 Integral LQR Controller Design

To ensure asymptotic tracking of the reference trajectory $r(t)$ and robust rejection of constant disturbances such as friction, the standard Linear Quadratic Regulator (LQR) is extended with integral action. While the standard LQR minimizes the cost $J = \int_0^\infty (\mathbf{x}^T \mathbf{Q} \mathbf{x} + u^T R u) dt$ to regulate states to the origin,

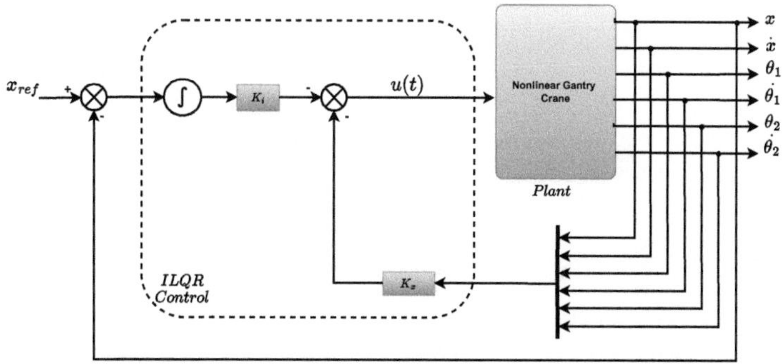

Fig. 2. ILQR conflagration with gantry caren system.

it inherently fails to eliminate the steady-state error $e(t) = r(t) - \mathbf{C}\mathbf{x}(t)$ in trajectory tracking applications (Fig. 2).

We introduce an integrator state x_I, defined by $\dot{x}_I = e(t)$. This augments the linearized state-space model derived in Sect. 4 into a composite state vector $\boldsymbol{\xi}(t) = [\mathbf{x}(t)^T, x_I(t)]^T \in \mathbb{R}^7$. The augmented open-loop dynamics are governed by:

$$\dot{\boldsymbol{\xi}}(t) = \bar{\mathbf{A}}\boldsymbol{\xi}(t) + \bar{\mathbf{B}}u(t) + \mathbf{E}_{ref}r(t) \tag{12}$$

where the augmented system matrices are constructed as:

$$\bar{\mathbf{A}} = \begin{bmatrix} \mathbf{A} & \mathbf{0}_{6\times 1} \\ -\mathbf{C} & 0 \end{bmatrix}, \quad \bar{\mathbf{B}} = \begin{bmatrix} \mathbf{B} \\ 0 \end{bmatrix} \tag{13}$$

with $\mathbf{C} = [1, 0, 0, 0, 0, 0]$ extracting the trolley position. The optimal control law $u(t) = -\mathbf{K}_{aug}\boldsymbol{\xi}(t)$ is derived by minimizing the augmented cost function $\bar{J}$:

$$\bar{J} = \int_0^{\infty} \left(\boldsymbol{\xi}^T \bar{\mathbf{Q}}\boldsymbol{\xi} + u^T R u\right) dt, \quad \text{with } \bar{\mathbf{Q}} = \text{diag}(\mathbf{Q}, Q_I) \tag{14}$$

where $\mathbf{Q}$ penalizes state transients and Q_I penalizes the integrated error. Solving the Algebraic Riccati Equation for the pair $(\bar{\mathbf{A}}, \bar{\mathbf{B}})$ yields the optimal gain vector $\mathbf{K}_{aug} = [\mathbf{K}_x, K_I]$. The resulting feedback law is:

$$u(t) = -\mathbf{K}_x\mathbf{x}(t) - K_I \int_0^t (r(\tau) - \mathbf{C}\mathbf{x}(\tau))d\tau \tag{15}$$

This formulation ensures the control effort adapts dynamically until the steady-state tracking error is driven to zero, providing robust performance despite the underactuated nature of the sway dynamics.

4 Optimization Algorithm for ILQR Tuning

To fully exploit the potential of the proposed Integral LQR controller, the selection of the weighting matrices Q and R is formulated as a global optimization problem. Traditional heuristic tuning often relies on trial-and-error, which is inefficient for the coupled dynamics of underactuated systems. In this work, we employ a Modified Sine Cosine Algorithm () to automate the tuning process, enhancing convergence speed and solution accuracy compared to the standard Sine Cosine Algorithm (SCA).

4.1 Standard Sine Cosine Algorithm

The Sine Cosine Algorithm (SCA) is a population-based meta-heuristic algorithm that navigates [12] the search space by fluctuating outwards (exploration) or inwards (exploitation) towards the best solution found so far. The position update equation for a search agent X_i in the t-th iteration is governed by:

$$X_i^{t+1} = \begin{cases} X_i^t + r_1 \sin(r_2)|r_3 P_i^t - X_i^t|, & r_4 < 0.5 \\ X_i^t + r_1 \cos(r_2)|r_3 P_i^t - X_i^t|, & r_4 \geq 0.5 \end{cases} \tag{16}$$

where P_i^t is the position of the target (best) solution, and r_2, r_3, r_4 are random variables. The parameter r_1 dictates the balance between diversification and intensification, linearly decreasing over iterations. While SCA is effective, its standard nonlinear search path often suffers from premature convergence when optimizing high-dimensional control parameters, leading to suboptimal gain selection.

4.2 Improved Sine Cosine Algorithm

To mitigate the risk of stagnation in local optima, the ISCA introduces a linear search path and an empirical parameter mechanism [13]. These enhancements modify the update rule to include historical information and a more direct trajectory toward the optimal region.

The primary modification replaces the standard absolute position update with a linear term that integrates an empirical best parameter, E_{best}, derived from previous iterations. The modified position update equation is defined as:

$$X_i^{t+1} = \begin{cases} r_1 X_i^t + c_1 \sin(r_2)(P_{best}^t - X_i^t) + c_2 \sin(r_2)(E_{best}^t - X_i^t), & r_3 < 0.5 \\ X_i^t + c_1 \cos(r_2)(P_{best}^t - X_i^t) + c_2 \cos(r_2)(E_{best}^t - X_i^t), & r_3 \geq 0.5 \end{cases} \tag{17}$$

where c_1 and c_2 are weighting constants for the global best and empirical best influence, respectively. The convergence factor r_1 is redefined to decay linearly from a user-defined maximum r_{max} to r_{min}:

$$r_1 = r_{max} - t\frac{r_{max} - r_{min}}{T_{max}} \tag{18}$$

This modification ensures a smoother transition from global exploration to local exploitation, significantly improving the algorithm's ability to fine-tune the LQR gains.

4.3 Objective Function Formulation

The optimization problem is encoded such that the decision variables correspond to the diagonal elements of the weighting matrices $\bar{Q}$ and R. Specifically, the search vector is defined as $\mathbf{\Psi} = [q_x, q_{\theta_1}, q_{\theta_2}, q_{\dot{x}}, q_{\dot{\theta}_1}, q_{\dot{\theta}_2}, q_I, r]$, representing the penalties for the states, integral error, and control effort.

The fitness of each candidate solution is evaluated using the Integral Time Absolute Error (ITAE) criterion, which provides a balanced penalty for long-duration transient errors. The cost function J is formulated to minimize tracking error and residual sway simultaneously:

$$J(\mathbf{\Psi}) = \int_0^{T_{sim}} t\left(|e_x(t)| + |\theta_1(t)| + |\theta_2(t)|\right) dt \tag{19}$$

where $e_x(t)$ is the cart tracking error, and $\theta_1(t), \theta_2(t)$ are the sway angles. The ISCA iteratively minimizes $J(\mathbf{\Psi})$ to yield the optimal gain vector $\mathbf{K}_{aug}$.

4.4 Implementation for Controller Tuning

The optimization process is executed with a population size of $N = 40$ over $T_{max} = 100$ generations. The complete algorithmic procedure for tuning the Integral LQR parameters is summarized as follows:

1. **Initialization:** Generate a population of 40 search agents $\mathbf{\Psi}_i$ $(i = 1, \ldots, 40)$ randomly within the defined bounds $[L_b, U_b]$ for the weighting matrices Q and R. Initialize E_{best} and set iteration counter $t = 0$.
2. **Evaluation:** For each agent $\mathbf{\Psi}_i$:
   (a) Construct the candidate matrices Q_i and R_i.
   (b) Solve the Algebraic Riccati Equation to obtain K_{aug}.
   (c) Simulate the closed-loop crane system with the derived gains.
   (d) Calculate the fitness value $J(\mathbf{\Psi}_i)$ using (21).
3. **Update Best Solutions:** Identify the current best agent P_{best}^t with the minimum fitness. Update the empirical best E_{best}^t based on historical performance.
4. **Parameter Update:** Update the convergence factor r_1 using (20) and update random variables r_2, r_3.
5. **Position Update:** Modify the position of each search agent using the ISCA update equation (19).
6. **Boundary Check:** Ensure all updated agents remain within the feasible search space $[L_b, U_b]$.
7. **Termination:** If $t < 100$, increment t and return to Step 2. Otherwise, terminate and output the global optimal vector $\mathbf{\Psi}_{opt}$.

5 Results and Discussion

To validate the efficacy and robustness of the proposed control strategy, a series of comprehensive simulations were conducted. The optimization process, was executed using MATLAB 2025a on a computer equipped with an Intel Core i5 processor and 20 GB of RAM. The gantry crane system itself was constructed in the SIMULINK environment, faithfully representing the complete nonlinear dynamic model derived in Sect. 2. All subsequent performance evaluations were carried out by interfacing the optimized controller with this high-fidelity nonlinear plant, ensuring a realistic assessment of the control law's effectiveness.

5.1 Simulation Setup and Parameters

The comparative analysis focuses on the performance of the Improved Sine Cosine Algorithm (ISCA) against the standard Sine Cosine Algorithm (SCA) for tuning the Integral LQR gains. Both algorithms were subjected to identical population sizes and iteration limits to ensure a fair evaluation of convergence speed and solution quality.

The simulation utilizes the specific physical parameters of the experimental gantry crane setup, which are detailed in Table 1. These parameters include significant viscous damping coefficients (f_x, f_l), reflecting the friction characteristics of the actual drive mechanism.

Table 1. System parameters of the gantry crane [5]

Variables	Symbol	Values	Unit
Trolley Mass	M	1.155	Kg
Hook Mass	m_1	0.20	Kg
Payload Mass	m_2	0.10	Kg
Cable Length (Nominal)	l_1	0.30	m
Cable Length (Range)	l_1	0.20–0.40	m
Sling Cable Length	l_2	0.20	m
Trolley Viscous Damping	f_x	82	Ns/m
Hoisting Viscous Damping	f_l	75	Ns/m
Gravitational Constant	g	9.81	m/s^2

To ensure a fair assessment of tracking precision under dynamic constraints, the desired cart trajectory $x_d(t)$ is defined as a cycloid curve. This trajectory profile is preferred in crane applications as it guarantees continuous velocity and acceleration, thereby minimizing the excitation of high-frequency structural modes. The reference trajectory is governed by the following equation:

$$x_d(t) = (x_f - x_0)\left\{\frac{t}{t_s} - \frac{1}{2\pi}\sin\left(2\pi\frac{t}{t_s}\right)\right\} + x_0 \tag{20}$$

where x_0 and x_f represent the initial and target positions, respectively, and t_s denotes the settling time of the motion profile.

Scenario 1: Nominal Operation. The first evaluation examines the system response under the nominal parameters ($m_2 = 0.30$ kg, $l_2 = 0.20$ m). The time-domain results for cart position and sway angles are presented in Fig. 3. The ISCA based controller yields a tightly regulated transient response. The cart tracks the cycloid reference with a settling time of 2.8 s, effectively eliminating steady-state error through the optimized integral action. Furthermore, the ISCA-tuned gains provide superior damping of the double-pendulum modes; the peak hook sway (θ_1) is attenuated relative to the SCA benchmark, and the payload sway (θ_2) settles to negligible amplitudes within 4 s.

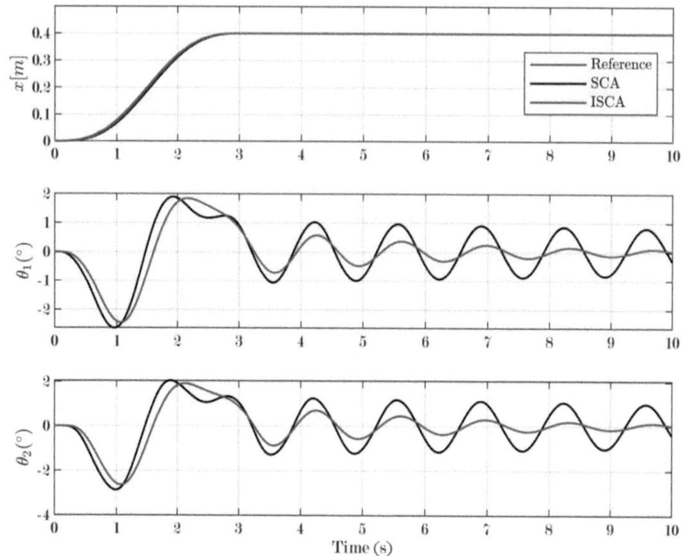

Fig. 3. System response under nominal parameters.

Scenario 2: Payload Mass Variation. To verify robustness against parameter variations, the payload mass m_2 was increased to 0.50 kg and the desired position x_d to 0.5 m. The system response is depicted in Fig. 4. Despite the significant inertial change, the ISCA-ILQR maintains robust tracking performance. The integral term (K_I) dynamically compensates for the increased load, preserving the zero steady-state error requirement. This confirms that the ISCA-optimized control law possesses a wider stability margin suitable for variable loading conditions.

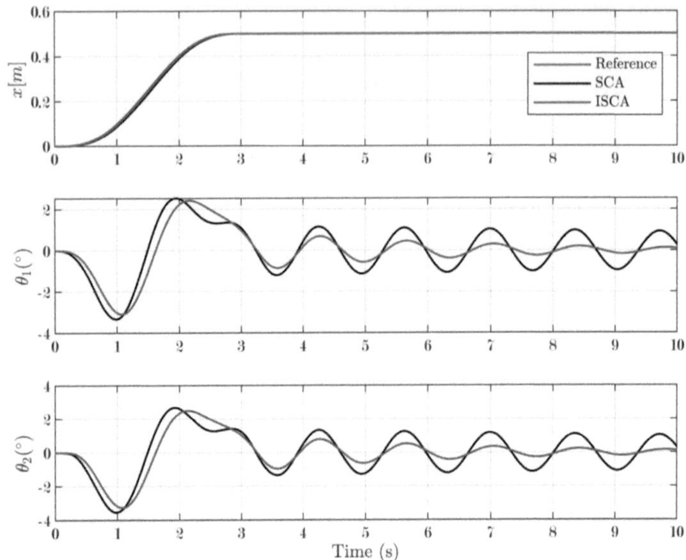

Fig. 4. System response under variations in desired position and payload mass.

Scenario 3: Combined Parameter Uncertainty. The final test introduces a compound disturbance by simultaneously extending the cable length to $l_2 = 0.40$ m and increasing the payload mass to $m_2 = 0.50$ kg. This scenario shifts the natural frequencies of both pendulums, rigorously testing the controller's resonance rejection capabilities. As shown in Fig. 5, the SCA-ILQR fails to adequately suppress the lower-frequency oscillations induced by the longer cable, resulting in persistent sway. In contrast, the ISCA-ILQR controller demonstrates exceptional resilience. The optimized gain structure effectively stiffens the system response against the parametric shift, containing the payload sway θ_2 within acceptable limits ($\pm 1°$) after 5 s.

The quantitative assessment of the control performance focuses specifically on the suppression of residual oscillations, which is the primary safety concern in crane operations. Table 2 summarizes the maximum sway amplitudes recorded for the hook ($\theta_{1,max}$) and the payload ($\theta_{2,max}$) across the three testing scenarios.

The results demonstrate the superior damping capability of the ISCA-ILQR controller. In the nominal scenario, the proposed method reduces the peak payload sway by approximately 12.5% compared to the standard SCA-ILQR. This advantage becomes even more pronounced under robust testing conditions. For instance, in the combined parameter variation scenario, the ISCA-ILQR limits the payload sway to 3.4°, whereas the SCA-ILQR allows it to reach 3.7°, indicating a significantly higher safety margin for the proposed approach.

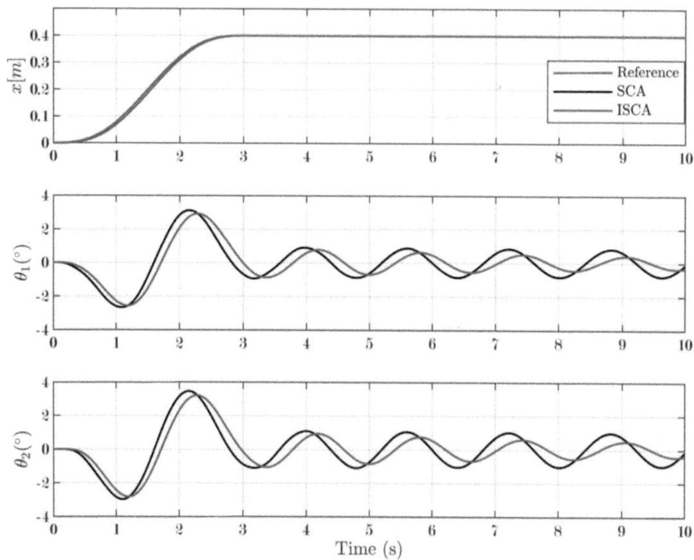

Fig. 5. System response under variations in cable length and payload mass.

Table 2. Comparative of maximum sway amplitudes

Scenario	Controller	$\theta_{1,max}$ (deg)	$\theta_{2,max}$ (deg)
Nominal	SCA-ILQR	3	3.2
	ISCA-ILQR	**2.4**	**2.8**
Payload Var.	SCA-ILQR	3.84	3.89
	ISCA-ILQR	**3.5**	**3.57**
Combined Var.	SCA-ILQR	3.4	3.68
	ISCA-ILQR	**3.2**	**3.37**

The quantitative outcomes of the simulated scenarios highlight a significant advancement over standard control methodologies prevalent in the literature. While traditional Linear Quadratic Regulator (LQR) implementations often suffer from steady-state errors when tracking time-varying cycloid references due to a lack of integral action, the proposed ILQR framework natively drives this tracking error to zero. Furthermore, optimizing these gains is notoriously difficult in high-dimensional spaces; standard meta-heuristics like the conventional Sine Cosine Algorithm (SCA) are prone to premature convergence. By introducing a linear search path and empirical best-history retention, the ISCA overcomes these local optima traps. This algorithmic superiority translates directly to the reported 12.5% reduction in peak payload sway under nominal conditions compared to standard SCA-tuned benchmarks.

The practical implications of these results are highly relevant to modern industrial applications. In severe combined uncertainty scenarios—such as a simultaneous 0.40 m cable extension and a 0.50 kg payload mass increase—the ISCA-ILQR controller successfully stiffened the system's response, strictly containing residual payload sway to within $\pm 1^\circ$ limits after just 5 s. In real-world maritime ports and construction sites, this rapid stabilization translates to significantly enhanced operational safety by mitigating the risk of collision caused by unsuppressed double-pendulum swing. Additionally, by minimizing the time operators must wait for residual oscillations to dampen before depositing a payload, this control strategy directly enables faster duty cycles and increased logistical throughput.

6 Conclusion

This study has successfully formulated and validated a hybrid control strategy for the stabilization of double-pendulum gantry crane systems. By augmenting the Linear Quadratic Regulator with integral action, the steady-state tracking errors characteristic of standard LQR implementations were effectively eliminated. The efficacy of this controller was maximized through the application of the Improved Sine Cosine Algorithm (ISCA), which leveraged a linearized search mechanism to bypass the premature convergence issues inherent in conventional meta-heuristics. Comparative simulations reveal that the ISCA-ILQR framework significantly enhances damping performance, achieving a 12.5% reduction in peak payload sway relative to the standard SCA-ILQR baseline. Furthermore, the control law displayed exceptional robustness against parametric volatility. In scenarios subjecting the system to simultaneous payload mass increases and cable length extensions, the controller maintained tight resonant suppression, restricting residual oscillations to within $\pm 1^\circ$. While this study establishes a robust theoretical and computational framework for the ISCA-ILQR controller, it is important to note that the current validation is simulation-based. This comprehensive nonlinear simulation phase serves as a necessary and safety-critical foundational step to verify algorithmic stability margins and parameter robustness prior to physical deployment. Consequently, our immediate future research will focus on transitioning this control strategy to experimental platforms to address real-world unmodeled dynamics, such as actuator saturation and complex stick-slip friction phenomena. Specifically, we intend to implement Hardware-in-the-Loop (HIL) testing and deploy the algorithm on a scaled physical gantry crane prototype. Furthermore, expanding the kinematic model to encompass three-dimensional spatial movements remains a priority to ensure the frameworkn's complete viability for complex industrial applications.

Acknowledgment. This work is supported by the Universiti Teknologi Malaysia under Vote no. 44H48 UTM-AI grant.

References

1. Hyla, P.: The crane control systems: a survey. In: 2012 17th International Conference on Methods & Models in Automation & Robotics (MMAR), pp. 505–509. IEEE (2012)
2. Abdel-Rahman, E.M., Nayfeh, A.H., Masoud, Z.N.: Dynamics and control of cranes: a review. J. Vib. Control **9**(7), 863–908 (2003)
3. Ramli, L., Mohamed, Z., Abdullahi, A.M., Jaafar, H.I., Lazim, I.M.: Control strategies for crane systems: a comprehensive review. Mech. Syst. Signal Process. **95**, 1–23 (2017)
4. Singhose, W., Porter, L., Kenison, M., Kriikku, E.: Effects of hoisting on the input shaping control of gantry cranes. Control. Eng. Pract. **8**(10), 1159–1165 (2000)
5. Huang, J., Liang, Z., Zang, Q.: Dynamics and swing control of double-pendulum bridge cranes with distributed-mass beams. Mech. Syst. Signal Process. **54**, 357–366 (2015)
6. Chen, Z.M., Meng, W.J., Zhao, M.H., Zhang, J.G.: Hybrid robust control for gantry crane system. Appl. Mech. Mater. **29**, 2082–2088 (2010)
7. Sun, N., Fang, Y., Chen, H.: Adaptive control of underactuated crane systems subject to bridge length limitation and parametric uncertainties. In: Proceedings of the 33rd Chinese Control Conference, pp. 3568–3573. IEEE (2014)
8. Corriga, G., Giua, A., Usai, G.: An implicit gain-scheduling controller for cranes. IEEE Trans. Control Syst. Technol. **6**(1), 15–20 (1998)
9. Adeli, M., Zarabadipour, H., Zarabadi, S.H., Shoorehdeli, M.A.: Anti-swing control for a double-pendulum-type overhead crane via parallel distributed fuzzy lqr controller combined with genetic fuzzy rule set selection. In: 2011 IEEE International Conference on Control System, Computing and Engineering, pp. 306–311. IEEE (2011)
10. Zawawi, M., Zamani, W.W., Ahmad, M., Saealal, M., Samin, R.: Feedback control schemes for gantry crane system incorporating payload. In: 2011 IEEE Symposium on Industrial Electronics and Applications, pp. 370–375. IEEE (2011)
11. Hussien, S.Y.S., Jaafar, H.I., Ghazali, R., Ghazaly, M.M., Johari, M.K.A., Ramli, L.: Modelling, simulation and validation for a multimode double-pendulum overhead crane. Appl. Model. Simul. **8**, 261–271 (2024)
12. Mirjalili, S.: Sca: a sine cosine algorithm for solving optimization problems. Knowl.-Based Syst. **96**, 120–133 (2016)
13. Wang, M., Lu, G.: A modified sine cosine algorithm for solving optimization problems. IEEE Access **9**, 27434–27450 (2021)

Preliminary Study of Corrosion-Induced Ultrasonic Signal Changes in Reinforced Concrete

Musa Iskandar Shah, Anita Ahmad(✉), Farah Aina Jamal Mohamad, and Noorhazirah Sunar

Faculty of Electrical Engineering, Universiti Teknologi Malaysia, 81310 UTM, Skudai, Johor, Malaysia
anita@utm.my

Abstract. Corrosion of steel reinforcement in reinforced concrete (RC) structures is a major factor compromising structural integrity and service life. While non-destructive evaluation (NDE) techniques such as ultrasonic testing are commonly used to detect cracks and voids, corrosion-specific detection remains underexplored, and simulation-based investigations are limited. This study presents a preliminary simulation-based analysis of the effect of corrosion layer on the ultrasonic wave propagation in RC structures. A two-dimensional model was developed in COMSOL Multiphysics® to simulate ultrasonic pulse propagation in both healthy and corroded configurations under identical conditions. The corrosion defect was represented by a rust layer surrounding the steel, introducing acoustic impedance mismatches with the surrounding materials. A modulated Gaussian ultrasonic pulse was applied at the transducer, and time-domain responses were analyzed to evaluate the effects of corrosion on wave amplitude. Simulation results indicate that the presence of corrosion significantly alters ultrasonic signal characteristics. The presence of the rust layer caused localized wave scattering and reflection, resulting in high-amplitude peaks and repeated defect-related features in the time-domain response. These changes demonstrate that ultrasonic waves are sensitive to corrosion-induced material degradation, suggesting potential for early detection of corrosion in RC structures.

Keywords: Corrosion detection · Finite element simulation · Non-destructive evaluation · Reinforced concrete · Ultrasonic testing

1 Introduction

Reinforced concrete (RC) is a cornerstone of modern infrastructure due to its high strength, durability, and cost-effectiveness. However, the long-term performance of RC structures is often compromised by the corrosion of embedded steel reinforcement. Corrosion induces expansion of the steel, cracking of the surrounding concrete, and delamination, which may cause a significant reduction in structural integrity and accelerate deterioration [1]–[3]. Early detection of corrosion is therefore critical for maintaining safety and prolonging service life.

N. A. Mohd Subha et al. (Eds.): SMS 2026, CCIS 2956, pp. 218–230, 2026.
https://doi.org/10.1007/978-981-92-0306-2_16

Non-destructive evaluation (NDE) techniques, including ultrasonic testing, acoustic emission, and electrical resistivity measurements, have been widely employed to monitor the health of RC structures [4]. These methods have been effective in detecting cracks, voids, and delamination within concrete. However, they often overlook corrosion-specific mechanisms [5]. Corrosion leads to the formation of rust products around steel reinforcement, introducing localized changes in stiffness and density that alter ultrasonic wave propagation in ways distinct from general defects.

Simulation-based studies provide a controlled and reproducible platform to investigate ultrasonic wave interaction with corrosion defects. While experimental studies provide practical insights, they are limited by accessibility, variability in corrosion development, and challenges in reproducing defect conditions. In contrast, numerical simulations allow systematic analysis of the effect of corrosion on ultrasonic signal characteristics such as amplitude and scattering patterns, which are critical for corrosion detection. Despite this potential, studies that combine corrosion-specific modeling with ultrasonic wave simulations remain limited, highlighting a clear gap in the current literature [6, 7].

This study aims to address this gap by investigating the effect of corrosion on ultrasonic signals in RC structures using a finite element simulation approach. Specifically, the objectives are to model ultrasonic wave propagation in reinforced concrete containing corroded steel, analyze the resulting changes in wave characteristics, and provide preliminary insights for designing ultrasonic NDE systems sensitive to corrosion. This study specifically models corrosion as a distinct material layer and investigates its direct influence on ultrasonic wave amplitude characteristics. The significance of this work lies in incorporating a corrosion-layer representation within a finite element ultrasonic model, introducing a baseline comparison framework, and quantifying corrosion-induced signal variation using relative amplitude metrics.

2 Theoretical Background

Corrosion of steel reinforcement is one of the most critical degradation mechanisms affecting the durability and service life of reinforced concrete structures. It typically initiates when chloride ions or carbonation penetrate the concrete cover, breaking down the passive oxide layer that protects steel. Once this layer is compromised, steel reacts with oxygen and moisture, forming rust products such as iron oxides and hydroxides [8]. The formation of these corrosion products leads to an increase in volume, generating internal tensile stresses that induce cracking, delamination, and spalling of the surrounding concrete, as illustrated in Fig. 1.

In addition to visible damage, corrosion causes localized reductions in stiffness, density, and bonding quality between steel and concrete, fundamentally altering the mechanical and acoustic properties of the composite material [9]. These corrosion-induced material changes directly influence the propagation of mechanical waves within RC structures. Variations in elastic modulus, density, and interfacial bonding create acoustic impedance mismatches that affect ultrasonic wave velocity, amplitude, scattering, and reflection behavior [10, 11].

As corrosion often develops internally before surface damage becomes apparent, non-destructive evaluation is required to detect and assess deterioration at an early stage.

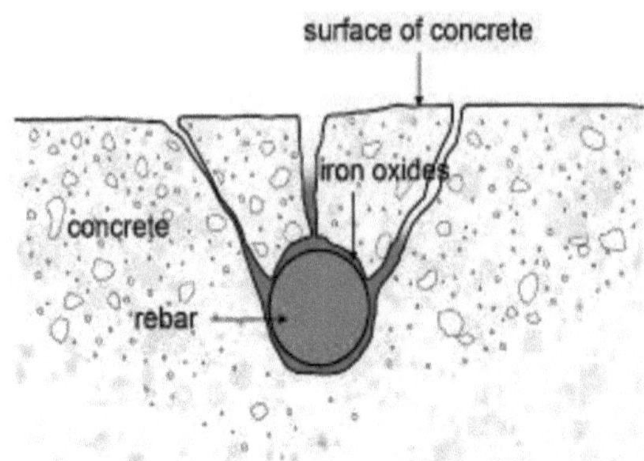

Fig. 1. Formation of corrosion in RC structures leads to other serious defects [12].

Among available NDE methods, ultrasonic techniques have demonstrated significant potential due to their sensitivity to changes in material properties and their ability to probe internal regions of concrete without causing damage.

One of the most common ultrasonic techniques for RC evaluation is ultrasonic pulse velocity (UPV), which evaluates the travel time of longitudinal waves through concrete [13]. Numerous studies have demonstrated that UPV decreases as corrosion progresses, correlating with steel mass loss and cracking of the concrete cover [14]. However, UPV provides a global, averaged measurement and exhibits limited sensitivity to localized corrosion. At advanced corrosion levels, velocity values may plateau, reducing the method's ability to quantify severe damage or distinguish between different corrosion severities [15]. These limitations indicate that while UPV is effective for general condition assessment, it may not adequately capture corrosion-specific mechanisms.

To enhance sensitivity to localized defects, guided ultrasonic wave techniques have been explored. Guided waves propagate along the embedded steel reinforcement and interact strongly with corrosion-induced defects and debonding at the steel–concrete interface. Experimental studies have shown that guided-wave signal amplitudes decrease during corrosion initiation due to increased energy dissipation, but may increase at later stages as debonding alters wave confinement [16]. Despite their potential, guided-wave responses are strongly influenced by concrete heterogeneity, boundary reflections, and propagation distance, often resulting in complex signal interpretation and significant energy loss in practical applications.

Beyond linear ultrasonic approaches, nonlinear ultrasonic methods have been explored to capture subtle material degradation associated with corrosion. These techniques analyze higher-order wave phenomena, such as harmonic generation and sideband peaks, which arise from microstructural changes and imperfect contacts within corroded regions. Nilsson et al. [17] demonstrated that nonlinear parameters increase significantly in corroded steel, enabling the detection of internal delamination that may not be evident in linear velocity-based measurements. Nevertheless, nonlinear responses can be affected by surface roughness and scattering effects, and in cases of extensive damage, the measured nonlinearity may decrease, complicating quantitative assessment.

Given the limitations of purely experimental approaches, numerical simulation has emerged as a valuable tool for investigating corrosion-induced ultrasonic responses in RC structures. Finite element modeling allows precise control over material properties, geometry, defect characteristics, and boundary conditions, enabling systematic study of wave–defect interactions. Zheng et al. [18] demonstrated that guided-wave simulations could accurately locate corrosion-related defects with errors below 2%, while reproducing theoretical wave velocities with high accuracy. These results highlight the capability of numerical models to isolate corrosion effects, evaluate signal sensitivity, and guide experimental design.

Despite extensive research on ultrasonic testing of RC structures, most existing studies focus on linear UPV measurements, guided-wave experiments affected by material heterogeneity, or nonlinear techniques with interpretation challenges. Simulation-based investigations that specifically examine corrosion-induced changes in ultrasonic signal characteristics, such as amplitude variation and scattering behavior, remain limited. Addressing this gap, the present study employs finite element simulations to investigate ultrasonic wave propagation in reinforced concrete containing corroded steel reinforcement. These theoretical considerations form the basis for the simulation methodology described in the following section.

3 Methodology

Fig. 2 presents the overall methodological framework adopted for corrosion detection in reinforced concrete. The framework comprises both numerical simulation and hardware-based experimental investigation. The simulation component is used to study ultrasonic wave propagation and corrosion-induced signal changes in a controlled environment, while the experimental component is intended for future validation of the numerical findings. This paper focuses exclusively on the simulation workflow, which forms the theoretical and computational foundation of the proposed ultrasonic corrosion detection approach.

The simulation was performed using COMSOL Multiphysics® version 6.0, a finite element-based platform for solving coupled multi-physics problems with high computational precision. COMSOL was selected due to its capability to model transient elastic wave propagation in heterogeneous media and to handle complex material interfaces in reinforced concrete containing corrosion defects. The simulation workflow implemented in COMSOL is illustrated in Fig. 3.

3.1 Physics Interface

Ultrasonic wave propagation within the reinforced concrete model was simulated using the Elastic Waves physics interface. This interface solves the governing equations of linear elastodynamics under the assumption of small deformations, making it suitable for ultrasonic non-destructive evaluation applications.

All materials in the model were defined as isotropic linear elastic materials and were characterized by their density (ρ), Young's modulus (E), and Poisson's ratio (ν). Under

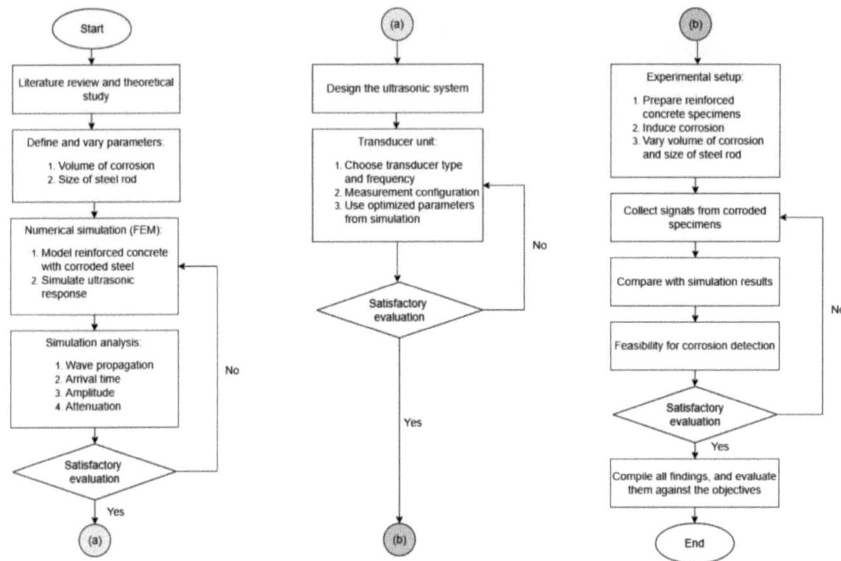

Fig. 2. Flowchart illustrating the overall methodology for corrosion detection in reinforced concrete.

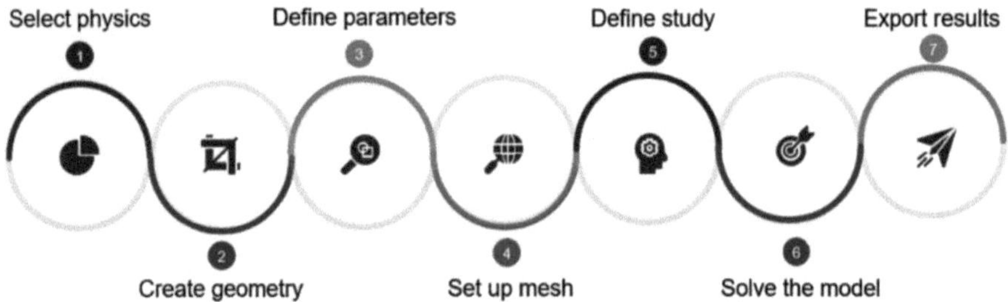

Fig. 3. Simulation workflow in COMSOL.

these assumptions, the wave propagation behavior is governed by the elastodynamic equation [19]:

$$\rho \frac{\partial^2 u}{\partial t^2} = \nabla.\sigma \tag{1}$$

where u is the displacement vector, and σ is the stress tensor. This equation represents the linear elastodynamic formula for isotropic elastic media and is widely used in ultrasonic wave propagation modeling [19]. This formula allows the propagation of both longitudinal (P) waves and shear (S) waves within the concrete medium, which are known to be sensitive to material discontinuities and interfacial defects, such as corrosion in reinforced concrete [20].

3.2 Geometrical Model

The geometrical configuration of the simulation model is shown in Fig. 4. The model consists of a two-dimensional concrete domain containing an embedded steel reinforcement bar surrounded by a corrosion layer. A two-dimensional representation was adopted in this preliminary study to reduce computational cost while capturing the essential wave-defect interaction mechanisms. This approach is commonly used in early-stage ultrasonic simulation studies before extending to three-dimensional studies [21].

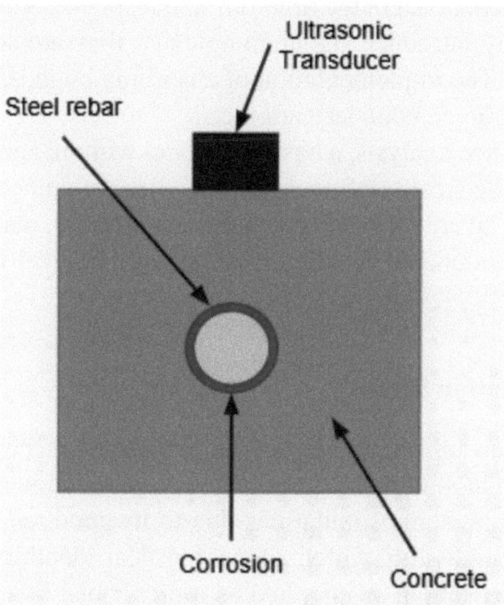

Fig. 4. The simulated concrete model.

The geometric parameters of the model are summarized in Table 1. The concrete domain has dimensions of 100 mm × 100 mm, containing a steel reinforcement bar with a diameter of 9 mm. Corrosion was modeled as a uniform layer with a thickness of 2 mm surrounding the steel bar. An ultrasonic transducer with dimensions of 20 mm × 20 mm was positioned at the concrete surface to generate and receive ultrasonic waves.

Table 1. Parameters of geometrical configuration.

Component	Dimension
Concrete	100 mm × 100 mm
Steel rebar	9 mm diameter
Corrosion layer	2 mm thickness
Ultrasonic transducer	20 mm × 20 mm

The corrosion defect was represented as a rust layer surrounding the reinforcement bar to simulate corrosion products formed due to steel degradation. Iron (III) oxide was selected to represent corrosion products due to its significantly lower stiffness compared to steel, which introduces a strong acoustic impedance mismatch at the steel-corrosion interface. This mismatch enhances ultrasonic wave scattering and reflection, and makes it suitable for investigating corrosion-induced signal changes [22, 23].

The outer boundaries of the concrete domain were assigned default elastic boundary conditions. Absorbing boundary conditions were not applied at this stage, resulting in partial wave reflections from the model boundaries. These reflections contribute to multiple wave arrivals observed in the time-domain response. Although such boundary-induced reflections may introduce signal complexity, they are acceptable for this preliminary investigation. The implementation of absorbing boundaries will be considered in future studies to minimize boundary artefacts.

To enable comparative analysis, a baseline model without corrosion was developed. In this configuration, the steel reinforcement was directly embedded in concrete without a surrounding rust layer. All other geometric parameters, material properties, excitation conditions, and boundary conditions remained identical to ensure a controlled comparison.

3.3 Ultrasonic Excitation Signal

The ultrasonic excitation applied at the transducer surface is schematically illustrated in Fig. 5. A modulated Gaussian pulse was used as the excitation signal. This waveform is commonly used in ultrasonic simulations due to its good temporal localization and controlled frequency bandwidth, which allows for clear identification of wave arrivals and reflections. The excitation was applied as a prescribed boundary condition at the transducer location to generate a transient ultrasonic wave propagating into the concrete domain.

3.4 Mesh Design

Mesh design plays a critical role in ensuring the accuracy and numerical stability of ultrasonic wave simulations. To accurately resolve wave propagation, the mesh must be sufficiently fine to capture the shortest wavelength present in the model, while remaining coarse enough to maintain computational efficiency.

In this study, the mesh size was determined based on the slowest wave speed in the materials, which corresponds to the shear wave velocity. The minimum wavelength was estimated using [24]:

$$\lambda_s = \frac{c_s}{f} \tag{2}$$

where c_s is the shear wave velocity, and f is the central frequency of the ultrasonic excitation. This relationship between wavelength, shear wave velocity, and excitation frequency is commonly adopted in finite element mesh design for ultrasonic simulations. The finite element mesh used in the simulation is shown in Fig. 6.

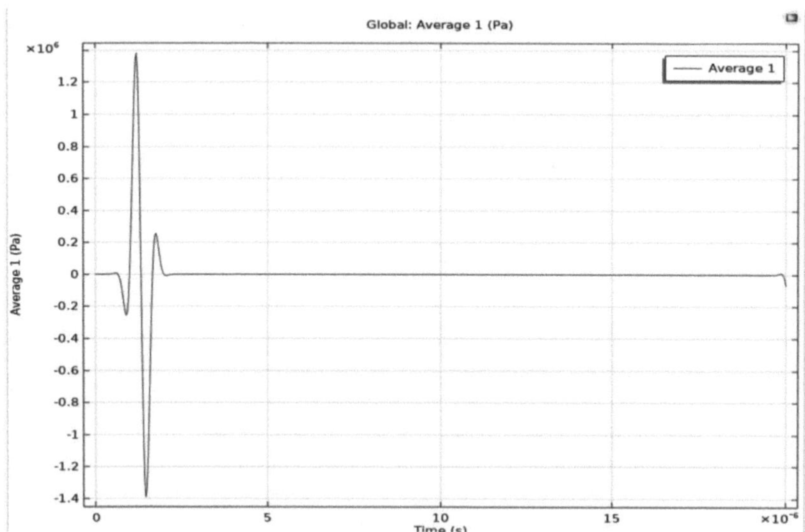

Fig. 5. Ultrasonic signal applied at the transducer.

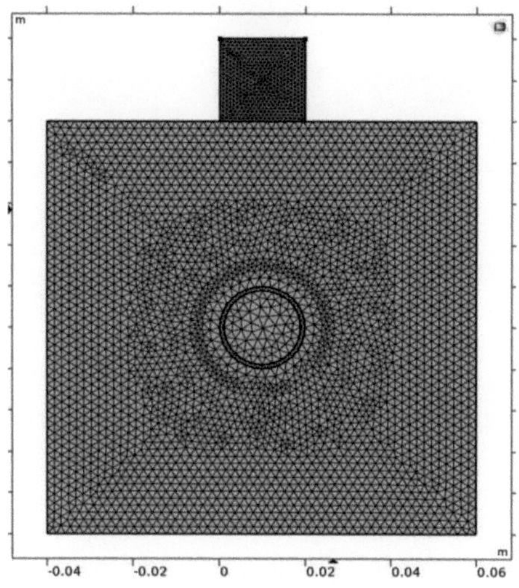

Fig. 6. Mesh design in the model.

3.5 Study Type

A time-dependent study was used to capture the transient pulse-echo response of the ultrasonic wave propagation within the reinforced concrete model. This approach enables the analysis of time-of-flight (ToF) and amplitude variations arising from wave interactions with the corrosion defect. These parameters are key indicators for defect detection in ultrasonic non-destructive evaluation.

4 Results and Discussion

Fig. 7 illustrates the simulated ultrasonic wave propagation within the reinforced concrete model without corrosion. The ultrasonic transducer positioned at the concrete surface emits a modulated Gaussian pulse that propagates downward into the concrete domain and interacts with the embedded steel reinforcement bar.

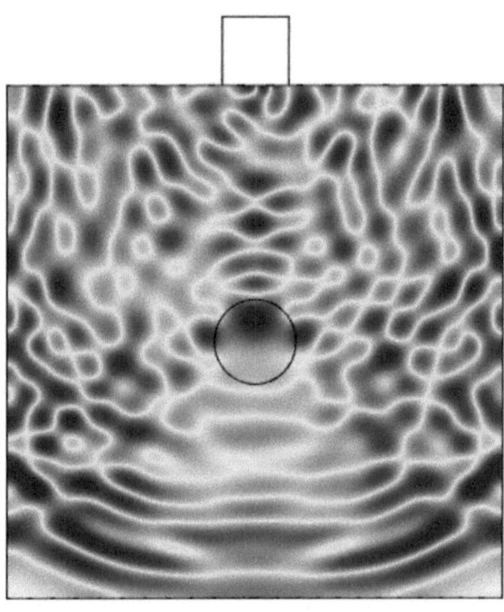

Fig. 7. Simulated ultrasonic wave propagation in reinforced concrete without corrosion.

In the healthy model, the wavefront propagates smoothly through the concrete domain. Partial reflection occurs at the concrete–steel interface due to acoustic impedance mismatch between the materials. The reflected waves return toward the surface and are captured by the transducer. The wavefield distribution remains relatively symmetric around the reinforcement, and no significant localized scattering is observed. The energy transfer across the steel–concrete interface appears stable, indicating intact bonding and uniform material continuity.

The corresponding time-domain signal for the healthy model presented in Fig. 8 exhibits an initial peak associated with the transmitted pulse and near-surface reflections, followed by a distinct reflection from the embedded steel reinforcement. The waveform shows relatively smooth oscillatory behavior with moderate amplitude levels and minimal irregular distortion.

Fig. 9 presents the simulated ultrasonic wave propagation in the reinforced concrete model containing a corrosion layer surrounding the steel reinforcement. The corrosion defect was modeled as a uniform rust layer, introducing additional material discontinuity between the steel and the surrounding concrete.

Compared to the healthy model, noticeable wave scattering and localized disturbances are observed around the corrosion region. The presence of steel–rust and rust–concrete interfaces creates multiple acoustic impedance mismatches, leading to partial reflection, transmission, and mode conversion of the incident ultrasonic waves. These

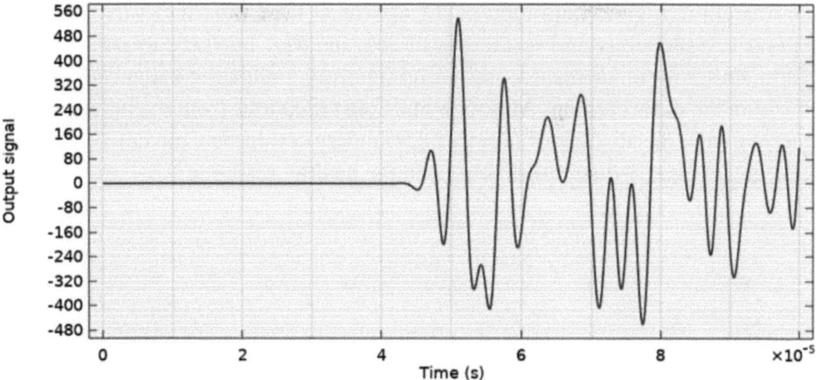

Fig. 8. Time-domain ultrasonic signal for healthy concrete.

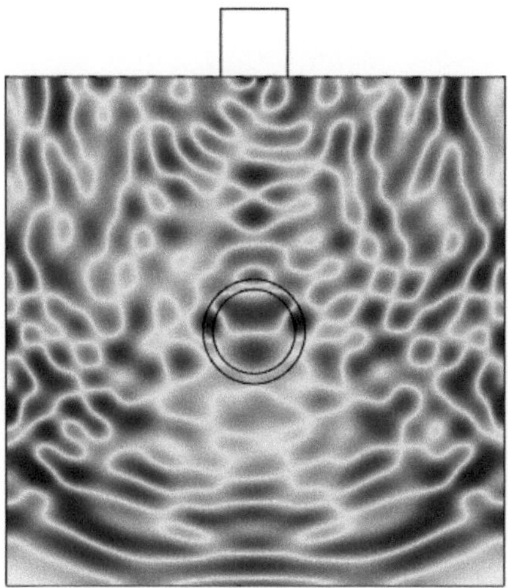

Fig. 9. Simulated ultrasonic wave propagation in reinforced concrete containing a corrosion defect.

interactions result in more complex wave patterns and localized energy redistribution around the reinforcement. Corrosion products generally exhibit lower stiffness and density compared to steel, which results in partial reflection and mode conversion of incident ultrasonic waves at the steel-corrosion interface [25]. As a consequence, localized wave disturbances are observed around the corrosion defect, indicating strong interaction between the propagating ultrasonic waves and the degraded reinforcement. Similar wavefield distortions have been reported in numerical and experimental studies of ultrasonic inspection in reinforced concrete, where defects such as corrosion, voids, or delamination cause scattering and reflection of elastic waves [26–28].

The time-domain signal obtained from the corroded model, as presented in Fig. 10, shows significant differences compared to the healthy case. In addition to the initial surface reflection, higher-amplitude peaks appear at later time intervals. These peaks correspond to reflections and scattering effects induced by the corrosion layer. The waveform also exhibits increased oscillation intensity, indicating enhanced wave interaction within the defect region. Although multiple reflections from the outer boundaries are present due to the absence of absorbing boundary conditions, the corrosion-induced features remain clearly distinguishable from the healthy response.

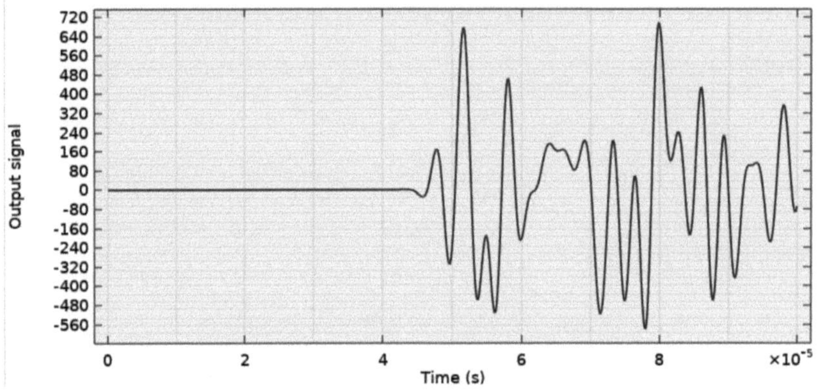

Fig. 10. Time-domain ultrasonic signal for corroded concrete.

To enable quantitative comparison, the peak amplitudes of the received time-domain signals were extracted from both healthy and corroded models. The relative amplitude variation was calculated as:

$$\Delta A\% = \frac{A_{corroded} - A_{healthy}}{A_{healthy}} \times 100\% \quad (3)$$

where $A_{corroded}$ and $A_{healthy}$ represent the maximum absolute peak amplitudes of the corroded and healthy models, respectively.

The results indicate that the presence of the corrosion layer produces a relative amplitude variation of approximately 25.9%. This confirms that corrosion significantly modifies ultrasonic reflection behavior. The additional impedance mismatches introduced by the rust layer enhance wave scattering and reflection, resulting in amplified defect-related signal components.

The present findings are consistent with previously reported numerical investigations. For example, Demi et al. [29] reported that corrosion damage significantly alters elastic wave propagation characteristics, including amplitude reduction and scattering effects. Similar wave–defect interaction mechanisms are observed in the current pulse-echo configuration, where the introduction of a corrosion layer generates additional impedance mismatches and measurable signal variation.

5 Conclusion

This paper presents a preliminary study on ultrasonic wave propagation for corrosion detection in reinforced concrete using a finite element simulation approach. The developed model captures the interaction between ultrasonic excitation, concrete, steel reinforcement, and corrosion-induced material degradation by analyzing the time-domain output of the ultrasonic response.

Simulation results demonstrate that corrosion significantly affects ultrasonic signal characteristics. Noticeable changes are observed in waveform amplitude and signal attenuation due to wave scattering and impedance mismatch introduced by the corrosion layer. These findings indicate that ultrasonic responses are sensitive to corrosion-related defects and can potentially be exploited for corrosion detection in reinforced concrete structures.

Based on the simulation outcomes, an ultrasonic testing system is proposed for corrosion detection, with key design parameters informed by the numerical results. Although this study focuses on simulation-based preliminary findings, experimental validation is necessary. Future work will involve hardware-based ultrasonic experiments to validate the simulated responses and assess the practical feasibility of the proposed approach.

Acknowledgement. This research was supported by MOE, FUNDER, UTM, NO VOTE (Q.J130000.3823.23.H04).

References

1. Yu, X., Yan, Z., Li, J.: Comprehensive study on corrosion induced structural deterioration of steel bars in reinforced concrete beams. J. Constr. Steel Res. **214**, 108504 (2024)
2. Taheri-Shakib, J., Al-Mayah, A.: Dynamics of localized accelerated corrosion in reinforced concrete: voids, corrosion products, and crack formation. Ceram. Int. **50**(22), 48755–48767 (2024)
3. Mahi, M.S.H., Ridoy, T.A.: Corrosion mechanisms in reinforced concrete: causes, effects, and sustainable mitigation strategies. Curr. Probl. Res. **1**(1), 52–66 (2025)
4. Mohammed, K.K., Tahir, H.S., Mohammed, A.S., Hassan, A.M.T.: A comprehensive review of structural health monitoring in reinforced concrete: advances in non-destructive testing (NDT), hybrid SHM systems, and conductive materials. Multiscale Sci. Eng., 1–25 (2025)
5. Bouhouche, I., et al.: Corrosion assessment and inspection in concrete corrosion. In: Corrosion in Concrete Structures: Integrating Advanced Technologies and Sustainable Practices, pp. 85–118 (2026)
6. Kim, J., Cho, Y., Lee, J., Kim, Y.: Defect detection and characterization in concrete based on FEM and ultrasonic techniques. Materials. **15**(22), 8160 (2022)
7. Hong, M., Lei, D., Hu, F., Chen, Z.: Assessment of void and crack defects in early-age concrete. J. Build. Eng. **70**, 106372 (2023)
8. Fuhaid, A.F.A., Niaz, A.: Carbonation and corrosion problems in reinforced concrete structures. Buildings. **12**(5), 586 (2022)
9. Robuschi, S.: Natural Corrosion in Reinforced Concrete Structures. Chalmers Tekniska Hogskola, Sweden (2021)

10. Biswas, R.K., Iwanami, M., Chijiwa, N., Nakayama, K.: Structural assessment of the coupled influence of corrosion damage and seismic force on the cyclic behaviour of RC columns. Constr. Build. Mater. **304**, 124706 (2021)
11. Rathod, V.T.: A review of acoustic impedance matching techniques for piezoelectric sensors and transducers. Sensors. **20**(14), 4051 (2020)
12. Koulouris, K., Apostolopoulos, C.: An experimental study on effects of corrosion and stirrups spacing on bond behavior of reinforced concrete. Meta. **10**(10), 1327 (2020)
13. Lin, Y.C., Lin, Y., Cheng, C.C.: A unified equation for prediction of concrete strength at various ages using the ultrasonic pulse velocity. Appl. Sci. **12**(17), 8416 (2022)
14. Othman, F.Z., Ayop, S.S.: Evaluation of corrosion in reinforced concrete: a review on the application of UPV method. Recent Trends Civ. Eng. Built Environ. **2**(1), 160–170 (2021)
15. Almashakbeh, Y., Saleh, E.: Evaluation of ultrasonic pulse velocity (UPV) for reinforced concrete corrosion. J. Appl. Eng. Sci. **20**(4), 1226–1233 (2022)
16. Habbaba, N., Mustapha, S., Lu, Y.: Early detection of corrosion in reinforced concrete using ultrasonic-guided waves. In: Life-Cycle of Structures and Infrastructure Systems, pp. 2379–2386. CRC Press (2023)
17. Nilsson, M., Ulriksen, P., Rydén, N.: Nonlinear ultrasonic characteristics of a corroded steel plate. Nondestr. Test. Eval. **38**(3), 456–479 (2023)
18. Zheng, Z., Lei, Y., Xue, X.: Numerical simulation of monitoring corrosion in reinforced concrete based on ultrasonic guided waves. Sci. World J. **2014**(1), 752494 (2014)
19. Spalvier, A., Sánchez, J., Pérez, N.: 3D Modeling of ultrasonic wave propagation in pervious concrete. J. Nondestruct. Eval. **44**(3), 104 (2025)
20. Angiulli, G., Burrascano, P., Ricci, M., Versaci, M.: Advances in the integration of artificial intelligence and ultrasonic techniques for monitoring concrete structures: a comprehensive review. J. Compos. Sci. **8**(12), 531 (2024)
21. Pereira, D., Le Duff, A., Painchaud-April, G., Belanger, P.: Simulation-based inversion for the characterization of adhesively bonded joints using ultrasonic guided waves. IEEE Trans. Ultrason. Ferroelectr. Freq. Control. **69**(7), 2400–2407 (2022)
22. Wieland, E., Miron, G.D., Ma, B., Geng, G., Lothenbach, B.: Speciation of iron(II/III) at the iron-cement interface: a review. Mater. Struct. **56**(2), 31 (2023)
23. Thirumalaiselvi, A., Sasmal, S.: Early detection and monitoring of corrosion in reinforced concrete structures using coupled time-frequency analysis. J. Mater. Civ. Eng. **36**(8), 04024207 (2024)
24. Hashemi, H.S., Mohammed, S.K., Zeng, Q., Azar, R.Z., Rohling, R.N., Salcudean, S.E.: 3-D ultrafast shear wave absolute Vibro-Elastography using a matrix Array transducer. IEEE Trans. Ultrason. Ferroelectr. Freq. Control. **70**(9), 1039–1053 (2023)
25. Alanazi, A.M.: Development of ultrasonic model-based iterative reconstruction technique for concrete structures with corroded rebars. Sci. Rep. **15**(1), 43566 (2025)
26. Ye, Y., Chai, H.K., Lee, F.W.: Assessing voided reinforced concrete by numerical modelling of impact-generated Rayleigh waves. Appl. Sci. **15**(7), 3635 (2025)
27. Barashok, K., Choi, Y., Choi, Y., Aslam, M., Lee, J.: Machine learning techniques in ultrasonics-based defect detection and material characterization: a comprehensive review. Adv. Mech. Eng. **17**(6) (2025)
28. Jain, H., Patankar, V.H.: Simulations and experimentation of ultrasonic wave propagation and flaw characterisation for underwater concrete structures. Nondestr. Test. Eval. **39**(6), 1581–1598 (2024)
29. Ai, D., Du, L., Li, H., Zhu, H.: Corrosion damage identification for reinforced concrete beam using embedded piezoelectric transducer: numerical simulation. Measurement. **192**, 110925 (2022)

Boolean-Algebraic Modeling and Analysis of NOMA Schemes in Two-User Interference Channels

Sock Theng Ooi[1(✉)], Marwan H. Azmi[2], and Razali Ngah[3]

[1] Space Engineering and Technology Division, Malaysian Space Agency (MYSA), Banting, Selangor, Malaysia
stooi@mysa.gov.my

[2] Wireless Communication Centre, Faculty of Artificial Intelligence, UTM Kuala Lumpur, 54100 Kuala Lumpur, Malaysia

[3] Wireless Communication Centre, Universiti Teknologi Malaysia, 81310 Johor Bahru, Johor, Malaysia

Abstract. This paper introduces a novel analytical framework based on Boolean algebra for modeling and analyzing Non-Orthogonal Multiple Access (NOMA) systems. Departing from conventional signal-to-interference-plus-noise ratio (SINR) based models, the proposed approach conceptualizes the Successive Interference Cancellation (SIC) process in a two-user interference channel as a probabilistic system of logical events. The core contribution is a systematic procedure that transforms the SIC decoding problem into a Boolean model, subsequently simplified and analyzed using techniques such as Karnaugh maps. This framework is demonstrated for systems characterized by 2, 4, and 5 elements, showcasing its capability to capture the fundamental logical interdependencies of SIC. The Boolean approach provides a new, complementary perspective for NOMA analysis, offering a structured methodology to evaluate decoding performance and system complexity in interference-limited environments.

Keywords: Successive Interference Cancellation (SIC) · Non-Orthogonal Multiple Access (NOMA) · Two-User Interference Channel (TU-IC)

1 Introduction

The relentless growth in wireless connectivity demands higher spectral efficiency, massive connectivity, and ultra-low latency for next-generation networks. NOMA has emerged as a promising technique to meet these demands [1–3]. SIC, employed at the receiver to sequentially decode and remove stronger users' signals, is crucial for enhancing spectral efficiency and user fairness [4,5]. However, practical SIC implementation faces significant challenges including imperfect interference cancellation, decoding order errors, and noise sensitivity [6,7].

N. A. Mohd Subha et al. (Eds.): SMS 2026, CCIS 2956, pp. 231–245, 2026.
https://doi.org/10.1007/978-981-92-0306-2_17

Conventional analytical frameworks for NOMA predominantly rely on stochastic geometry or signal-to-interference-plus-noise ratio (SINR) formulations to evaluate performance metrics such as outage probability and ergodic capacity [1,8]. While these methods provide valuable insights into average system performance, they often abstract away the logical interdependencies between fundamental decoding events in two-user scenarios. Boolean algebra offers a compelling alternative framework, representing system behavior through logical relationships among discrete events such as decoding success or failure. Techniques like the Karnaugh map (K-map) enable systematic simplification of these Boolean expressions, providing a foundational model for the SIC process.

This paper proposes a novel Boolean-algebraic framework to model and analyse the NOMA systems. The core approach transforms the sequential NOMA detection process into a probabilistic Boolean system. Beginning with a two-user interference channel (TU-IC) as baseline, modeled as a probabilistic system with two events, the framework is generalized to more complex interactions represented by 4 and 5 elements. This approach provides a new probabilistic-logical perspective that complements conventional analytical models [9].

The principal contributions of this work are summarized as follows:,

(i) Systematic procedure to model SIC scheme in TU-IC using probability system with two events.
(ii) The decoding capability algorithm of SIC is further proposed by exploiting Boolean Algebra in order to lighten the system's complexity and also reveals its dynamic system properties which could anticipate changes over time, enable the system to change the decision flexibility.
(iii) Numerical simulation is conducted to compare the performance among the proposed schemes and also to their existing benchmarked schemes.
(iv) All proposed cases are outperformed existing benchmarked schemes before received SNR 50dB in both single and total achievable rates.

The remainder of this work is organized as follows. Section 2 introduces system channel for TU-IC in downlink transmission. Section 3 presents systematic procedure to model the SIC schemes of the TU-IC using Boolean Function with 2 elements. Section 4 extends this framework to Boolean functions with higher elements. Section 5 provides performance evaluation results, and Sect. 6 concludes the paper.

2 System Model

We consider the TU-IC illustrated in Fig. 1. The received signals Y_i, for $i \in \{1,2\}$, of the TU-IC can be written by

$$Y_1 = h_{11}X_1 + h_{21}X_2 + N_1 \tag{1}$$
$$Y_2 = h_{22}X_2 + h_{12}X_1 + N_2, \tag{2}$$

where X_i is the transmitted signal from user i subject to power constraint P_i. The channel realization is modelled as the independent and identically distributed

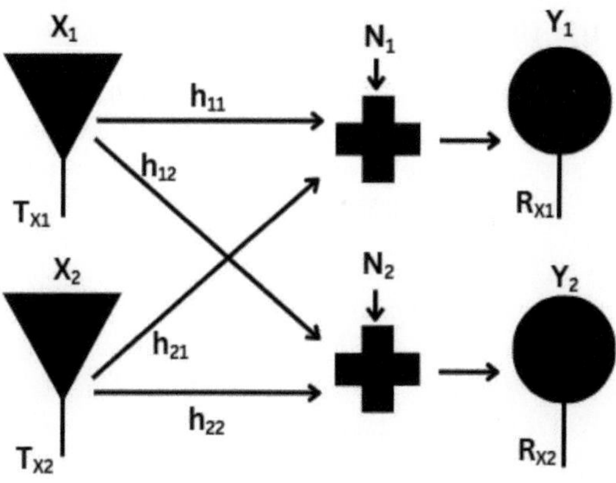

Fig. 1. Two-user interference channel (TU-IC).

(i.i.d.) complex Gaussian random variable (RV) $h_{ij} \sim \mathcal{CN}(0, \sigma_{ij})$, and h_{ij} denotes the channel realization from node i to node j. The noise is also a complex Gaussian RV with zero mean and unit variance, $N \sim \mathcal{CN}(0, 1)$. For user i and $i \neq j$, we define $\text{SNR}_i = \frac{|h_{ii}{}^2|P_i}{N_i}$, $\text{INR}_i = \frac{|h_{ji}{}^2|P_j}{N_i}$, $\text{SNIR}_i = \frac{|h_{ii}{}^2|P_i}{|h_{ji}{}^2|P_j + N_i}$, and $\text{ISNR}_i = \frac{|h_{ji}{}^2|P_j}{|h_{ii}{}^2|P_i + N_i}$. Perfect global channel knowledge is assumed to be available at all nodes and the bandwidth is normalized to one. It is worth emphasizing that this simple TU-IC channel setting captures the essential features of SIC decoding and the methodology underlying the proposed model is in fact very general and broadly applicable to a more complex interference channel.

In the baseline formulation, perfect global channel state information and perfect SIC are assumed to be available at all nodes. This idealized setting serves as a foundational abstraction that enables the rigorous construction of the proposed Boolean-algebraic framework and facilitates the explicit characterization of logical dependencies among decoding events. Such assumptions allow the analytical structure of the interference-coupled system to be isolated from implementation-specific perturbations.

3 Modelling NOMA Schemes of TU-IC Using Boolean Function With 2 Elements

The NOMA scheme under SIC decoding in a TU-IC is modeled as a probabilistic system with 2-events [9]. The first event represents successful SIC decoding, while the second corresponds to failed SIC decoding where interference is treated as noise, leading to reduced achievable rates. This formulation transforms the achievable rate analysis into a two-event probability problem. The following subsections detail the systematic modeling procedure.

3.1 Step 1: Define the Proposition Statement

A probability system event occurs when its proposition statement's condition is satisfied. For the 2-events NOMA model, we begin by defining the proposition statement with the set of channel conditions that determine the occurrence of the two events, given by Proposition 1.

Proposition 1. *In TU-IC, user i can perform SIC decoding when its interference channel is sufficiently strong allowing it to firstly decode the interference signal message from the other user j by treating its own received signal message as noise. This SIC decoding can happen for user 1 when the following channel condition* $ISNR_1 > SNR_2$ *is true, while for user 2 is when* $ISNR_2 > SNR_1$.

3.2 Step 2: Identify the Elements to Partition the Basic Space of the 2-Events Probability System

Based on the Proposition 1, two channel conditions determine whether the user is able to perform SIC decoding, i.e. $\mathrm{ISNIR}_1 > \mathrm{SNR}_2$ for user 1, and $\mathrm{ISNIR}_2 > \mathrm{SNR}_1$ for user 2. These conditions are used as the elements to partition the basic space of the developed 2-events probability system. For $k = \{1, 2\}$, the two elements ξ_k with their corresponding channel conditions can be mathematically written as [9]

$$\begin{aligned} \xi_1 : &\quad \mathrm{ISNR}_1 > \mathrm{SNR}_2 \\ \xi_2 : &\quad \mathrm{ISNR}_2 > \mathrm{SNR}_1 \end{aligned} \tag{3}$$

3.3 Step 3: Generate the Input Variables of the Boolean Function

Next, we generate the input variables for the Boolean expression of the 2-events probability system. Each element in (3) can be seen as a set of question with either true or false as an answer. These two answers can be represented by using a binary random variable (RV), with values '1' and '0' correspond to the case whenever the channel condition is true and false, respectively. For elements ξ_k with channel condition K, mapping the answer of elements to a binary RV can be performed using indicator function [9]

$$\mathrm{I}(\xi_k) = \begin{cases} 0, \text{for } \xi_k : K^c \\ 1, \text{for } \xi_k : K, \end{cases} \tag{4}$$

where K^c is the complement condition of K. Using (4), the two binary RVs, representing the answers to the defined elements of (3), are given by

$$A = \mathrm{I}(\xi_1) \tag{5}$$

$$B = \mathrm{I}(\xi_2) \tag{6}$$

Note that a pair of answer sets A and B generates four partitions of A^cB^c, A^cB, AB^c and AB, which can be arranged into 4 distinct binary sequences of 00, 01, 10 and 11 called minterms m_l for $l = 1, 2, 3, 4$. In general, a class of N distinct binary answer sets generates a partition of 2^N members. A very important point when partitioning the probability space S into 2^N is that some of the members of the generated partition may be empty. Finally, the binary RVs of A and B act as the input variables for the Boolean function of the proposed 2-events probability system.

3.4 Step 4: Generate the Output Variables of the Boolean Function

Next step is to define the output variables for the Boolean function of the transformed 2-events probability systems. Here the number of output variables depends on the number of users. For the studied TU-IC, there are two output variables defined as O_1 for user 1 and O_2 for user 2. Note that extension to a more complex interference channel with more users is straightforward. Each output variable is used to represent the two events of the probability system, with value '1' indicates the state where the user can perform SIC decoding, while value '0' means no SIC decoding by the user.

3.5 Step 5: Map the Input Variables to Output Variables

We can now map each of the possible minterms m_l (input variables) to one of the two events (output variables) for each user. With two input variables (A and B), and two output variables (O_1 and O_2), the mapping of the input to output variable can be easily depicted by using the truth table in Table 1. For example, minterm $m_3 = AB$, with $A = 1$ and $B = 1$ in Table 1, indicates that both channel conditions are true, which translated to the case where both user 1 and 2 are able to perform SIC decoding, i.e. O_1=1 and O_2=1.

Table 1. The truth table for system of two events with 2 elements.

Minterm	Input		Output	
	A	B	O_1	O_2
m_0	0	0	0	0
m_1	0	1	0	1
m_2	1	0	1	0
m_3	1	1	1	1

3.6 Step 6: Produce the Boolean Function for the 2-Events Probability System

Once the mapping of input to output variables is completed, the Boolean expression for the two-events system can be derived using the Canonical Sum of product

(SOP) form below:

$$F_{O_i}(A,B) = \sum_{l=0}^{2^N-1} O_i.m_l. \tag{7}$$

The Canonical SOP Boolean expression in (7) can be further simplified using law of Boolean algebra or using simplification methods like McCluskey algorithm [10], and K-map [11]. By using (7) and the Boolean inverse law of $U + \bar{U} = 1$, the Boolean expression for the 2-events system in Table 1 can be simplified as

$$\begin{aligned} F_{O_1}(A,B) &= \; A\bar{B} + AB = A, \\ F_{O_2}(A,B) &= \; \bar{A}B + AB = B. \end{aligned} \tag{8}$$

From (8), user 1 can perform SIC decoding when $A = 1$, i.e. the condition $\text{ISNR}_1 > \text{SNR}_2$ is true. Similarly, user 2 can perform SIC decoding when$B = 1$, i.e. the condition $\text{ISNR}_2 > \text{SNR}_1$ is true. Using the Boolean function F_{O_i} in (8), the achievable rate for NOMA schemes using Boolean function with two variables A and B can be expressed by

$$\text{R}_i = \begin{cases} \log_2(1+\text{SNR}_i), & \text{if } F_{O_i}(A,B) = 1; \\ \log_2(1+\text{SNIR}_i), & \text{if } F_{O_i}(A,B) = 0. \end{cases} \tag{9}$$

4 NOMA Schemes Using Boolean Function with Higher Elements

In this section, we further derive a 2-events probability system with higher elements for the SIC schemes of the TU-IC by applying similar procedure steps in Sect. 3.

4.1 Boolean Function with 4 Elements

To analyse the achievable rate of SIC schemes in TU-IC in more detail, the probability system with 2 elements presented Sect. 3 can be extended further using higher 4 elements, defined by the following proposition statement:

Proposition 2. *In the TU-IC, user i can perform SIC decoding when its interference channel is strong enough to firstly decode user j's signal while treating its own as noise. Two thresholds govern SIC applicability at user 1: first, when user 2 can also perform SIC, achieving rate $\log_2(1 + SNR_2)$; second, when user 2 cannot perform SIC, limited to $\log_2(1+SNIR_2)$. Thus, user 1 can employ SIC when its $ISNR_1$ satisfies:*

(i) $ISNR_1 > SNR_2$, when SIC also happen at user 2,
(ii) $ISNR_1 > SNIR_2$, when no SIC happen at user 2.

Similarly, SIC conditions for user 2 are:

(i) $ISNR_2 > SNR_1$, when SIC also happen at user 1

(ii) $\mathit{ISNR}_2 > \mathit{SNIR}_1$*, when no SIC happen at user* 1.

Based on Proposition 2, 4 elements are defined to represent all channel conditions.

$$\begin{aligned} \xi_1 &: \quad \mathrm{ISNR}_1 > \mathrm{SNR}_2 \\ \xi_2 &: \quad \mathrm{ISNR}_2 > \mathrm{SNR}_1 \\ \xi_3 &: \quad \mathrm{ISNR}_1 > \mathrm{SNIR}_2 \\ \xi_4 &: \quad \mathrm{ISNR}_2 > \mathrm{SNIR}_1 \end{aligned} \tag{10}$$

By mapping the 4 elements in (10) using binary designator into four binary RVs of $A = \mathrm{I}(\xi_1)$, $B = \mathrm{I}(\xi_2)$, $C = \mathrm{I}(\xi_3)$ and $D = \mathrm{I}(\xi_4)$, we define the binary variables A, B, C and D as the input of the Boolean function for the studied NOMA schemes of TU-IC. The set of the generated four RVs produces 16 partitions or minterms m_l for $l = \{0, 1, .., 15\}$. From these 16 partitions, Theorem 1 provides the list of partitions with empty member, i.e. these input combinations can't happen.

Theorem 1. *For input variables of A, B, C and D as defined from the elements in* (10)*, the generated partition of minterms m_4, m_6, m_8, m_9, m_{12}, m_{13} and m_{14} are empty.*

Proof. The elements of ξ_1 and ξ_3 in (10) defined by variables A and C, respectively, are dependant to each other. Since $\mathrm{SNR}_2 > \mathrm{SNIR}_2$, when element ξ_1 is true, i.e. A=1, the element of ξ_3 is also true, i.e. C=1. As a result, the input combination of A=1 and C=0 will not happen in practice. Similar reason applies to elements of ξ_2 and ξ_4 defined by variables B and D, respectively, where input combination of B=1 and D=0 is not valid since $\mathrm{SNR}_1 > \mathrm{SNIR}_1$. These invalid input combinations of $\{B = 1, D = 0\}$ and/or $\{A = 1, C = 0\}$ exist in the generated partitions of minterms m_4, m_6, m_8, m_9, m_{12}, m_{13} and m_{14}. Since these input combination is not happening in practice, the partition of minterms m_4, m_6, m_8, m_9, m_{12}, m_{13} and m_{14} are empty.

The list of input conditions in Theorem 1 are never realized in practice, and shall be designated using don't care condition in digital logic design [11]. There is another interesting combination of output variables O_1 and O_2 for the partition of minterm m_3, given by the following theorem.

Theorem 2. *For minterm m_3, the values of output variables are either $\{O_1 = 1, O_2 = 0\}$, or $\{O_1 = 0, O_2 = 1\}$.*

Proof. The input combinations of minterm m_3 are $A = 0$, $B = 0$, $C = 1$ and $D = 1$, which translate into the satisfaction of only 2 elements of ξ_3 and ξ_4 in (10), i.e. both users satisfy the channel condition of $\mathrm{ISNIR}_i > \mathrm{SNIR}_j$. For this particular channel combination, only one user can perform SIC decoding and achieves the rate of $\log_2(1 + \mathrm{SNR})$, which happen only if the other user does not perform SIC decoding and operates at the lower rate of $\log_2(1 + \mathrm{SNIR})$. As a consequence, the output combinations for m_3 are either $\{O_1 = 1, O_2 = 0\}$, or $\{O_1 = 0, O_2 = 1\}$.

Using the input variables of $A, B, CandD$, and the output variables of O_1 for user 1 and O_2 for user 2, Table 2 presents the truth table that maps between the input and the output variables for the derived 2-event probability system with 4 elements. Based on Theorem 2, we set the output of $O_1 = 1$ and $O_2 = 0$ for minterm m_3. By using the K-Map minimization technique, the Boolean expressions for the 2-events system in Table 2 are given by (Figs. 2 and 3)

$$\begin{aligned} F_{O_1}(A,B,C,D) = &\ A + \bar{B}C, \\ F_{O_2}(A,B,C,D) = &\ B + \bar{C}D \end{aligned} \tag{11}$$

and the achievable rate for user i can be expressed using (11) by

$$\mathrm{R}_i = \begin{cases} \log_2(1 + \mathrm{SNR}_i), & \text{if } F_{O_i}(A,B,C,D) = 1; \\ \log_2(1 + \mathrm{SNIR}_i), & \text{if } F_{O_i}(A,B,C,D) = 0. \end{cases} \tag{12}$$

Table 2. The truth table for 2-events system comprises with 4 elements. Following Theorem 2, output $O_1 = 1$ and $O_2 = 0$ are selected for minterm m_3, indicating user 1 can perform SIC. Symbol 'X' denotes don't care condition for empty partitions from Theorem 1.

Minterm	Input				Output	
	A	B	C	D	O_1	O_2
m_0	0	0	0	0	0	0
m_1	0	0	0	1	0	1
m_2	0	0	1	0	1	0
m_3	0	0	1	1	1	0
m_4	0	1	0	0	X	X
m_5	0	1	0	1	0	1
m_6	0	1	1	0	X	X
m_7	0	1	1	1	0	1
m_8	1	0	0	0	X	X
m_9	1	0	0	1	X	X
m_{10}	1	0	1	0	1	0
m_{11}	1	0	1	1	1	0
m_{12}	1	1	0	0	X	X
m_{13}	1	1	0	1	X	X
m_{14}	1	1	1	0	X	X
m_{15}	1	1	1	1	1	1

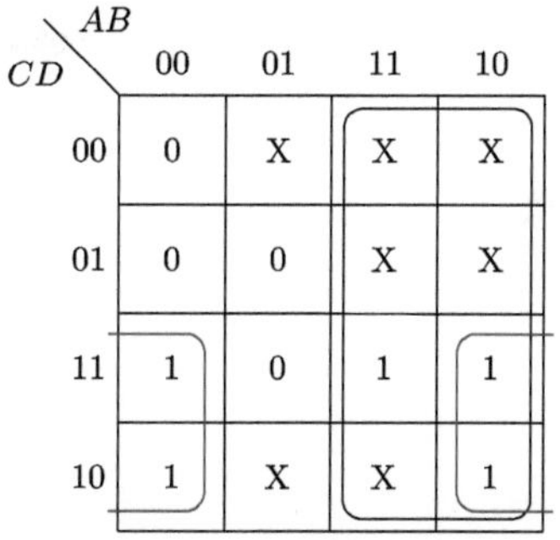

CD \ AB	00	01	11	10
00	0	X	X	X
01	0	0	X	X
11	1	0	1	1
10	1	X	X	1

Fig. 2. The K-map for the output O_1 from Table 2. The black and blue circles produce Boolean terms of A and $\bar{B}C$, respectively.

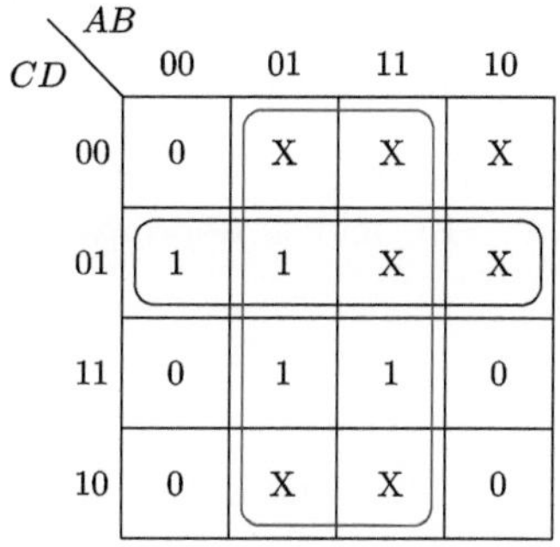

CD \ AB	00	01	11	10
00	0	X	X	X
01	1	1	X	X
11	0	1	1	0
10	0	X	X	0

Fig. 3. The K-map for the output O_2 from Table 2. The black and blue circles produce Boolean terms of $\bar{C}D$ and B, respectively.

4.2 Boolean Function with 5 Elements

In Subsect. 4.1, we select user 1 instead of user 2 to perform SIC decoding for partition minterm m_3 based on Theorem 2. Now let add one more element to the previous 4 elements in (10). We define the fifth element as

$$\xi_5 : \quad \mathrm{SNR}_1 > \mathrm{SNR}_2 \tag{13}$$

and map this element to an input variable $S = \mathrm{I}(\xi_5)$, which can be used as a selector to determine on which user should perform SIC decoding for minterm m_3, summarized in the following proposition

Proposition 3. *For partition of minterm m_3 in Theorem 2, the user with stronger SNR should be selected to perform the SIC decoding in order to maximize the overall achievable rate of the NOMA schemes in TU-IC user.*

Based on Proposition 3, user 1 will perform SIC decoding when $S = 1$, while user 2 when $S = 0$. The truth table of the probability system with 5 elements is given in Table 3, where we can see that when the channel of user 2 is stronger denoted by input combinations of minterm $m_3 = \bar{S}\bar{A}\bar{B}CD = 00011$, user 2 is selected to perform SIC represented by the output $O_1 = 0$ and $O_2 = 1$. As for user 1, this condition is represented using input combinations of minterm $m_{19} = S\bar{A}\bar{B}CD = 10011$ (Figs. 4 and 5).

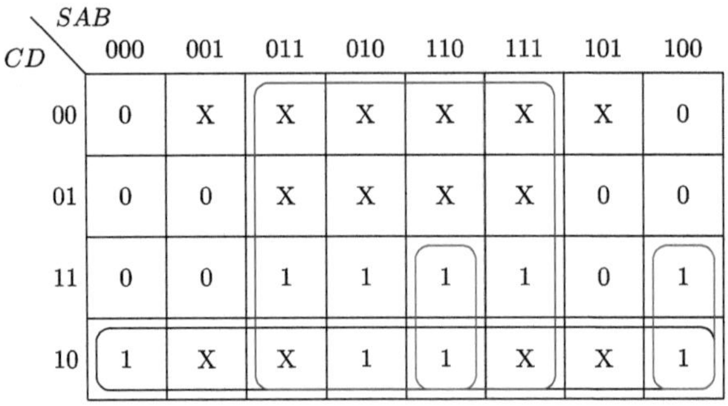

CD \ SAB	000	001	011	010	110	111	101	100
00	0	X	X	X	X	X	X	0
01	0	0	X	X	X	X	0	0
11	0	0	1	1	1	1	0	1
10	1	X	X	1	1	X	X	1

Fig. 4. The K-map for the output O_1 from Table 3. The black, red and blue circles produce Boolean terms of $C\bar{D}$, $S\bar{B}C$ and A, respectively.

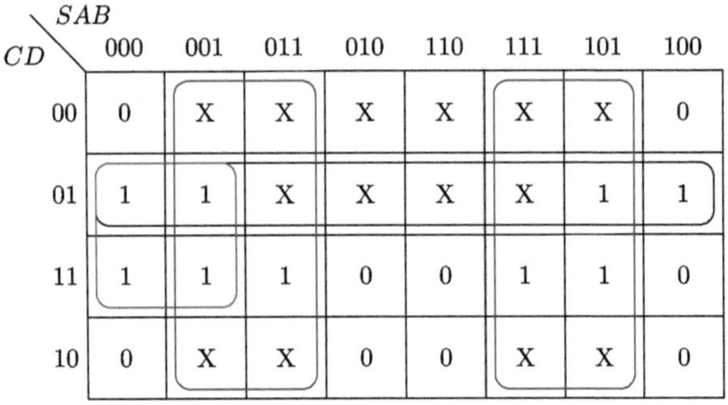

CD \ SAB	000	001	011	010	110	111	101	100
00	0	X	X	X	X	X	X	0
01	1	1	X	X	X	X	1	1
11	1	1	1	0	0	1	1	0
10	0	X	X	0	0	X	X	0

Fig. 5. The K-map for the output O_2 from Table 3. The black, red and blue circles produce Boolean terms of $\bar{C}D$, $\bar{S}\bar{A}D$ and B, respectively.

By using the K-Map minimization technique, the Boolean expressions for the 2-events system in Table 3 is given by

$$\begin{aligned} F_{O_1}(S,A,B,C,D) = &\quad A + C\bar{D} + S\bar{B}C, \\ F_{O_2}(S,A,B,C,D) = &\quad B + \bar{C}D + \bar{S}\bar{A}D \end{aligned} \tag{14}$$

and the achievable rate for user i under SIC schemes can be expressed using (11) by

$$\mathrm{R}_i = \begin{cases} \log_2(1+\mathrm{SNR}_i), & \text{if } F_{O_i}(S,A,B,C,D) = 1; \\ \log_2(1+\mathrm{SNIR}_i), & \text{if } F_{O_i}(S,A,B,C,D) = 0. \end{cases} \tag{15}$$

Table 3. The truth table for system of 2-events with 5 elements.

Minterm	Input					Output	
	S	A	B	C	D	O_1	O_2
m_0	0	0	0	0	0	0	0
m_1	0	0	0	0	1	0	1
m_2	0	0	0	1	0	1	0
m_3	0	0	0	1	1	0	1
m_4	0	0	1	0	0	X	X
m_5	0	0	1	0	1	0	1
m_6	0	0	1	1	0	X	X
m_7	0	0	1	1	1	0	1
m_8	0	1	0	0	0	X	X
m_9	0	1	0	0	1	X	X
m_{10}	0	1	0	1	0	1	0
m_{11}	0	1	0	1	1	1	0
m_{12}	0	1	1	0	0	X	X
m_{13}	0	1	1	0	1	X	X
m_{14}	0	1	1	1	0	X	X
m_{15}	0	1	1	1	1	1	1
m_{16}	1	0	0	0	0	0	0
m_{17}	1	0	0	0	1	0	1
m_{18}	1	0	0	1	0	1	0
m_{19}	1	0	0	1	1	1	0
m_{20}	1	0	1	0	0	X	X
m_{21}	1	0	1	0	1	0	1
m_{22}	1	0	1	1	0	X	X
m_{23}	1	0	1	1	1	0	1
m_{24}	1	1	0	0	0	X	X
m_{25}	1	1	0	0	1	X	X
m_{26}	1	1	0	1	0	1	0
m_{27}	1	1	0	1	1	1	0
m_{28}	1	1	1	0	0	X	X
m_{29}	1	1	1	0	1	X	X
m_{30}	1	1	1	1	0	X	X
m_{31}	1	1	1	1	1	1	1

5 Numerical Result

This section presents Monte-Carlo simulation results demonstrating the achievable rate for the proposed Boolean probability system model of the TU-IC. The main ideas and results are presented in the simplest possible setting, we consider the symmetric TU-IC setting with $h_{1,1} = h_{2,2}$, $h_{1,2} = h_{2,1}$ and $P_1 = P_2$. Figures 6 and 7 show the achievable rate of user 1 and 2, respectively, for the Boolean probability system model of 2, 4 and 5 elements in (9), (12) and (15) as a function of SNR with fixed INR of 30dB. For performance benchmarking purpose, the achievable rate of, the rate of $\log_2(1 + \text{SNR}_i)$ representing the point-to-point (PTP) channel with no interference is used as the upper bound.

Generally, the achievable rate performance in Figs. 6 and 7 can be divided into three regions:

(i) Region SNR is extremely weaker than SNR: User performance approaches the upper bound of PTP, indicating user 1 is able to apply SIC.
(ii) Region SNR approaches the value of INR: User performance starts decreasing as user SIC might not be applied. The gap between the performance of PTP and TU-IC is starting to be existed.
(iii) Region SNR is stronger than INR: User performance is increased however user is unable to perform SIC. Though the achievable rate linearly increases with the increase of SNR but there is a wide gap between the performance of PTP and TU-IC.

Figures. 6 and 7 show single achievable rate at user 1 and user 2 attained by four schemes PTP, 2 elements, 4 elements and 5 elements. Under symmetry channel condition, the results obtained by all schemes are almost the same, except for 4 elements where user 1 achieve higher rate when compared to user 2. The rate of user 1 increases up to SNR=25dB before it starts to decrease while user 2 only able to achieve up to 12dB. The reason behind is due to the implication of Theorem 2 where under 4 elements scenario, only user 1 can perform SIC when minterm m_3 is happened as the values of output variables are $\{O_1 = 1, O_2 = 0\}$. Vice versa, same results will be obtained with alternate between user 1 and user 2 if output variables are $\{O_1 = 0, O_2 = 1\}$ in Theorem 2 for minterm m_3. Next, we focus on fairness performance where the Proposition 3 is implemented by including fifth Boolean variable as an indicator, offering opportunity for either user 1 or use 2 to perform SIC based on their superior SNR value. Followed by this, the system could fairly choose which user should perform SIC, creating balance number of '1' in both output variables. Therefore, user 1 and user 2 attain similar achievable rates in Figs. 6 and 7 since the TU-IC setting is symmetry.

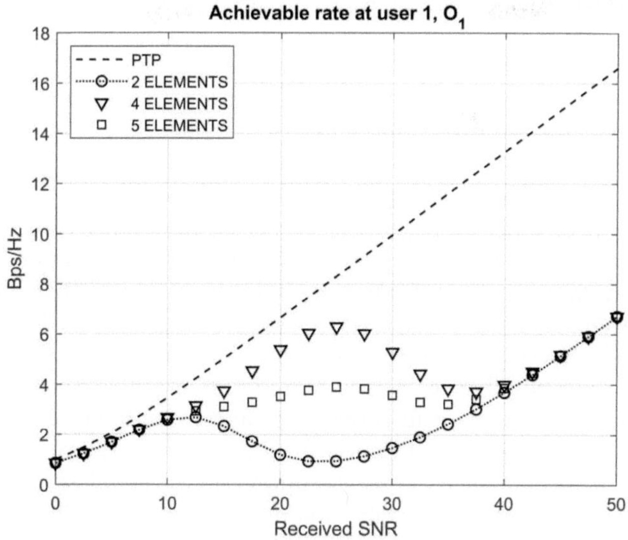

Fig. 6. The rate vs. the received SNR at INR=30dB for user 1 when modelling NOMA scheme with 2, 4 and 5 elements.

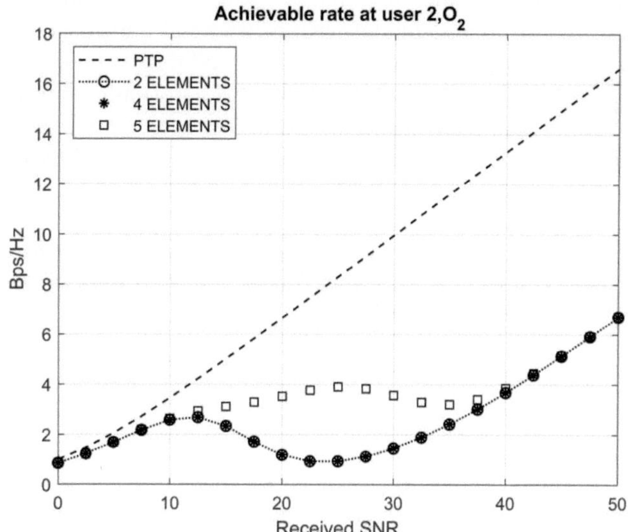

Fig. 7. The rate vs. the received SNR at INR=30dB for user 2 when modelling NOMA scheme with 2, 4 and 5 elements.

6 Conclusions

This paper proposed a new Boolean Algebra method to model and analyse the NOMA scheme under SIC decoding. The procedure to transform NOMA scheme into a probability system with two events are presented. Using the setting of the NOMA scheme in TU-IC, we showed in detail the way to analyse the NOMA scheme. Note that the procedure proposed here is general and it can be applied to other settings as well. Although the proposed framework reduces analytical complexity and introduces a complementary Boolean-algebraic perspective for NOMA system analysis, several inherent limitations remain that motivate further investigation. The current baseline model is formulated under the assumption of idealized Successive Interference Cancellation (SIC), whereas practical SIC realization is generally constrained by residual interference effects, decoding order uncertainties, hardware imperfections, and noise-induced perturbations. Extending the Boolean-probabilistic representation to incorporate imperfect SIC therefore constitutes a primary theoretical direction, where logical event probabilities can be reformulated to capture non-deterministic cancellation behaviour.

In addition, the present numerical evaluation is restricted to a symmetric TU-IC configuration. A more general characterization of system behaviour can be achieved by extending the analysis to asymmetric channel conditions, which may reveal structural variations in interference interaction patterns under heterogeneous propagation environments. Finally, future research aims to explore the applicability of the proposed model within practical aerospace communication scenarios. In particular, the Boolean-algebraic NOMA scheme will be investigated in a dual ground-station setting where two stations operate concurrently over a shared frequency band. Such an extension is expected to provide deeper insight into adaptive interference mitigation mechanisms and dynamic decoding strategies under time-varying channel conditions.

Acknowledgements. This work has been supported by MYSA which is funded by Program Pembangunan Inovasi Teknologi Angkasa Melalui Kerjasama Antarabangsa under project code P30006030040001.

References

1. Khan, W.U., Li, X., Ihsan, A., Khan, M.A., Menon, V.G., Ahmed, M.: NOMA-enabled optimization framework for next-generation small-cell IoV networks under imperfect SIC decoding. IEEE Trans. Intell. Transp. Syst. **23**(11), 22442–22451 (2021)
2. Ding, Z., Schober, R., Poor, H.V.: Unveiling the importance of SIC in NOMA systems–Part 1: State of the art and recent findings. IEEE Commun. Lett. **24**(11), 2373–2377 (2020)
3. Yue, X., Qin, Z., Liu, Y., Kang, S., Chen, Y.: A unified framework for non-orthogonal multiple access. IEEE Trans. Commun. **66**(11), 5346–5359 (2018)

4. Al-Dweik, A., Iraqi, Y., Park, K.-H., Al-Jarrah, M., Alsusa, E., Alouini, M.-S.: Efficient NOMA design without channel phase information using amplitude-coherent detection. IEEE Trans. Commun. **70**(1), 245–263 (2021)
5. Beddiaf, S., Khelil, A., Khennoufa, F., Kara, F., Kaya, H., Li, X., Rabie, K., Yanikomeroglu, H.: A unified performance analysis of cooperative NOMA with practical constraints: Hardware impairment, imperfect SIC and CSI. IEEE Access **10**, 132931–132948 (2022)
6. Singh, S., Bansal, M.: Performance analysis of NOMA-based AF cooperative overlay system with imperfect CSI and SIC. IEEE Access **9**, 40263–40273 (2021)
7. Kavitha, K., Vappangi, S.: Performance analysis of novel user pairing-based hybrid NOMA system with fixed/optimal power allocation strategy. IEEE Access **11**, 22442–22451 (2023)
8. Bisen, S., Bhatia, V., Brida, P.: Successive interference cancellation with multiple feedback in NOMA-enabled massive IoT network. EURASIP J. Wirel. Commun. Netw. **2024**(1), 71 (2024)
9. Pfeiffer, P.E.: Concepts of Probability Theory, 2nd edn. Dover Publications, New York (1978)
10. McCluskey, E.J.: Minimization of Boolean functions. The Bell System Technical Journal **35**(6), 1417–1444 (1956)
11. Karnaugh, M.: The map method for synthesis of combinational logic circuits. Transactions of the American Institute of Electrical Engineers, Part I: Communication and Electronics **72**(5), 593–599 (1953)

Neuro-Evolutionary Heuristics for Euclidean Traveling Salesman: A Computational Intelligence Benchmark

Mayank Jha(✉)

Amazon Robotics, North Reading, USA
mayankjh@amazon.com

Abstract. This paper studies the analysis of the Euclidean Traveling Salesman Problem (ETSP), a variant of the Traveling Salesman Problem (TSP), through application of computational intelligence techniques. The TSP is a classical NP-hard combinatorial optimization problem. The ETSP specifically considers cities as points in Euclidean space where distances satisfy the triangle inequality. Heuristic frameworks for algorithmic performance evaluation of genetic algorithm, greedy constructive method, local search operators using artificial intelligence on synthetic and real-life geographic data. The results demonstrate that neuro-evolutionary strategies can yield tours that are close to optimal, particularly as the dimension of the problem increases. The balance between quality of solution, temporal complexity and scalability in an artificial intelligence optimisation framework are further explored. The open-source implementation serves as a consistent platform for advancing intelligent routing in automated logistics and spatial reasoning.

Keywords: Traveling salesman problem · Computational intelligence · Genetic algorithms · Local search · Combinatorial optimization · Heuristic benchmarking

1 Introduction

For a long time now, the Traveling Salesman Problem (TSP) has been studied widely, more than any other combinatorial optimization problem. The TSP falls into the NP-hard (Non-deterministic Polynomial-time hard) category of computational problems, meaning that no known algorithm can solve all instances optimally in polynomial time. The Euclidean Traveling Salesman Problem (ETSP) is a variant where cities are represented as points in Euclidean space, and the distance between cities satisfies the triangle inequality. The ETSP has many applications including robotic path planning, circuit board drilling, package delivery, and logistics optimization. The relevance of ETSP extends beyond theoretical interest due to its practical applications in real-world routing and scheduling problems.

N. A. Mohd Subha et al. (Eds.): SMS 2026, CCIS 2956, pp. 246–258, 2026.
https://doi.org/10.1007/978-981-92-0306-2_18

The emergence of innovative searching approaches designed to optimize the search space marked important developments in modern computational intelligence. Metaheuristic algorithms provide satisfactory solutions within reasonable computation time by guiding subordinate heuristics toward promising regions of the search space. The difficulty of finding exact solutions stems from the combinatorial explosion of possible tours, which grows factorially with problem size. Genetic algorithm (GA) techniques are commonly combined with local search heuristics such as 2-opt and its structural variants to balance exploration and exploitation in the search process.

This paper examines the performance of three classical heuristics on synthetic and actual ETSP-based instances: the greedy algorithm, 2-opt local-search, and genetic algorithm. The evaluation assesses the effectiveness and scalability of these heuristic procedures with respect to both solution quality and runtime performance. The study provides a repeatable experimental setup with standardized performance measures across several datasets, enabling systematic comparison of heuristic tradeoffs.

Previous examinations of TSP heuristics have primarily used synthetic datasets with limited problem sizes. This study extends the analysis to include both synthetic instances and standard benchmark datasets from the TSP Library (TSPLIB). The paper is organized as follows: Sect. 2 summarizes related work on heuristics for TSP. Section 3 discusses the relevance and positioning of the proposed framework. Section 4 provides a detailed account of the algorithms and experimental setup. Section 5 describes the datasets and performance metrics. Section 6 presents the empirical findings. Finally, Sect. 7 concludes with future research directions.

2 Related Work

Intelligent decision-making and optimization in complex digital systems have been widely studied across artificial intelligence, distributed systems, software engineering, and computational intelligence. Recent research emphasizes the integration of data-driven inference, scalable computation, and adaptive control to address challenges arising from system heterogeneity, dynamic workloads, and real-time constraints.

Unified frameworks for intelligent optimization increasingly combine algorithmic reasoning with adaptive data modeling to support decision-making in large-scale environments [1,2]. Data-driven optimization architectures that exploit heterogeneous data streams have demonstrated strong potential in improving stability, scalability, and resource efficiency in complex digital infrastructures. In parallel, distributed AI training systems have evolved beyond rigid synchronization models. Adaptive staleness control and parameter-server orchestration mechanisms dynamically balance throughput and convergence, improving training efficiency under heterogeneous compute conditions [3].

Vision-based learning systems represent another important class of intelligent decision frameworks, particularly in cyber-physical and smart-city domains.

Deep learning architectures leveraging large-scale visual data have shown effectiveness in predicting dynamic urban flow patterns and supporting real-time automation, albeit with trade-offs between accuracy, latency, and deployment scalability. Complementary to vision-driven intelligence, semantic retrieval and text analytics pipelines enhance decision support by transforming unstructured textual data into structured representations. Hybrid retrieval methods that combine classical information retrieval with neural language models significantly improve efficiency and semantic robustness in question answering and large-scale document similarity analysis [4,5].

Beyond system-level optimization, organizational and managerial dimensions play a critical role in operational effectiveness. Studies on Agile and DevOps environments highlight the importance of continuous monitoring, feedback loops, and metric-driven optimization to sustain team performance in distributed software systems [6]. Similarly, research on generative AI adoption in software product management underscores the need for governance structures, ethical guardrails, and lifecycle integration to ensure responsible and scalable deployment of intelligent technologies [7].

Security and trustworthiness remain fundamental concerns in intelligent digital systems. Formal behavioral models addressing speculative execution vulnerabilities in managed runtime environments contribute to safer decision execution by enabling verification of information flow and mitigation correctness [8]. At the algorithmic level, advances in computational intelligence, including evolutionary heuristics, hybrid graph encodings, interpretable AI models, and autonomous navigation systems, demonstrate the effectiveness of adaptive learning and optimization techniques across combinatorial optimization, network representation learning, robotics, and 3D perception tasks [9,10].

Despite these advances, existing approaches often remain siloed, focusing on either algorithmic optimization, distributed execution, domain-specific intelligence, or governance considerations [11,12]. This fragmentation motivates the need for a unified computational framework capable of integrating intelligent inference, scalable execution, and adaptive decision optimization across heterogeneous digital systems.

3 Relevance and Positioning of the Proposed Framework

The proposed computational framework directly addresses the limitations identified in prior work by offering a unified model for intelligent data-driven decision optimization across complex digital environments. Unlike domain-specific or single-layer optimization approaches, the framework is designed as a general-purpose foundation that integrates algorithmic intelligence, adaptive inference, and scalable computation.

From a systems perspective, the framework aligns with advances in adaptive distributed learning and feedback-driven optimization by abstracting decision-making as a control process operating over heterogeneous data streams. While prior studies focus on specific training pipelines or communication architectures, the proposed framework generalizes these mechanisms into a broader computational paradigm applicable to predictive control, intelligent automation, and large-scale analytics.

In data-intensive and perception-driven domains, the framework supports vision-based prediction systems and semantic analytics pipelines under a unified optimization strategy [4,5]. This enables coordinated intelligence across visual, textual, and structured data sources for robust decision support.

The framework further complements organizational and managerial optimization models by providing a technical substrate for continuous monitoring, feedback, and performance improvement in software-intensive environments [6]. When combined with governance-aware adoption strategies, it supports responsible scaling of intelligent systems while addressing ethical, accountability, and risk considerations [7].

Finally, by emphasizing robustness, interpretability, and secure execution, the framework remains compatible with formal verification approaches and interpretable AI paradigms. Its extensibility also enables integration with adaptive intelligence techniques in robotics, graph-based learning, and cyber-physical systems, reinforcing its applicability across diverse computational domains.

Overall, the framework serves as a unifying abstraction that consolidates fragmented advances in intelligent computing into a coherent, scalable, and extensible decision optimisation paradigm.

4 Methodology

The ETSP is solved with the help of the 2-opt local search heuristic algorithm, a greedy constructive algorithm, and a genetic algorithm. In the below section, the description of every algorithm is provided along their theoretical complexity and implementation details. The design of all algorithms makes it reproducible and extendable.

4.1 Greedy Algorithm

The algorithm grows the tour by always going to a nearest unvisited city to the current location. Beginning from an arbitrarily selected city, the algorithm continuously adds the nearest unvisited city to the tour, until all cities have been visited, following which it returns to the starting city. This method ensures a tour that is connected and runs in$O(n^2)$ time, where n is the number of cities. Despite its simplicity, the greedy method is known to produce tours within $15 - -20\%$ of the Held-Karp lower bound on average. The algorithm can be enhanced using multi-start techniques, which conduct multiple runs with different commencement cities and keep the best result. It implements this enhancement with 10 random restarts per dataset.

4.2 2-opt Local Search

The 2-opt method optimizes an existing tour by taking two edges and changing the connection of the two edges if and only if the cost of the tour improves with the change. This process goes on until no more improving swaps can be found. The algorithm investigates $O(n^2)$ The number of possible swaps in each iteration is usually quite small. The initial tour effectiveness of this local search procedure relies heavily on the quality of the initial tour; hence, it is often employed as post-processing after a constructive heuristic. An efficient version is developed that makes use of candidate lists and early termination criteria to reduce overhead. The algorithm ceases execution at two points. The first is when there exists no improving swap after examining all possible edge pairs. The second point is when the algorithm has reached a limit of 50 cycles.

4.3 Genetic Algorithm

Genetic algorithm is a population-based evolutionary algorithm for ETSP. Every person represents a different sequence of cities, expressed as a number. The complete distance of the tour is used as fitness function. The ordered crossover (OX) combines parent tours for selection with probability 0.8 while swapping with 0.2.

Selection is via tournament selection of size two, elitism is used to keep the best individual through generations. The process is repeated for at most 500 generations or if there is no improvement after 50 generations. In the case of datasets with under 50 cities, the population size is 100. Moreover, the other instances, the population size is 200. Per generation, the time complexity is $O(p \cdot n)$, where p is the population size.

4.4 Experimental Setup

The algorithms employed Python 3.9 for numerical calculations, utilizing NumPy, while Matplotlib served as the visualisation tool. The experiments were conducted on an Intel Core i7-10700K CPU system with configuration: 32 GB RAM, Windows 10 OS. All algorithms for all datasets were executed ten times with different random seeds, and average performance metrics were stored. The time module of Python was used to record execution duration, not including file I/O, to microsecond precision. The solution quality is reported as the percentage above the known optimal tour length (or the best-found solution in unknown-optimum datasets). A paired t-test was used to check the significance at $\alpha = 0.05$.

4.5 Overall Computational Framework Architecture

Figure 1 presents the architectural workflow of the proposed computational intelligence benchmarking framework.

The framework consists of five major modules:

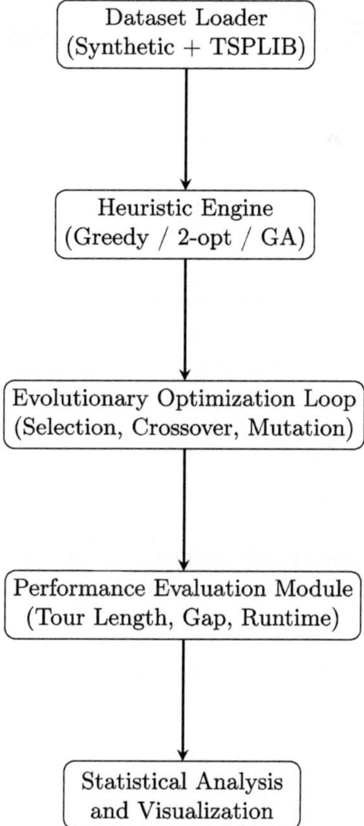

Fig. 1. Block diagram of the computational intelligence benchmarking framework.

- Dataset Loader (synthetic and TSPLIB instances)
- Heuristic Engine (Greedy, 2-opt, Genetic Algorithm)
- Evolutionary Optimisation Loop (for GA)
- Performance Evaluation Module
- Statistical Analysis and Visualisation Unit

Data flows sequentially from dataset preprocessing to heuristic execution, followed by metric extraction and statistical validation. The modular separation ensures reproducibility, scalability, and extensibility of the benchmarking pipeline.

4.6 Neuro-Evolutionary Interpretation and Justification

The adopted scheme is based on a classical Genetic Algorithm (GA). However, it adheres to the more general neuro-evolutionary computing paradigm. Neuroevolution refers to applying evolutionary search processes to structured representations with adaptive behaviours. In the proposed work, each chromosome represent a structured spatial decision policy over permutations of the cities, and the use of evolutionary operators iteratively optimises this representation by making use of fitness-driven adaptation.

In contrast with the gradient-based neural optimisation, the GA functions as a black-box adaptive search across a high-dimensional combinatorial manifold. The use of population-based evolutionary dynamics, along with elitism-preserving and stochastic recombination simulate an adaptive selection pressure. This process is analogous to the evolution of weights in neuro-evolutionary systems.

Also, the ordered crossover mechanism protects positional building blocks (schema protection), which is similar in concept to the protection of functional subnetworks in topology-evolving neuro-evolution frameworks. The mutation operator alters solutions at a given rate in controlled amounts.

For this reason, although the implementation does not cause neural network weights to evolve, it would comply with the basic principle of neuro-evolution, which is to evolve structured representations of solutions through adaptive selection and variation. Thus, "evolutionary" refers to the adaptive element (or population-based evolutionary intelligence). It is inherently embedded in the optimisation framework and not the actual neural network topology evolution.

5 Datasets and Performance Metrics

To test the algorithms, three datasets are used, which include two real geographical instances obtained from the TSP Library (TSPLIB), a standard benchmark repository for combinatorial optimisation problems. Each dataset presents unique difficulties regarding spatial arrangement, overall density, and efficient tour organisation. Table 1 details relevant characteristics and counts.

5.1 Datasets

- **P15:** To verify algorithm correctness and baseline performance, a small synthetic instance with 15 cities. The data set contains randomly distributed points in a unit square and is used for preliminary validation.
- **ATT48:** One of the best-known 48-city instances representing cities in the USA, with known optimal tour length of 10628. This dataset has a non-uniform spatial distribution, with clustering into some regions to make the algorithm work on varying densities.
- **R200:**A dataset containing two hundred uniformly distributed points in $[0, 4000]^2$ was generated to test scalability. The uniform distribution ensures that algorithms cannot take advantage of spatial patterns and serves to stress test general-purpose heuristics.
- **SGB128:** Another data set from TSPLIB with 128 cities in the United States, for intermediate scale validation. This dataset is created to bridge small and large instances.

5.2 Performance Metrics

Multiple performance metrics are employed to comprehensively evaluate algorithm behavior:

- **Tour Length:** The total Euclidean distance of the computed tour, measured using double precision floating-point arithmetic.
- **Optimality Gap:** Percentage deviation from the known optimal or best-found solution, calculated as $\frac{L-L^*}{L^*} \times 100\%$, where L is the algorithm's tour length and L^* is the optimal/best-known length.
- **Runtime:** It averaged over 10 independent runs and measures the time taken from algorithm initialisation until the solution output in seconds.
- **Scalability Coefficient:** The regression slope on log-scales estimates how fast runtime increases with increasing problem size.
- **Convergence Rate:** A measure of the number of generations or iterations used to reach within 5% of the final solution demonstrates the efficiency of the algorithm in exploration.
- **Solution Stability:** Coefficient of variation of tour length over multiple runs, which measures the robustness of the algorithm to random initialisation.

Table 1. Dataset Characteristics and Statistical Properties

Dataset	Cities	Optimal Length	Area	Point Density
P15	15	291.0	1.0	15.0
ATT48	48	10628	10600	0.0045
SGB128	128	50600	25000	0.0051
R200	200	Unknown	1.6×10^7	1.25×10^{-5}

6 Results and Discussion

The experimental results measuring three algorithms on the selected datasets are discussed in this section. Unless otherwise stated, differences of interest are statistically significant at the 95% confidence level.

6.1 Solution Quality Analysis

Table 2 presents the average tour length and optimality gap for the respective algorithms. Among all datasets, the GA method is able to achieve the lowest optimality gap. Particularly, for R200, GA's upper bound is significantly better than Greedy and 2-opt. The greedy algorithm works fine for small instances, but fails for larger ones as it does not consider the global picture. The initial greedy tours can be improved by the 2-opt algorithm, which gets stuck in local optima.

The results of statistical analysis infer that the GA solutions are significantly better than the greedy and 2-opt for all datasets with a $p < 0.01$. As the problem size increases, the improvement margin, in percentage, increased from 0.9% for P15 to 19.9% for R200 greedy baseline. The 2-opt algorithm demonstrates varying performances. It achieves a significant improvement over greedy for ATT48 (11.8% reduction in the optimality gap), while the results for R200 (12.2% reduction) are not as dramatic.

6.2 Runtime Performance and Scalability

The runtime of each algorithm on a log scale is illustrated in Fig. 2. Out of the three algorithms, Greedy is the fastest, while 2-opt has a slow performance. The genetic algorithm is the slowest among the three of them due to population-based search with iterative processing. Though, the GA has a time complexity which grows approximately linearly with the problem size (regression shows an $O(n^{1.2})$ performance), suitable for very large instances where quality is preferred over speed. The 2-opt algorithm has a quadratic growth rate of $O(n^{2.1})$, restricting it to small instances.

Table 2 thorough runtime examination showcases that although GA requires three to five times longer than greedy for small instances, it requires a similar amount of time for large instances, almost 1.5 to 2 times longer. The 2-opt algorithm exhibits an extremely high variance of running time based on the quality of the initial solution, and its worst-case scenario will behave like $O(n^3)$ or cubic time in nature when the initial tour being optimised is a poorly initialised tour.

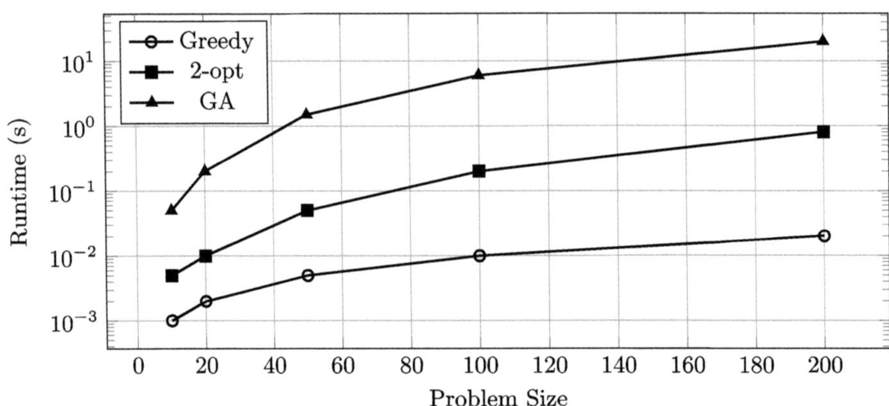

Fig. 2. Runtime scaling comparison across algorithms (log scale).

6.3 Scalability and Robustness Analysis

As shown in Table 2, the genetic algorithm preserves a low optimality gap even on large instances, while a greedy algorithm and 2-opt drop sharply. This demonstrates the power of GA in high-dimensional search spaces. The GA gives the best tours, but at the expense of most runtime. Nonetheless, it has more favourable scaling properties than the other approaches.

The robustness of the algorithms, as measured by the coefficient of variation across multiple runs, shows that the GA has the lowest variability from 1.2% to 3.5%, followed by 2-opt from 2.8% to 5.1% and the greedy, which has the greatest variability between 4.5% and 7.2%. GA's evolution-based methods produce consistent outcomes as a result of a random initialisation. Thus, it is useful in cases needing predictability.

Table 2. Scalability analysis: optimality gap vs. problem size

Algorithm	P15 (%)	ATT48 (%)	SGB128 (%)	R200 (Relative Gap)
Greedy	8.3	14.3	13.1	1.00
2-opt	4.1	2.5	3.5	0.81
Genetic	3.1	0.02	1.00	0.76

6.4 Visualisation of Generated Tours

The algorithms' generated tours on the ATT48 data set are presented in Fig. 3. It is also observed that the GA tour's path is closer to the optimum path. The borders of the path are smooth and do not intersect each other. The pathway is organized properly in space. Conversely, a greedy tour has characteristic "nearest-neighbour" artefacts such as sharp angles, edge crossings and backtracking. The solution reached with the greedy technique does not match that of the one described above, as it represents a local optimum. The evolutionary algorithm does not yield these edges as part of its outcome.

The examination of algorithmic signatures showed that greedy tours were myopic with a lot of double-backing. The 2-opt tours were locally optimised with fewer crossings but a few long edges. The GA tours were globally optimised with balanced edge lengths and good spatial coverage. These observations indicate that as the greedy algorithm makes local decisions, these patterns lead to local optima. Similarly, 2-opt is a neighbourhood search with a limited neighbourhood, while GA performs a population-based global search.

6.5 Algorithmic Insights and Practical Implications

Although there have been quite a few studies which have employed evolutionary algorithms for the capacitated vehicle routing problem with promising results,

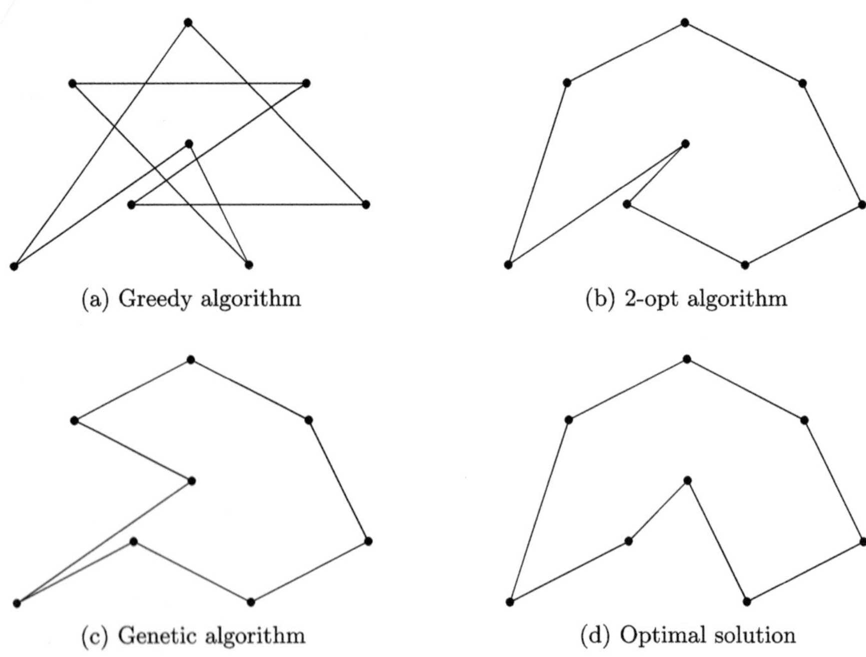

Fig. 3. Simplified 8-city illustration showing tour differences between algorithms. The genetic algorithm produces solutions closest to optimal with smooth non-crossing edges, whereas greedy solutions exhibit crossings and inefficiencies.

there is still a lack of research involving real-time applications. Though theoretical principles are known, not much is done to develop a real-time algorithm that could ease operational stress on live systems for customers and dispatch managers.

The choice of algorithm should be guided by the application requirements. The greedy algorithm performs well, even when not optimal, when quick solution generation is preferred. The 2-opt algorithm is suitable for offline optimisation problems with moderate quality requirements. The genetic algorithm tends to yield better solution quality when sufficient time is available. However, the merit of the genetic algorithm can frequently overshadow that of an application-specific algorithm, even if the latter would deliver a superior solution quality.

7 Conclusion and Future Work

The paper will review three heuristic algorithms for the two-dimensional TSP. The solution obtained from a genetic algorithm beats the other two algorithms. A greedy heuristic is often chosen as the baseline algorithm in many applications because it is simple and fast. The experimental results for large-sized problems indicate that 2-opt takes much less time and outperforms greedy. Tours achieved using two optimisation techniques are found to be around 5 to 25% more expensive than the genetic algorithm tours. Out of the three algorithms, the genetic algorithm provides the best quality solution, although it requires high computational cost.

One key finding is that GA has optimality gaps (lower bound from CPLEX) of less than 3% even for 200-city instances. In addition, GA achieves better solution quality (15 to 20% improvement) than the classical metaheuristics. A second key observation concerns runtime scaling: although GA takes longer to complete than classical heuristics, its time growth with increasing problem size remains consistently better. Hence, for large instances, GA is preferred to classical heuristics.

Future work will explore several directions. Hybrid GA techniques, which combine evolutionary search with local refinements, will be investigated for quicker runtimes and similar quality solutions. Parallel and distributed implementations of neuro-evolutionary heuristics will be explored to achieve scalability through population partitioning and concurrent evaluation. Variants of dynamic TSP with time constraints will eventually be incorporated into the benchmark. Adaptive parameter tuning methods designed according to problem-specific characteristics will be included. Finally, these approaches will be combined to enable training sophisticated routing policies end-to-end, similarly to other works that make use of deep reinforcement learning frameworks.

The empirical framework developed here offers a basis for future research studies. Additionally, the open-source implementation could help other researchers reproduce this work or expand upon it.

References

1. Bagheri, M., Bagheritaba, M., Alizadeh, S., Parizi, M.S., Matoufinia, P., Luo, Y.: AI-driven decision-making in healthcare information systems: a comprehensive review (2024)
2. Wei, J., et al.: Decision-making in the age of AI: a review of theoretical frameworks, computational tools, and human-machine collaboration. Contemp. Math. **6**, 2089–2112 (2025)
3. Razdan, V.: LinkShard: Communication-Centric Parameter Serving for Distributed AI. (2026). https://doi.org/10.31224/6165
4. Koilada, D.K.: Hybrid semantic retrieval: augmenting weighted TF–IDF with BERT for enhanced question answering. In: Proceedings of the International Conference on Sustainable Innovation with Artificial Intelligence and Machine Learning 2025 (ICSIAIML 2025), pp. 359–365 (2026). https://doi.org/10.2991/978-94-6463-948-3_26
5. Razdan, V.: Vector semantics at scale: an AI pipeline for financial text similarity. 29 January 2026, PREPRINT (Version 1) available at Research Square. https://doi.org/10.21203/rs.3.rs-8696862/v1
6. Sibaram Prasad, P.: Optimizing performance in agile and DevOps teams. In: 2025 8th International Conference of Computer and Informatics Engineering (IC2IE), pp. 1–4 (2025). https://doi.org/10.1109/IC2IE67206.2025.11283346

7. Mahajan, G.: Operationalizing generative AI in software product management: a review of managerial use-cases, governance, and ethical guardrails (2026). https://doi.org/10.31224/6204
8. Shikha, P.: Java platform security: a behavioral model for speculative execution vulnerabilities. In: 2025 International Conference on Intelligent Communication Networks and Computational Techniques (ICICNCT), pp. 1–7 (2025). https://doi.org/10.1109/ICICNCT66124.2025.11233140
9. Kadiyala, L., Sajja, R., Sermet, Y., Muste, M., Demir, I.: AI-driven decision-making for water resource planning and hazard mitigation using automated multi-agents. J. Hydroinformatics (2025)
10. Khakifirooz, M., Fathi, M., Dolgui, A.: Theory of AI-driven scheduling (TAIS): a service-oriented scheduling framework by integrating theory of constraints and AI. Int. J. Prod. Res. 1–35 (2024)
11. Anthis, Z., Mavrikis, M.: In search of answers: supporting educators to understand AI through heuristic thinking. In: International Conference on Artificial Intelligence in Education, pp. 119–133. Springer Nature Switzerland, Cham, July 2025
12. Kumar, P.S.: Ai-driven decision support system for intuitionistic fuzzy assignment problems. In: Using Traditional Design Methods to Enhance AI-Driven Decision Making, pp. 352–398. IGI Global Scientific Publishing (2024)

An Extended Feedforward Neural Network for Solving Caputo-Fabrizio Fractional Differential Equations

Mohd Rashid Admon[1], Norazak Senu[2(✉)], Noorehan Yaacob[1], Mohd Ariff Admon[1], Mohamad Shahiir Saidin[1], and Hanis Nasir[1]

[1] Department of Mathematical Sciences, Faculty of Science, Universiti Teknologi Malaysia, Johor Bahru, Johor, Malaysia

[2] Institute for Mathematical Research, Universiti Putra Malaysia, Selangor, Malaysia

norazaksenu@upm.edu.my

Abstract. The contemporary application of artificial neural network (ANN) has transcended traditional computer science problems, making significant inroads into the field of mathematics. A notable variant, the feedforward neural network (FNN), has emerged as a powerful tool for approximating solutions to fractional differential equations (FDEs). However, researchers have often limited themselves to ANN with a single hidden layer. Recent advancements in various real-world applications have demonstrated the superior performance achieved by incorporating multiple hidden layers. This study focuses on investigating the accuracy of Caputo-Fabrizio FDEs (CFFDEs) by extending the hidden layers in feedforward ANN from one to two, resulting an extended feedforward neural network (EFNN). At first stage, the method formulation involving type of CFFDEs, construction of approximation solution and approximation of Caputo- Fabrizio derivative is derived. Then, the vectorization is implemented for forward propagation, approximate solution and error function. In the last stage, the learning solver that use BroydenFletcher-GoldfarbShanno (BFGS) algorithm is described. The effectiveness of the EFNN-BFGS approach is validated through several CFFDEs, including a practical application to tumor-immune dynamics. Numerical results demonstrate that the proposed scheme offers improved accuracy and outperforms existing methods.

Keywords: Artificial Neural Network · Extended feedforward neural network · Hidden layer · BroydenFletcherGoldfarbShanno algorithm · Caputo-Fabrizio fractional differential equation

1 Introduction

Fractional calculus has evolved from a purely theoretical concept into a powerful tool for real-world applications. Originating from the 1695 correspondence between Leibniz and L'Hopital, it generalizes classical differentiation and integration to arbitrary orders [1,2]. Its key advantage lies in capturing memory

N. A. Mohd Subha et al. (Eds.): SMS 2026, CCIS 2956, pp. 259–276, 2026.
https://doi.org/10.1007/978-981-92-0306-2_19

and nonlocal effects through fractional operators [3]. This capability has been successfully applied in diverse areas, including economics [4], various applied sciences [5], robotics [6], agriculture [7], and epidemic modeling [8]. Fractional differential equations (FDEs) provide a natural framework for modeling systems with memory and hereditary properties. Unlike classical differential equations, FDEs incorporate the entire past history of a system [1], enabling more accurate representations of complex phenomena. However, their analytical solutions are often unavailable or expressed in complicated special functions, making practical implementation difficult. Traditional numerical methods, such as predictorcorrector and finite difference schemes, have shown effectiveness but may face high computational cost and scalability issues. Recent advances in machine learning, particularly neural networkbased approaches, offer flexible and efficient alternatives. These methods approximate nonlinear mappings without explicit discretization and have demonstrated strong performance in solving FDE-based models across scientific and engineering applications [9–11].

In recent years, artificial neural networks (ANN) have gained significant attention for their strong function-approximation capability, particularly in solving ordinary differential equations (ODEs) and FDEs. ANN generate continuous approximations whose accuracy depends on tunable parameters (weights and biases), optimized by minimizing a residual function derived from the governing equations. Lagaris et al. [12] pioneered the use of feedforward neural networks for differential equations by constructing a differentiable trial solution that satisfies initial or boundary conditions, with parameters trained using the Broyden-FletcherGoldfarbShanno (BFGS) algorithm. A major advantage of this approach is the creation of continuous solutions, unlike classical discrete schemes such as finite difference method (FDM) and finite element method (FEM). Pakdaman et al. [13] extended this framework to linear and nonlinear FDEs by incorporating the Caputo fractional derivative, approximated numerically via the trapezoidal rule. Raja et al. [14] aimed to enhance the speed and accuracy of solving FDEs in Caputo sense based on the previous work done by Raja et al. [15] through different implementation of optimization algorithm. The optimization technique used is constrained one, known as interior-point algorithm that has good efficiency and fast convergence.

The use of multiple hidden layers in ANN, known as deep neural network (DNN), has raised questions about the necessity of deep architectures for solving differential equations. Dufera [16] applied deep feedforward networks with Adam optimization and vectorization to efficiently solve integer-order systems. Panghal and Kumar [17] studied up to three hidden layers for delay differential equations, observing improved accuracy with two layers but deterioration with three. Inspired by these works, Admon et al. [18] implemented deep FNN for FDEs in Caputo sense, showing that Adam with suitable nodes and iterations outperformed gradient descent and momentum methods. Their vectorized approach was later extended to nonlinear Riccati systems for both ODEs and FDEs, yielding solutions consistent with exact results [19].

Motivated by findings that increasing the number of hidden layers from one to two can enhance accuracy in solving differential equations [16–18], this paper extends the existing work of Pakdaman et al. [13] by employing a two hidden layer feedforward neural network with BFGS for solving fractional differential equations that having the latest fractional derivate operator known as Caputo-Fabrizion derivative. The proposed scheme is referred to as EFNNBFGS. The CaputoFabrizio derivative is adopted instead of the classical Caputo derivative due to the comparatively limited exploration of alternative fractional operators in previous studies. Introduced by Caputo and Fabrizio [20], this derivative features a non-singular kernel, offering a more general formulation and avoiding the singularity issues commonly encountered at the endpoints in numerical implementations of the classical Caputo derivative.

The article is formulated as follows: In Sect. 2, some definitions were highlighted. In Sect. 3, the detailed of the scheme is presented through the architecture of EFNN, method formulation, vectorization and BFGS learning solver. The main results are discussed in Sect. 4, where the proposed EFNN-BFGS is applied to several CFFDEs. Finally, the findings were conclude in Sect. 6.

2 Preliminaries and Notation

This section introduces several definitions that will be used throughout the paper, following [1,21].

Definition 1. *Let $f \in \mathbb{R}[a,b]$. Then, Caputo fractional derivative with order $\beta > 0$ are defined as*

$$\begin{aligned}({}_a^C D_x^\beta f)(x) &= {}_a J_x^{m-\beta}[f^{(m)}(x)] \\ &= \frac{1}{\Gamma(m-\beta)} \int_a^x (x-t)^{m-\beta-1} f^{(m)}(t)dt, \quad x > a, \end{aligned} \tag{1}$$

where m is positive integer satisfying $m - 1 < \beta \leq m$.

Caputo and Fabrizio [20] then proposed a fractional derivative with a non-singular exponentially decaying kernel, referred to as the CaputoFabrizio (CF) fractional derivative. Let $H^1(a,b) = \{g|g \in L^2(a,b) \text{ and } g' \in L^2(a,b)\}$ where $L^2(a,b)$ is the space of square integrable functions on interval (a,b). Then, CF fractional derivative defined as follows:

Definition 2. *Let $g(t) \in H^1(a,b)$ and $\beta \in (0,1)$. Then, the Caputo-Fabrizio fractional derivative with order β is defined as*

$${}_a^{CF} D_x^\beta g(x) = \frac{M(\beta)}{1-\beta} \int_a^x g'(t) e^{\left(-\beta \frac{x-t}{1-\beta}\right)} dt, \tag{2}$$

where $M(\beta)$ is normalization function such that $M(0) = M(1) = 1$.

3 Methodology

3.1 Extended Feedforward Neural Network Architecture

The architecture of EFNN is an extension of one hidden layer FNN proposed by Pakdaman et al. [13]. Consider a schematic diagram of EFNN as shown in the Fig. 1. It can be regard as 1-2-1 EFNN indicating that the architecture of EFNN has 1 input layer, 2 hidden layers and 1 output layer.

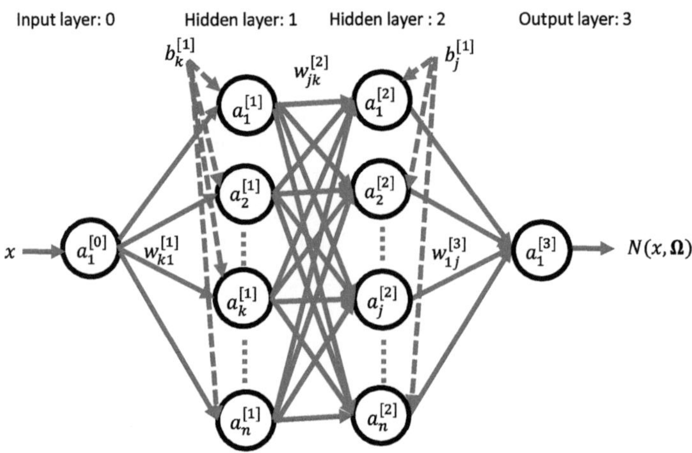

Fig. 1. Schematic diagram of EFNN.

The output of EFNN can be obtained through the process known as forward propagation. In this process, the value at each node in EFNN will go through two phase of modification values. In the first phase, the multiplication of weights and the node in the previous layer plus with bias is calculated. Then, the values obtained at this phase is further substitute in activation function, φ. This value is then become input to the node in the next layer. In addition, the activation functions use in this network are the hyperbolic tangent function in first hidden layer and second hidden layer, while the output layer use linear activation function. The mathematical formulation for EFNN can be summarized as follows:

- In the input layer:

$$a_1^{[0]} = x. \tag{3}$$

- In the first hidden layer:

$$s_k^{[1]} = w_{k1}^{[1]} a_1^{[0]} + b_k^{[1]}, \tag{4}$$

$$a_k^{[1]} = \varphi(s_k^{[1]}). \tag{5}$$

– In the second hidden layer:

$$s_j^{[2]} = \sum_{k=1}^{n} w_{jk}^{[2]} a_k^{[1]} + b_j^{[2]}, \tag{6}$$

$$a_j^{[2]} = \varphi(s_j^{[2]}). \tag{7}$$

– In the output layer:

$$s_1^{[3]} = \sum_{j=1}^{n} w_{1j}^{[3]} a_j^{[2]}, \tag{8}$$

$$a_1^{[3]} = s_1^{[3]}. \tag{9}$$

3.2 Method Formulation

Before going further, the problem of Caputo-Fabrizio fractional differential equations (CFFDEs) need to be defined. Consider a form of CFFDEs that will be solved is as follows:

$$\left({}_{0}^{CF}D_x^{\beta} y\right)(x) = F\Big(x, y(x)\Big), \quad x \in [0, x_f], \tag{10}$$

with initial condition $y(0) = a$ and $0 < \beta \leq 1$. ${}_{0}^{CF}D_x^{\beta}$ denote the Caputo-Fabrizio with order β. From the Definition 3, the normalization constant $M(\beta)$ are assumed to be equal 1 throughout this work. The function F is a continuous function that can be linear and nonlinear. Next, it turns to define the form of approximate solution. The mathematical formulation of approximate solution of CFFDEs in Eq. (10) can be constructed as:

$$y_T(x, \mathbf{\Omega}) = a + xN(x, \mathbf{\Omega}). \tag{11}$$

The first term in Eq. (11) satisfies initial condition while the second term consists the EFNN output, $N(x, \mathbf{\Omega})$. Substituting the approximate solution in Eq. (11) into original problem of CFFDE in Eq. (10) will result in

$$\left({}_{0}^{CF}D_x^{\beta} y_T\right)(x, \mathbf{\Omega}) = F\Big(x, y_T(x, \mathbf{\Omega})\Big). \tag{12}$$

It is clear to see that $y_T(x, \mathbf{\Omega})$ satisfies initial condition. Thus, instead of solving problem in Eq. (10), now it turns to solve Eq. (12). Now, it is ready to define the error function. But first, the discretization of the domain $[0, x_f]$ are required. By discretizing the domain $[0, x_f]$ into ρ subinterval with equal step size $h = \dfrac{x_f}{\rho}$, the error function can be defined as

$$E(\mathbf{\Omega}) = \sum_{d=1}^{m_t} \left[\left({}_{0}^{CF}D_x^{\beta} y_T\right)\left(x^{(d)}, \mathbf{\Omega}\right) - F\left(x^{(d)}, y_T\left(x^{(d)}, \mathbf{\Omega}\right)\right)\right]^2. \tag{13}$$

During training process, this error function need to be minimized its value to obtain optimally trained of network parameters, $\boldsymbol{\Omega}$ in EFNN. To perform the computation of this error function, the approximation of Caputo-Fabrizio fractional derivative are needed based on the following theorem:

Theorem 1. *Consider interval $[0, x_f]$ that subdivided into ρ subinterval of equal width $h = \frac{x_f}{\rho}$ to generate $\rho+1$ training points. For all discretized training points in the interval $[0, x_f]$, let $x^{(1)} = 0 < x^{(2)} = h < x^{(3)} = 2h < ... < x^{(m_t-1)} = \rho h < x^{(m_t)} = (\rho+1)h$. Then, for each training points $x^{(d)}$ where $d = 1, 2, ...m_t$, discretizing the interval $[0, x^{(d)}]$ into τ subinterval where $h = \frac{x^{(d)}}{\tau}$, then the approximation of Caputo-Fabrizio fractional derivative at training point $x^{(d)}$ is given by*

$$\left({}_{0}^{CF}D_{x}^{\beta}y_T\right)\left(x^{(d)}, \boldsymbol{\Omega}\right)$$

$$= -\frac{M(\beta)}{h_d\beta}\left\{\begin{array}{l} \left(y_T\left(x_d^{(2)}, \boldsymbol{\Omega}\right) - y_T\left(x^{(1)}, \boldsymbol{\Omega}\right)\right)\left(e^{\frac{-\beta}{1-\beta}(\tau+1)h_d} - e^{\frac{-\beta}{1-\beta}\tau h_d}\right) \\ + \sum_{r=2}^{d-1}\left[\left(y_T\left(x_d^{(r+1)}, \boldsymbol{\Omega}\right) - y_T\left(x_d^{(r)}, \boldsymbol{\Omega}\right)\right)\right. \\ \left.\times\left(e^{\frac{-\beta}{1-\beta}(\tau-r+2)h_d} - e^{\frac{-\beta}{1-\beta}(\tau-r+1)h_d}\right)\right] \\ + \left(y_T\left(x^{(d)}, \boldsymbol{\Omega}\right) - y_T\left(x_d^{(d-1)}, \boldsymbol{\Omega}\right)\right)\left(e^{\frac{-\beta}{1-\beta}h_d} - 1\right) \end{array}\right\} \tag{14}$$

3.3 Vectorization

In order to compute efficiently the forward propagation in (3)–(9), approximate solution in (11) and error function in (13) can be further converted into vectorized form.

i. Vectorization of forward propagation

   The forward propagation shown in (3)–(9) is computed for single input of x. At dth training point of x, these process can be shown in condensed form as follows:

Input layer to first hidden layer:
The mathematical expression at each node can be written as

$$\begin{bmatrix} s_1^{[1](d)} \\ s_2^{[1](d)} \\ \vdots \\ s_n^{[1](d)} \end{bmatrix} = \begin{bmatrix} w_{11}^{[1]} \\ w_{21}^{[1]} \\ \vdots \\ w_{n1}^{[1]} \end{bmatrix} a_1^{[0](d)} + \begin{bmatrix} b_1^{[1]} \\ b_2^{[1]} \\ \vdots \\ b_n^{[1]} \end{bmatrix} \tag{15}$$

where in condensed form is

$$\mathbf{s}^{[1](d)} = \mathbf{w}^{[1]} a_1^{[0](d)} + \mathbf{b}^{[1]}, \tag{16}$$

that finally yield

$$\mathbf{a}^{[1](d)} = \varphi\left(\mathbf{s}^{[1](d)}\right). \tag{17}$$

Then, the value obtained in (17) become input to the node in the next layer.

First hidden layer to second hidden layer:

The mathematical expression at each node can be written as

$$\begin{bmatrix} s_1^{[2](d)} \\ s_2^{[2](d)} \\ \vdots \\ s_n^{[2](d)} \end{bmatrix} = \begin{bmatrix} w_{11}^{[2]} & w_{12}^{[2]} & \cdots & w_{1n}^{[2]} \\ w_{21}^{[2]} & w_{22}^{[2]} & \cdots & w_{2n}^{[2]} \\ \vdots & \vdots & \ddots & \vdots \\ w_{n1}^{[2]} & w_{n2}^{[2]} & \cdots & w_{nn}^{[2]} \end{bmatrix} \begin{bmatrix} a_1^{[1](d)} \\ a_2^{[1](d)} \\ \vdots \\ a_n^{[1](d)} \end{bmatrix} + \begin{bmatrix} b_1^{[2]} \\ b_2^{[2]} \\ \vdots \\ b_n^{[2]} \end{bmatrix} \tag{18}$$

where in condensed form is

$$\mathbf{s}^{[2](d)} = \mathbf{W}^{[2]} \mathbf{a}^{[1](d)} + \mathbf{b}^{[2]}, \tag{19}$$

that finally yield

$$\mathbf{a}^{[2](d)} = \varphi\left(\mathbf{s}^{[2](d)}\right). \tag{20}$$

Second hidden layer to output layer:

The mathematical expression at each node can be written as

$$s_1^{[3](d)} = \begin{bmatrix} w_{11}^{[3]} & w_{12}^{[3]} & \cdots & w_{1j}^{[3]} & \cdots & w_{1n}^{[3]} \end{bmatrix} \begin{bmatrix} a_1^{[2](d)} \\ a_2^{[2](d)} \\ \vdots \\ a_n^{[2](d)} \end{bmatrix} \tag{21}$$

where in condensed form is

$$s^{[3](d)} = \mathbf{w}^{[3]}\mathbf{a}^{[2](d)}, \tag{22}$$

and straight away it can shows that

$$a^{[3](d)} = s^{[3](d)}. \tag{23}$$

In order to perform forward propagation efficiently, it need to be computed for all values of training points at one time. This can be done through vectorization process. Since there is m_t training points such that $X = \left[x^{(1)}, x^{(2)}, ..., x^{(m_t)}\right]$ for $d = 1, 2, ..., m_t$, then by stacking process (16)–(17) will result as

$$S^{[1]} = \mathbf{w}^{[1]}A^{[0]} + \mathbf{b}^{[1]}, \tag{24}$$

$$A^{[1]} = \varphi\left(S^{[1]}\right) \in \mathbb{R}^{n\times m_t}, \tag{25}$$

where

$$\begin{aligned} A^{[0]} &= \left[a^{[0](1)}, a^{[0](2)}, \ldots, a^{[0](m_t)}\right], \\ &= \left[x^{(1)}, x^{(2)}, \ldots, x^{(m_t)}\right], \\ &= X \in \mathbb{R}^{1\times m_t}. \\ S^{[1]} &= \begin{bmatrix} | & | & & | \\ \mathbf{s}^{[1](1)} & \mathbf{s}^{[1](2)} & \ldots & \mathbf{s}^{[1](m_t)} \\ | & | & & | \end{bmatrix} \in \mathbb{R}^{n\times m_t}, \\ \mathbf{w}^{[1]} &\in \mathbb{R}^{n\times 1}, \\ \mathbf{b}^{[1]} &\in \mathbb{R}^{n\times 1}. \end{aligned}$$

The same thing done for (19)–(20) that yields

$$S^{[2]} = \mathbf{W}^{[2]}A^{[1]} + \mathbf{b}^{[2]}, \tag{26}$$

$$A^{[2]} = \varphi\left(S^{[2]}\right) \in \mathbb{R}^{n\times m_t}, \tag{27}$$

where

$$\begin{aligned} A^{[1]} &= \begin{bmatrix} | & | & & | \\ \mathbf{a}^{[1](1)} & \mathbf{a}^{[1](2)} & \ldots & \mathbf{a}^{[1](m_t)} \\ | & | & & | \end{bmatrix} \in \mathbb{R}^{n\times m_t}, \\ S^{[2]} &= \begin{bmatrix} | & | & & | \\ \mathbf{s}^{[2](1)} & \mathbf{s}^{[2](2)} & \ldots & \mathbf{s}^{[2](m_t)} \\ | & | & & | \end{bmatrix} \in \mathbb{R}^{n\times m_t}, \\ \mathbf{W}^{[2]} &\in \mathbb{R}^{n\times n}, \\ \mathbf{b}^{[2]} &\in \mathbb{R}^{n\times 1}. \end{aligned}$$

and finally the vectorized form for (22)–(23) is

$$S^{[3]} = \mathbf{w}^{[3]} A^{[2]}, \tag{28}$$

$$A^{[3]} = S^{[3]} \in \mathbb{R}^{1\times m_t}, \tag{29}$$

where

$$A^{[2]} = \begin{bmatrix} | & | & & | \\ \mathbf{a}^{[2](1)} & \mathbf{a}^{[2](2)} & \dots & \mathbf{a}^{[2](m_t)} \\ | & | & & | \end{bmatrix} \in \mathbb{R}^{n\times m_t},$$

$$S^{[3]} = \begin{bmatrix} s^{[3](1)} & s^{[3](2)} & \dots & s^{[3](m_t)} \end{bmatrix} \in \mathbb{R}^{1\times m_t},$$

$$\mathbf{w}^{[3]} \in \mathbb{R}^{1\times n}.$$

As can be seen, there is inconsistent in dimension during addition in (24) and (26). This shorthand notation is allowed in describing these process, which also can be known as broadcasting [22]. Thus, the vectorized version of the FNN output is $N(X, \mathbf{\Omega}) = A^{[3]} \in \mathbb{R}^{1\times m_t}$ where

$$\begin{aligned} \mathbf{\Omega} &= \left[\mathbf{w}^{[1]}, \mathbf{W}^{[2]}, \mathbf{w}^{[3]}, \mathbf{b}^{[1]}, \mathbf{b}^{[2]}\right], \\ &= \left[w_{11}^{[1]}, w_{12}^{[1]}, ..., w_{1n}^{[3]}, b_1^{[1]}, b_2^{[1]} ..., b_n^{[1]}, b_1^{[2]}, b_2^{[2]} ..., b_n^{[2]}\right] \in s_1. \end{aligned}$$

where $s_1 = \mathbb{R}^{(n^2+4n)\times 1}$.

ii. *Approximate solution in vectorized form*

It is straighforward to show that vectorized form of approximate solution in (11) is given as

$$\underbrace{y_T(X, \mathbf{\Omega})}_{\mathbb{R}^{1\times m_t}} = a + \underbrace{X \odot N(X, \mathbf{\Omega})}_{\mathbb{R}^{1\times m_t}}. \tag{30}$$

respectively, where $\odot$ is Hadamard product or elementwise multiplication and broadcasting is implemented to (30).

iii. *Error function in vectorized form*

The error function defined in (13) can be written in vectorized form as:

$$E = ||\overline{G} - \overline{F}||^2, \tag{31}$$

$$= (\overline{G} - \overline{F})(\overline{G} - \overline{F})^T. \tag{32}$$

where $\overline{G} = \left[\left({}_0^{CF}D_x^{\beta} y_T\right)\left(x^{(1)}, \mathbf{\Omega}\right), \left({}_0^{CF}D_x^{\beta} y_T\right)\left(x^{(2)}, \mathbf{\Omega}\right), ..., \left({}_0^{CF}D_x^{\beta} y_T\right)\left(x^{(m_t)}, \mathbf{\Omega}\right)\right] \in \mathbb{R}^{1\times m_t}$ and $\overline{F} = \left[F\left(x^{(1)}, y_T\left(x^{(1)}, \mathbf{\Omega}\right)\right), F\left(x^{(2)}, y_T\left(x^{(2)}, \mathbf{\Omega}\right)\right), ..., F\left(x^{(d)}, y_T\left(x^{(d)}, \mathbf{\Omega}\right)\right)\right] \in \mathbb{R}^{1\times m_t}$.

3.4 Learning Solver: Broyden-Fletcher-Goldfarb-Shanno Method

The Broyden-Fletcher-Goldfarb-Shanno (BFGS) method is an effective method for solving unconstrained optimization. It can be implemented in solving CFFDEs by minimizing the error function defined in (13) as

$$\min_{\boldsymbol{\Omega}} E(\boldsymbol{\Omega})$$

through adjusting the network parameters, $\boldsymbol{\Omega}$ until reached an optimal ones, $\boldsymbol{\Omega}^*$. BFGS is one of the quasi-Newton method, under second-order optimization method which provide information of Hessian matrix or surface of error function to reach an optimal faster than its counterpart such as among variants in first-order optimization. Instead of applying the formula explicitly, BFGS algorithm can be perform through built-in solver in Matlab software known as `fminunc`. This input argument can be summarized in Table 1. Note that the initial values of network parameters are set to be random numbers taken from uniform random distribution for $\mathbf{w}^{[1]}$, $\mathbf{w}^{[3]}$, $\mathbf{W}^{[2]}$. The remaining network parameters which are biases, are all set initially to zero.

Table 1. Settings for argument in `optimset` used in EFNN-BFGS scheme.

Argument	Setting
Algorithm	'quasi-newton'
HessianApproximation	'bfgs'
StepTolerance	'1×10^{-8}'
MaxIterations	'1000'
MaxFunctionEvaluations	'1×10^{7}'
OutputFcn	'OutputFcn'

4 Numerical Results

In this section, EFNN-BFGS scheme is implemented on solving linear and nonlinear CFFDEs. There are three examples discussed, wherein the first two examples involve a single equation of CFFDEs, and the last example involves a system of CFFDEs, that is an application problems known as the tumor-immune interaction.

Example 1. [23]. Consider linear CFFDE

$$\left({}_0^{CF}D_x^{\beta}y\right)(x) - h(x) = 2y(x), \quad x \in [0,1], \tag{33}$$

where

$$h(x) = \frac{(\beta - \beta^2 - 1)e^{\frac{\beta x}{\beta - 1}} - (1 + 2\beta(-1 + \beta))(-1 + 2\beta x) - \beta((1 - 3\beta + 4\beta^2)\cos x + \beta \sin x)}{\beta + 2\beta^2(\beta - 1)},$$

with the initial condition $y(0) = 1$. The exact solution for this example is $y(x) = x + \cos x$. According to Eq. (11), the trial solution for this example can be constructed as

$$y_T(x, \mathbf{\Omega}) = 1 + xN(x, \mathbf{\Omega}). \tag{34}$$

The numerical results along with the absolute error for both cases are tabulated in Table 2. The results show that the proposed scheme provides satisfactory results for this problem in terms of accuracy and computational time. This can be seen through the absolute error (AE) where for the cases of $\beta = 0.75$, the range of absolute error obtained are from 10^{-4} to 10^{-6} with a runtime of 7.91 s and the cases of $\beta = 0.95$ are from 10^{-3} to 10^{-5} with a runtime of 11.59 s.

Table 2. Numerical solution and absolute error obtained by EFNN-BFGS for Example 1 with $\beta = 0.75$ and $\beta = 0.95$.

		$\beta = 0.75$		$\beta = 0.95$	
x	Exact	EFNN-BFGS	AE	EFNN-BFGS	AE
0	0	0	0	0	0
0.1	1.095004	1.095061	5.68×10^{-5}	1.094957	4.76×10^{-5}
0.2	1.180067	1.180117	5.04×10^{-5}	1.180005	6.14×10^{-5}
0.3	1.255336	1.255343	6.55×10^{-6}	1.255258	7.83×10^{-5}
0.4	1.321061	1.321010	5.09×10^{-5}	1.320938	1.23×10^{-4}
0.5	1.377583	1.377479	1.03×10^{-4}	1.377374	2.09×10^{-4}
0.6	1.425336	1.425193	1.42×10^{-4}	1.424999	3.37×10^{-4}
0.7	1.464842	1.464665	1.77×10^{-4}	1.464336	5.06×10^{-4}
0.8	1.496707	1.496462	2.45×10^{-4}	1.495988	7.19×10^{-4}
0.9	1.521610	1.521195	4.15×10^{-4}	1.520617	9.93×10^{-4}
1	1.540302	1.539499	8.03×10^{-4}	1.538930	1.37×10^{-3}
CPU Time(s)		7.91		11.59	

To check the convergence plot of network parameters and error function for these cases throughout the iteration, Fig. 2(a) and Fig. 2(b) show the behaviour of network parameters and error function for the case of $\beta = 0.75$ and $\beta = 0.95$ respectively. For $\beta = 0.75$, all network parameters start to maintain at specific values or converge after 25 iterations and ended the training process in less than 35 iterations. While for $\beta = 0.95$, all network parameters start to maintain

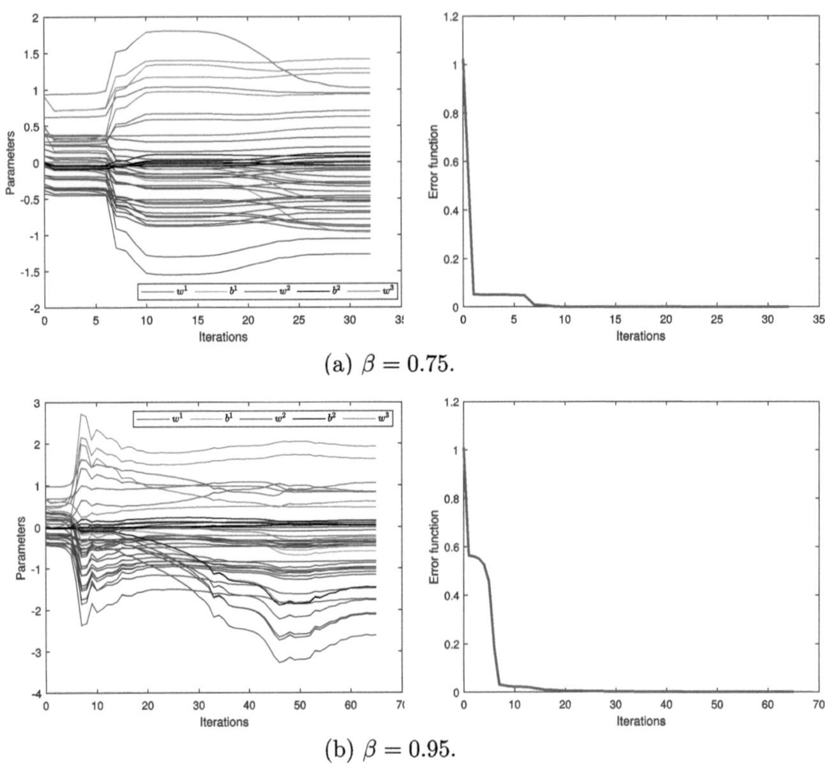

(a) $\beta = 0.75$.

(b) $\beta = 0.95$.

Fig. 2. Convergence of network parameters and error function of EFNN-BFGS with different values of β for Example 1.

Table 3. Comparison of L^2-errors between EFNN-BFGS and NSFR [24] for Example 1 when $\beta = 0.9$.

	L^2- error		
EFNN-BFGS	$h = 0.1$	$h = 0.0333$	$h = 0.02$
	9.7645×10^{-3}	1.9581×10^{-3}	5.8307×10^{-4}
NSFR	$h = 0.1$	$h = 0.01$	$h = 0.001$
	2.6850×10^{-1}	1.0907×10^{-2}	6.9686×10^{-4}

Table 4. Comparison of MAE between EFNN-BFGS and MQ-RBF [23] for Example 1 with different values of β.

Method	MAE	
	$\beta = 0.1$	$\beta = 0.5$
EFNN-BFGS	8.9621×10^{-5}	9.4736×10^{-4}
MQ-RBF ($N = 6$)	4.3360×10^{-4}	8.9050×10^{-3}

at specific values or converge towards the end of the training process (60–70 iterations).

For both cases, the error function are decreasing towards zero after some points and maintain its behaviour until iteration ended. The proposed scheme is then compared with two methods which are Numerical Fractional Simpsons Rule (NSFR) [24] and Multiquadric Radial Basis Function (MQ-RBF) [23] in Table 3 and Table 4 respectively. In comparison with NSFR, Table 3 shows that the proposed scheme require less step size to provide better L^2-errors for Problem 1 when $\beta = 0.9$. In the initial comparison, both scheme use same step size $h = 0.1$ and EFNN-BFGS provide better accuracy than NSFR. In the next comparison, EFNN-BFGS require only $h = 0.0333$ to give better L^2-errors compared with NSFR that use more step size, $h = 0.01$. The step size of proposed scheme is then decreased to $h = 0.02$, which is better enough to get the same accuracy with NSFR that require very small step size, $h = 0.0001$. In comparison with MQ-RBF, the same style of analysis has been done like the previous comparison. But in this case, the comparison is conducted based on MAE on solving Example 1 with two different values of β which are $\beta = 0.1$ and $\beta = 0.5$. With same number of taining points, $m_t = 6$ (MQ-RBF indicate by $N = 6$), Table 4 shows EFNN-BFGS are provide better MAE for both cases. Note that for this example, the number of nodes use in EFNN-BFGS for all cases is 5.

Example 2. [25]. Consider nonlinear CFFDE

$$\left({}_0^{CF}D_x^{\beta}y\right)(x) + y^2(x) = p(x), \quad x \in [0, x_f], \tag{35}$$

where

$$p(x) = M(\beta)\left(\frac{3}{\beta}x^2 - \frac{6(1-\beta)}{\beta^2}x + \frac{6(1-\beta)^2}{\beta^3} - \frac{6(1-\beta)^2}{\beta^3}e^{-\frac{\beta}{1-\beta}x}\right),$$

with the initial condition $y(0) = 0$. The exact solution for this problem is $y(x) = x^3$. Set $M(\beta) = 1$. Based on Eq. (11), the designated trial solution is

$$y_T(x, \mathbf{\Omega}) = xN(x, \mathbf{\Omega}). \tag{36}$$

In this example, the nonlinear CFFDE is solved with different values of $\beta = 0.5$ and $\beta = 0.7$ using EFNN-BFGS with 31 number of training points. The numerical results along with the absolute error for both cases are tabulated in Table 5. For both cases, the numerical solutions indicate that the provided solutions are accurate up to four to fifth decimals places. From runtime for the both cases, there is no significant differences in terms of computational time for solving this problem with different values of β.

To check the convergence plot of network parameters and error function for these cases throughout the iteration, Fig. 3(a) show the behaviour of network parameters and error function for the case of $\beta = 0.5$ and Fig. 3(b) shows the behaviour of network parameters and error function for the case of $\beta = 0.7$. For $\beta = 0.5$, some of the network parameters start to converge after 100 iterations

Table 5. Numerical solution and absolute error of EFNN-BFGS for Example 2 with $\beta = 0.5$ and $\beta = 0.7$

		$\beta = 0.5$		$\beta = 0.7$	
x	Exact	EFNN-BFGS	AE	EFNN-BFGS	AE
0	0	0	0	0	0
0.1	0.001	0.000808	1.92×10^{-4}	0.000872	1.28×10^{-4}
0.2	0.008	0.008018	1.82×10^{-5}	0.008095	9.48×10^{-5}
0.3	0.027	0.027097	9.74×10^{-5}	0.0270	3.00×10^{-5}
0.4	0.064	0.064026	2.62×10^{-5}	0.063913	8.70×10^{-5}
0.5	0.125	0.124959	4.15×10^{-5}	0.124970	2.99×10^{-5}
0.6	0.216	0.215978	2.23×10^{-5}	0.216103	1.03×10^{-4}
0.7	0.343	0.343035	3.48×10^{-5}	0.343126	1.26×10^{-4}
0.8	0.512	0.512063	6.29×10^{-5}	0.512092	9.20×10^{-5}
0.9	0.729	0.729099	9.92×10^{-5}	0.729220	2.20×10^{-4}
1	1	1.000070	7.00×10^{-5}	1.000227	2.27×10^{-4}
CPU Time(s)		7.51		10.39	

and ended the training process in less than 140 iteration. The other network parameters do not exhibit consistency in maintaining the same value even after the iterations have been ended. In the case $\beta = 0.7$, most of the network parameters are seem to adjust the network parameters in a small amount of value even though the iteration has been ended by the solver. For both cases, the error function are decreasing towards zero after some points and maintain its behaviour until iteration ended. Note that for this example, the number of nodes use in EFNN-BFGS for all cases is 5.

Example 3. [26]. Consider nonlinear system of ODE that involving the interaction between tumor and immune interaction is represented by

$$\frac{dT_I}{dt} = 2a_4 T_M - (d_2 + a_1)T_I, \tag{37}$$

$$\frac{dT_M}{dt} = a_1 T_I - d_3 T_M - a_4 T_M \tag{38}$$

with the initial condition $T_I(0) = 1.3$ and $T_M(0) = 1.2$. T_I denoted the population of tumor cells during interphase, T_M population of tumor cells during mitosis. The proportion of cell death was denoted by the term $d_2 T_I$ and $d_3 T_M$ while $a_4 T_M$ and $a_1 T_I$ represent the rates of cell reproduction. The system of ODE was analyzed by the author in [26] to investigate the interaction of tumor population at two different cell cycle stages T_M and T_I for $t > 0$ when there is absence of immune and drug.

In this work, the classical integer-order derivative in the system is replaced by the CaputoFabrizio fractional derivative to examine the resulting effects on

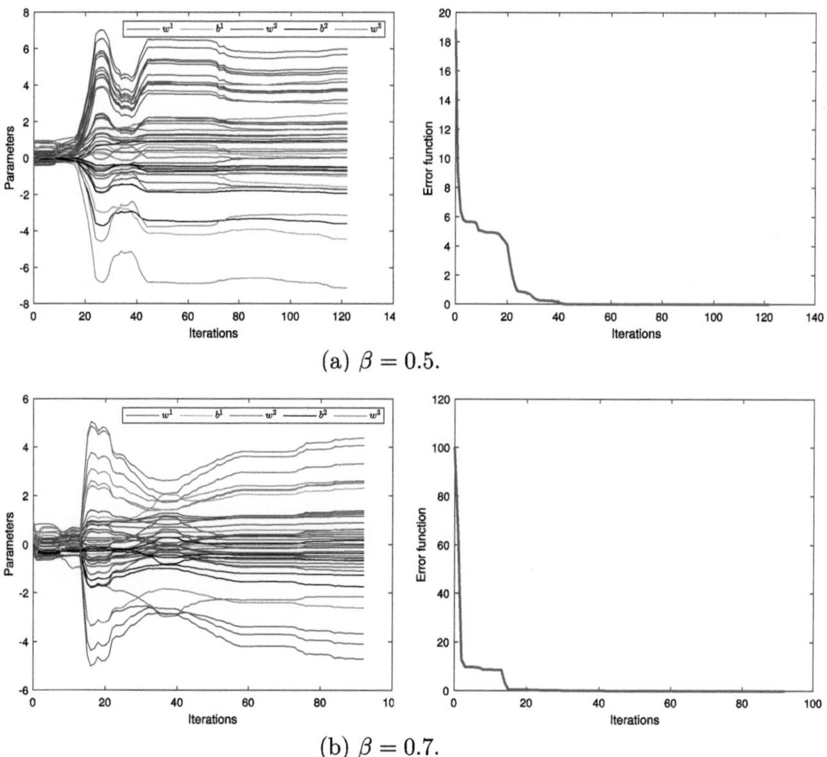

Fig. 3. Convergence of network parameters and error function of EFNN-BFGS with different values of β for Example 2.

the system dynamics. Thus, the new CFFDE system is

$$\left({}_{0}^{CF}D_t^{\beta}T_I\right)(t) = 2a_4T_M - (d_2 + a_1)T_I, \tag{39}$$

$$\left({}_{0}^{CF}D_t^{\beta}T_M\right)(t) = a_1T_I - d_3T_M - a_4T_M. \tag{40}$$

Based on Eq. (11), the designated trial solution for both system in Eq. (37)–(38) and Eq. (39)–(40) is

$$y_{T_I}(t, \mathbf{\Omega}) = 1.3 + tN(t, \mathbf{\Omega}), \tag{41}$$

$$y_{T_M}(t, \mathbf{\Omega}) = 1.2 + tN(t, \mathbf{\Omega}). \tag{42}$$

In the first stage, the EFNNBFGS scheme is implemented on the original system given in Eq. (37)–(38) over the domain $t \in [0, 10]$ and the obtained results are validated against those reported in [26] using the classical fourth-order RungeKutta (RK4) method. The EFNNBFGS framework successfully approximates the system behavior, with the computed solutions closely matching the

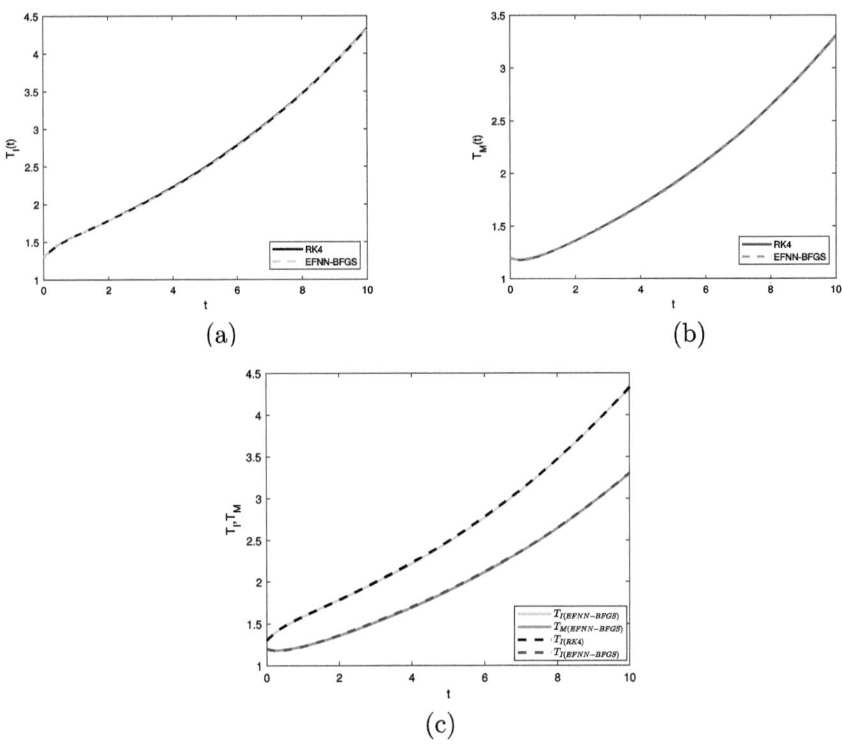

Fig. 4. Numerical solution for Example 3 for system with integer-order models.

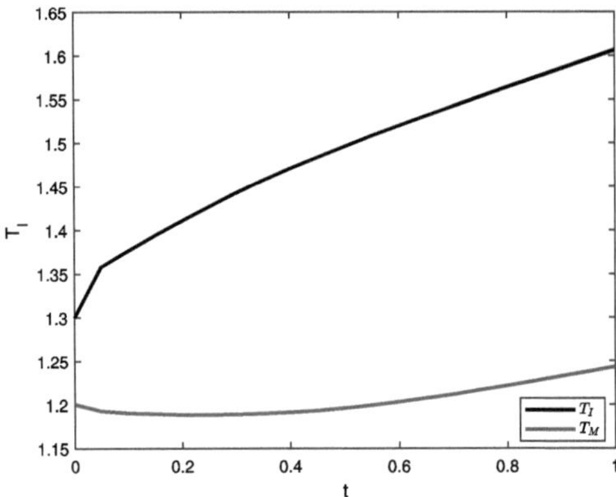

Fig. 5. Numerical solution for Example 3 for system of CFFDE with $\beta = 0.99$.

RK4 benchmark, as illustrated in Fig. 4(a)–(c). The model parameters are set to $a_1 = 1.0, a_4 = 0.8, d_2 = 0.11$ and $d_3 = 0.4$, corresponding to the stability region identified in [26], where tumor populations increase in both the interphase and metaphase stages. The EFNNBFGS results are consistent with these theoretical predictions. While the CPU time to solve this cases is 34.55 s. In the second stage, the EFNNBFGS scheme is implemented on the original system given in Eq. (39)–(40) over the domain $t \in [0, 1]$ with $\beta = 0.99$. Figure (5) shows the numerical solution of T_I and T_M over time and roughly the behaviour follows the original system of ODE. Note that for this example, the number of nodes use in EFNN-BFGS for both system is 3. The CPU time taken for solving this problem is 47.36 s.

5 Conclusion

In the present study, an extended feedforward neural network with BFGS has been developed to solve CFFDEs. Based on the first two examples, it can be concluded that the proposed EFNNBFGS scheme developed in this study is capable of producing highly accurate results with fewer than 100 iterations in most cases. The computational time varies across cases, depending on the number of training points and the total iterations required for convergence. In terms of parameter convergence, not all network parameters reach full convergence. Nevertheless, the proposed method is able to produce accurate approximations. Moreover, the EFNNBFGS scheme remains effective to larger domains. Through the last example, EFNNBFGS demonstrates the applicability of this newly extended scheme in solving real-world problems.

Acknowledgements. This study was supported by the Geran Dana Penyelidik Berpotensi Dana DTD (Vot No. 4J724) awarded by Universiti Teknologi Malaysia.

References

1. Podlubny, I.: Fractional differential equations: an introduction to fractional derivatives, fractional differential equations, to methods of their solution and some of their applications. Elsevier (1998)
2. Diethelm, K.: The Analysis of Fractional Differential Equations: An Application-Oriented Exposition Using Differential Operators of Caputo Type. Lecture Notes in Mathematics. Springer, Heidelberg (2010)
3. Tarasov, V.E.: No nonlocality. no fractional derivative. Commun. Nonlinear Sci. Numerical Simul. **62**, 157–163 (2018)
4. Tarasov, V.E.: Mathematical economics: application of fractional calculus. Mathematics **8**(5), 660 (2020)
5. Arora, S., Mathur, T., Agarwal, S., Tiwari, K., Gupta, P.: Applications of fractional calculus in computer vision: a survey. Neurocomputing **489**, 407–428 (2022)
6. Singh, A.P., Bingi, K.: Applications of fractional-order calculus in robotics. Fractal Fractional **8**, 403 (2024)
7. Tapadar, A., Roy, D., Adhikary, A.: Ripe stage detection and optimal ripe hours prediction of a banana using modified fractional order colpitts oscillator. IEEE Trans, Instrumentation Measure (2025)

8. Nisar, K.S., Farman, M., Ghannam, M., Hincal, E., Sambas, A.: A case study of fractional-order varicella virus model to nonlinear dynamics strategy for control and prevalence. Nonlinear Engineering **14**(1), 20240072 (2025)
9. Wang, X., Yang, T., Maeder, M., Marburg, S.: A novel fractional calculus modeling and physics-informed machine learning study on dynamic performance of hybrid flax/basalt fiber-reinforced composite. Nonlinear Dyn. **112**(22), 19917–19937 (2024)
10. Wang, J., Liu, Y., Yan, L., Han, K., Feng, L., Zhang,R.: Fractional sub-equation neural networks (fsenns) method for exact solutions of space–time fractional partial differential equations. Chaos: Interdisciplinary J. Nonlinear Sci. **35**(4) (2025)
11. Dabiri, D., Saadat, M., Mangal, D., Jamali, S.: Fractional rheology-informed neural networks for data-driven identification of viscoelastic constitutive models. Rheol. Acta **62**(10), 557–568 (2023)
12. Lagaris, I.E., Likas, A., Fotiadis, D.I.: Artificial neural networks for solving ordinary and partial differential equations. IEEE Trans. Neural Networks **9**(5), 987–1000 (1998)
13. Pakdaman, M., Ahmadian, A., Effati, S., Salahshour, S., Baleanu, D.: Solving differential equations of fractional order using an optimization technique based on training artificial neural network. Appl. Math. Comput. **293**, 81–95 (2017)
14. Raja, M.A.Z., Samar, R, Manzar, M.A., Shah, S.M.: Design of unsupervised fractional neural network model optimized with interior point algorithm for solving bagley–torvik equation. Math. Comput. Simul. **132**, 139–158 (2017)
15. Raja, M.A.Z., Khan, J.A., Qureshi, I.M.: Solution of fractional order system of bagley-torvik equation using evolutionary computational intelligence. Math, Problems Eng (2011)
16. Dufera, T.T.: Deep neural network for system of ordinary differential equations: vectorized algorithm and simulation. Mach. Learn. Appl. **5**, 100058 (2021)
17. Panghal, S., Kumar, M.: Neural network method: delay and system of delay differential equations. Eng. Comput., 1–10 (2021)
18. Admon, M.R., Senu, N., Ahmadian, A., Majid, Z.A., Salahshour, S.: A new efficient algorithm based on feedforward neural network for solving differential equations of fractional order. Commun. Nonlinear Sci. Numerical Simul. **117**, 106968 (2023)
19. Admon, M.R., Senu, N., Ahmadian, A., Majid, Z.A.: An advanced scheme based on artificial intelligence technique for solving nonlinear riccati systems. Comput. Appl. Math. **43**(6), 362 (2024)
20. Caputo, M., Fabrizio, M.: A new definition of fractional derivative without singular kernel. Progress Fractional Differentiation Appl. **1**(2), 73–85 (2015)
21. Kilbas, A.A., Srivastava, H.M., Trujillo, J.J.: Theory and applications of fractional differential equations, vol. 204. Elsevier (2006)
22. Goodfellow, I., Bengio, Y., Courville, A.: Deep learning. MIT press (2016)
23. Kazemi, B.F., Jafari, H.: Error estimate of the MQ-rbf collocation method for fractional differential equations with caputo-fabrizio derivative. Math. Sci. **11**(4), 297–305 (2017)
24. Arshad, S., Saleem, I., Akgül, A., Huang, J., Tang, Y., Eldin, S.M.: A Novel Numerical Method for Solving the Caputo-Fabrizio Fractional Differential Equation. AIMS Math **8**, 9535–9556 (2023)
25. Cao, J., Wang, Z., Chuanju, X.: A high-order scheme for fractional ordinary differential equations with the caputo-fabrizio derivative. Commun. Appl. Math. Comput. **2**(2), 179–199 (2020)
26. Admon, M.R., Maan, N.: Modelling tumor growth with immune response and drug using ordinary differential equations. J. Teknologi (Sci. Eng.) **79**(5) (2017)

Machine Learning Predictive Analytics for Sustainable Management of Consumer Response Data in the Automotive Sector

Samud Jaidina(✉), Nor Asmaa Alyaa, and Wong Kuan Yew

Faculty of Mechanical Engineering, Universiti Teknologi Malaysia, Johor, Malaysia
samud@graduate.utm.my

Abstract. The automotive sector increasingly relies on consumer-generated data, such as customer complaints and market claims, to support quality assurance and post-market decision-making. However, the growing volume, complexity, and unstructured nature of consumer responses pose significant challenges to sustainable data governance, particularly in ensuring timely analysis, consistent prioritization, and accountable decision-making. This study proposes a machine learning-based predictive analytics framework to support sustainable data governance of consumer responses in an automotive manufacturing context. The framework integrates data preprocessing, feature learning, and feature selection to enhance data quality and governance readiness. Unsupervised clustering techniques are applied to identify latent patterns within consumer response data and to distinguish crisis-level issues that require immediate attention, while supervised classification models are developed to automate the categorization of incoming claims. The predictive models achieve an accuracy exceeding 90%, demonstrating their effectiveness in supporting reliable and scalable analytics. To enhance transparency and operational usability, the analytical outcomes are visualized through a business intelligence dashboard developed using Power BI, enabling real-time monitoring and cross-functional decision support. The results indicate that the proposed approach improves responsiveness, accountability, and sustainability in managing consumer response data. This study contributes to the adoption of data-driven governance practices in the automotive sector and supports Industry 4.0 initiatives through the integration of predictive analytics and business intelligence.

Keywords: Predictive Model. Quality Issue Classification. Manufacturing Industry. Crisis Detection. Machine Learning

1 Introduction

The automotive manufacturing industry has undergone rapid transformation due to increasing vehicle complexity, stricter quality requirements, and higher customer expectations for reliability and safety [1, 2]. Modern vehicles integrate advanced electronics, sensors, and interconnected mechanical systems, generating large volumes of data throughout the product lifecycle [3]. Among these sources, consumer-generated data such as customer complaints, warranty claims, and service feedback are important for post-market quality monitoring and continuous improvement [4, 5].

N. A. Mohd Subha et al. (Eds.): SMS 2026, CCIS 2956, pp. 277–293, 2026.
https://doi.org/10.1007/978-981-92-0306-2_20

Effective governance of consumer response data is necessary to ensure data quality, consistency, and accountability across organizations [6]. However, many manufacturers still rely on manual inspection or rule-based filtration of market claims, which limits scalability and delays decision-making when handling large volumes of heterogeneous data [7]. These limitations reduce the ability to respond quickly to customer concerns and prioritize critical quality issues [8].

Consumer response data often contain high dimensionality, textual variability, missing values, and inconsistent terminology across customer and dealer reports [9, 10]. These challenges complicate traditional filtration processes and weaken expert-driven decision-making [11]. Consequently, governance objectives such as transparency, repeatability, and fairness may be compromised when different stakeholders interpret the same data differently [12]. Delayed identification of high-risk quality issues can increase warranty costs, prolong corrective actions, and create potential safety risks [13]. Therefore, analytical approaches are required to transform raw consumer response data into actionable insights for effective decision-making.

Despite recent advances in data analytics and machine learning, many studies still focus on structured manufacturing or sensor-based datasets, with limited attention given to post-market consumer response data that are often noisy and unstructured. Furthermore, the lack of integration between analytical models and practical decision-support tools, such as business intelligence dashboards, limits the practical use of predictive analytics in industrial environments.

The purpose of this study is to analyze consumer response data in the automotive manufacturing sector using machine learning-based predictive analytics to support sustainable data governance and quality issue management. Specifically, the study aims to identify significant attributes from consumer response data, develop predictive models for issue filtration and prioritization, and evaluate the performance of the proposed approach against existing practices. The study involves obtaining real-world consumer response data, performing preprocessing and feature selection, developing machine learning models for clustering and classification, evaluating model performance, and analyzing the implications for data governance and decision-making in the automotive sector.

The remainder of this paper is organized as follows. The next section reviews related studies on predictive analytics and machine learning in manufacturing contexts. The subsequent section describes the research methodology, followed by the results and discussion. Finally, the paper concludes with key findings, implications, limitations, and recommendations for future research.

2 Methodology

This study adopts a predictive analytics–based research framework to automate quality issue filtration in an automotive manufacturing environment. Predictive analytics and machine learning techniques have been widely applied to improve quality management, defect prioritization, and decision-making in manufacturing systems [1, 2]. The proposed methodology integrates data collection, data preprocessing, exploratory data analysis, feature learning and selection, clustering, supervised classification, and visualization to transform raw market claim data into actionable quality insights. The overall research workflow is illustrated in Fig. 1.

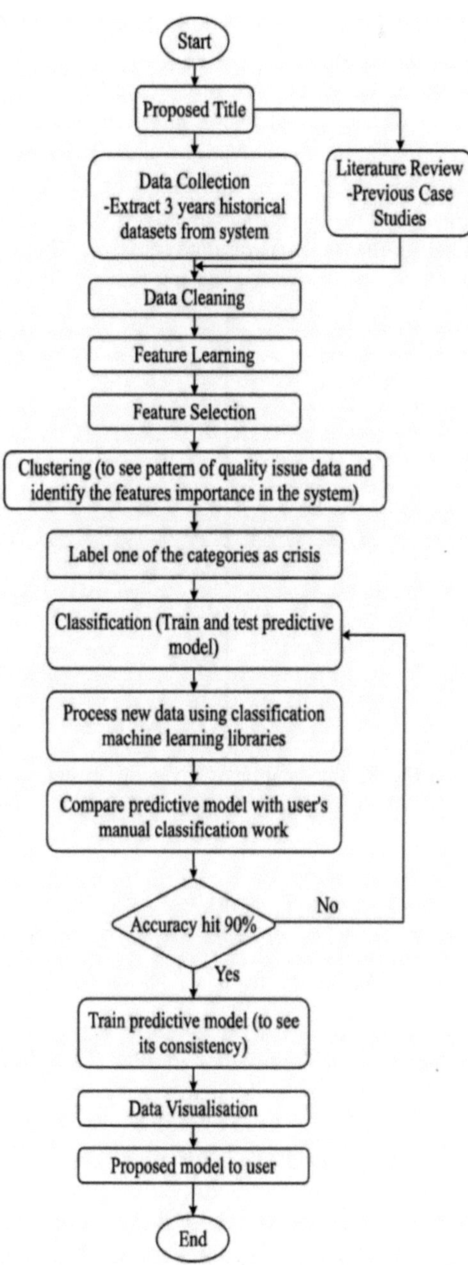

Fig. 1. Flow chart of the research methodology for predictive quality issue filtration.

2.1 Data Preprocessing

After data collection, preprocessing is performed to improve data quality, particularly through data cleaning. Data cleaning involves correcting or removing inaccurate, damaged, duplicated, or missing information from the dataset. One common method is data dropping, which removes rows containing missing values. However, since no duplicated records were found in the collected dataset, this method was not applied.

Instead, missing data were handled using data replacement to maintain all observations. In the dataset, the columns *Part Number* and *Part Description* contained 181 missing values as shown in Fig. 2 (left). These missing values were replaced with the label *"Unknown"* (Fig. 3). After the replacement process, the dataset contained 1132 non-null records, indicating that no missing data remained (Fig. 2 right)

```
RangeIndex: 1132 entries, 0 to 1131
Data columns (total 12 columns):
 #   Column             Non-Null Count
---  ------             --------------
 0   Sale               1132 non-null
 1   Model              1132 non-null
 2   Build Date         1132 non-null
 3   Sold Date          1132 non-null
 4   Repair Date        1132 non-null
 5   Km                 1132 non-null
 6   Month Used         1132 non-null
 7   Part Number        951 non-null
 8   Part Description   951 non-null
 9   Customer Note      1132 non-null
 10  Dealer Note        1132 non-null
 11  Dealer Solution    1132 non-null
```

```
RangeIndex: 1132 entries, 0 to 1131
Data columns (total 12 columns):
 #   Column             Non-Null Count
---  ------             --------------
 0   Sale               1132 non-null
 1   Model              1132 non-null
 2   Build Date         1132 non-null
 3   Sold Date          1132 non-null
 4   Repair Date        1132 non-null
 5   Km                 1132 non-null
 6   Month Used         1132 non-null
 7   Part Number        1132 non-null
 8   Part Description   1132 non-null
 9   Customer Note      1132 non-null
 10  Dealer Note        1132 non-null
 11  Dealer Solution    1132 non-null
```

Fig. 2. Information of data in the system

```
data['Part Number'] = data['Part Number'].fillna('Unknown')
data['Part Description'] = data['Part Description'].fillna('Unknown')
```

Fig. 3. Missing data replacement

During data exploration, it is found that in 'Part Description', the term "NON DEFINI' is referring to battery replacement in 'Dealer Note' and 'Dealer Solution', therefore Python mapping function is used to replace the needless term (Fig. 4). Mapping function is defined as a set of entries that lets users search for a particular key and get its value in return and a wide variety of items may be used as the keys in this function. It is used to eliminate noisy data that will cause the predictive model to be inaccurate. This mapping function is also applied to column 'Customer Note', 'Dealer Note' and 'Dealer Solution' as there are a lot of inconsistent data to be cleaned.

From the CSV file, the date shown is in string format and it is known that performing actions on a string instead of a Date Time item is extremely difficult. For this case,

```
part_mapping={
    'NON DEFINI':'12V BATTERY',
}
data['Part Description']=data['Part Description'].replace(part_mapping)
```

Fig. 4. Mapping Function

the Python Date Time item can be created from a string Date Time using the Pandas to_datetime() function. Then, the date time item will be transformed to numeric representation. The .toordinal() function can covert the date time into an integer indicating the number of days with ordinal 1 for January first of year 1. In data analytics and predictive modeling, label encoding is a method for transforming categorical variables into numerical representation. It is very helpful when developing algorithms since the majority of machine learning frameworks can only work with numerical inputs. Therefore, before go into feature learning and feature selection, the data need to be encoded using LabelEncoder() function to ease the next steps.

2.2 Feature Learning and Feature Selection

Since the data is unlabeled to target output, unsupervised feature learning is done to examine the connections between the data elements. In this project, heatmap is used for feature learning as it is excellent at displaying the correlation matrix between several variables, which makes it simple to quickly discover variables that are significantly or negatively connected. The heatmap layer uses a gradient hue, ranging from blue to red, to show the corresponding percentage of interactions. White denotes the absence of correlations; red denotes significant correlations and blue denotes correlations that are adverse.

The stronger the correlation between the features when the color is darker to red. Next, variance inflation factor (VIF) is used to select features, thus increasing the reliability of the predictive model. Multicollinearity is clearly indicated by a high VIF score. The results show that a predictor is highly correlated with other predictors when they surpass specific limits, which may have an impact on the model's accuracy, stability and dependability. Therefore, the features with VIF scores above 10 are removed as there is high multicollinearity in the features. After reducing the features, clustering combining with filter method, F-test is done to see the features importance. Then, the top 3 features with the F-scores more than 1000 are selected to test and train the predictive model.

2.3 Exploratory Data Analysis (EDA)

During feature learning and selection, heatmap and clustering algorithms are implemented as exploratory data analysis (EDA) tools to examine correlations within quality issue data and identify feature importance. Heatmap visualizes relationships between variables using color gradients, allowing patterns, trends, and correlations in complex datasets to be interpreted quickly.

Clustering groups unlabeled data into distinct clusters based on similar characteristics. The elbow method is first applied to determine the optimal number of clusters, *k*, by analysing the within-cluster sum of squares (WCSS). The optimal *k* value is identified at the elbow point of the graph. The clustering results are then visualized using scatter plots to illustrate data patterns. In this study, one cluster is labelled as a crisis category to represent quality issues that require higher priority and immediate attention.

2.4 Predictive Model Development

In the predictive model development stage, a classification algorithm is first applied to classify the initial dataset into identified clusters. The purpose of classification is to divide dataset components into several groups based on shared characteristics, making it easier to organize information and predict behavior. The predictive model is first trained and tested using existing data to classify quality issue behavior. Subsequently, the model processes new incoming data using classification machine learning libraries. Classification algorithms belong to supervised machine learning techniques, which rely on past data to generate predictions by forming categories based on shared features.

Evaluating the performance of the predictive model is important to ensure it behaves reliably across different types of data. In practice, comparable, measurable, and repeatable results are required, especially for industries with strict quality standards. For classification algorithms, the classifier score (clf.score) is commonly used to measure the average accuracy based on the provided attribute labels and dataset. The best score is 1.0, indicating perfect accuracy. In this project, the classification report function, which includes the classifier score, is used to evaluate the model performance (Fig. 5). The report presents precision, recall, f1-score, support, accuracy, macro average, and weighted average.

Classification Report:

	precision	recall	f1-score	support
0	0.99	1.00	1.00	117
1	1.00	1.00	1.00	4
2	1.00	1.00	1.00	6
3	0.91	0.91	0.91	11
4	0.99	0.99	0.99	86
5	0.00	0.00	0.00	1
6	1.00	1.00	1.00	2
accuracy			0.99	227
macro avg	0.84	0.84	0.84	227
weighted avg	0.98	0.99	0.98	227

Fig. 5. Example of classification report

Precision represents the proportion of correct positive predictions among all predicted positives, while recall represents the proportion of correct positive predictions among all actual positives. The f1-score is the weighted mean of precision and recall, calculated as 2 × (Precision × Recall) / (Precision + Recall). A score closer to 1 indicates better model performance. The support column indicates the number of actual instances of each category in the dataset. Besides overall accuracy, the report also provides the macro average, which is the unweighted mean and may penalize the model for poor performance in minority classes. The weighted average addresses class imbalance by considering the number of instances in each class. In this project, the predictive model aims to achieve an accuracy of 90%. If this target is reached, the model will be retrained using a different dataset to ensure consistency.

2.5 Data Visualization

Data visualization is a technique of representing data with graphical components such as maps, graphs and charts. The tools for data visualization enhance and simplify the process of visual communication for precision and detail. To comprehend and spot trends in data sets, data scientists have accessibility to libraries that use tools or coding languages like Python. The data scientists can operate more productively by using visualization tools that code information using designs, hues, graphs and lines. In this project, data visualization is essential to display the important information generated by the predictive model. The quality and manufacturing department can easily access the information to deal with the quality issues.

3 Result and Discussion

3.1 Data Preprocessing Workflow

Since the data in the system consists of a lot of inconsistent data and missing data, the data preprocessing step is first started with data cleaning. To encounter the missing data, the blanks found in columns 'Part Number' and 'Part Description' are filled with the term 'Unknown'. For inconsistent data, they are replaced with appropriate terms using mapping function in the columns 'Part Description', 'Customer Note', 'Dealer Note' and 'Dealer Solution'. As a result, the cleaned data have 1132 non-null counts and Fig. 6. represents the first 5 rows of the cleaned data. However, it can be seen that the data except for 'Km' and 'Month Used' are not in numerical items. Since the clustering and classification only can apply with numerical inputs, the data need to be transformed into numerical representation. The 'Repair Date' is transformed to Python Data Time item using the Pandas to_datetime() function and the date time item is then transformed to numeric representation using .toordinal() function which indicating the number of days with proleptic Gregorian ordinal 1 for January first of year 0001. For example, second January of 2024 is transformed to ordinal date of 738887 (Fig. 6).

Next, for columns 'Part Number', 'Part Description', 'Customer Note', 'Dealer Note' and 'Dealer Solution', LabelEncoder() function is used to convert the categorical variables into numerical representation as shown in Figs. 7 and 8. The dictionaries of the encoded mapping for each categorical variables will be shown in Appendix C.

Repair Date	Km	Month Used	Part Number	Part Description	Customer Note	Dealer Note	Dealer Solution
2024-01-02	16221	21	1629085380	12V BATTERY	Car starting problem	Battery to be replaced	Replace battery
2024-01-02	44967	19	Unknown	Unknown	Multiple warning lights on	Battery to be replaced	Replace battery
2024-01-02	19932	18	9830210580	TURBO REGULATN SOLENOID VALVE	Alert engine light on	Solenoid valve defect	Replace solenoid valve
2024-01-02	15153	15	98393686ZD	MULTIFUNCTION SCREEN PANEL	Multifunction screen appearance fault	Multifunction screen panel poor fitting	Replace multifunction screen panel
2024-01-02	33154	15	1629085380	12V BATTERY	Car starting problem	Battery to be replaced	Replace battery

Fig. 6. Example of classification report

Repair Date	Km	Month Used	Part Number	Part Description	Customer Note	Dealer Note
2024-01-02	16221	21	1629085380	12V BATTERY	Car starting problem	Battery to be replaced
2024-01-02	44967	19	Unknown	Unknown	Multiple warning lights on	Battery to be replaced
2024-01-02	19932	18	9830210580	TURBO REGULATN SOLENOID VALVE	Alert engine light on	Solenoid valve defect
2024-01-02	15153	15	98393686ZD	MULTIFUNCTION SCREEN PANEL	Multifunction screen appearance fault	Multifunction screen panel poor fitting
2024-01-02	33154	15	1629085380	12V BATTERY	Car starting problem	Battery to be replaced

Fig. 7. Example of date ordinal

Dealer Solution	Repair Date Ordinal	Part Number Encoded	Part Description Encoded	Customer Note Encoded	Dealer Note Encoded	Dealer Solution Encoded
Replace battery	738887	11	3	25	31	50
Replace battery	738887	108	135	88	31	50
Replace solenoid valve	738887	83	134	9	222	182
Replace multifunction screen panel	738887	103	87	85	145	141
Replace battery	738887	11	3	25	31	50

Fig. 8. Example of label encoded data

3.2 Feature Learning for Governed Predictive Analytics

After data preprocessing, the feature learning is done by constructing a heatmap to determine the correlation between each feature. When the color of cell is closer to red in the heatmap, meaning that the features are having a strong correlation. Based on Fig. 9, 'Dealer Note' and 'Dealer Solution' hit the highest score of 0.77, indicating that these two features are strongly related. For best illustration we can see in the Fig. 17(page 16), the correlation can be clearly seen that when 'Dealer Note' states 'Battery to be replaced', the solution will be replacing the battery. When there is defect on particular part, the solution will be replacing the part. 'Customer Note' and 'Dealer Note' also have a high score of 0.7 where the customer describes the problem happen in the car and the dealer determines which part is defect. For 'Part Number' and 'Part Description', they are strongly correlated as the part number represent the specific part. However, their score is slightly lower as some parts may have different part number, for example the number 5600VF, 1629085380 and 9678006180 are all represent 12 V battery. For the negative scores in the heatmap, the features are hardly related to each other, such as the significant negative score (−0.26) by the feature 'Repair Date' and 'Month Used'. The repair date cannot indicate the month used in field of the car. Overall, it can be seen that in the heatmap, the features 'Part Number', 'Part Description', 'Customer Note', 'Dealer Note' and 'Dealer Solution' are correlated to each other.

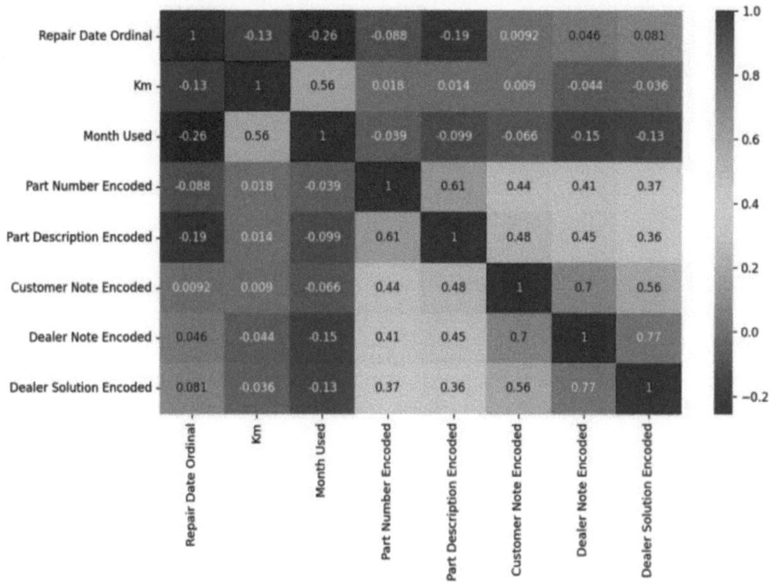

Fig. 9. Heatmap

After learning the relationship between the features, the VIF method is used to reduce the unnecessary features with multicollinearity. VIF can detect multicollinearity between variables even the correlations are low. Other method such as pairwise correlations also can used for reducing features, however pairwise correlations only can detect multicollinearity for variables that are in direct linear relationships, it is challenging when there are complex interactions among variables. According to Fig. 10, the features

'Repair Date' and 'Month Used' have a very high VIF score which exceeds optimum VIF score of 10, indicating that these two features have significant multicollinearity and need to be reduced. The other six features are selected to proceed to next step.

	feature	VIF
0	Repair Date Ordinal	12.153843
1	Km	5.134176
2	Month Used	12.611332
3	Part Number Encoded	4.198088
4	Part Description Encoded	3.474849
5	Customer Note Encoded	4.877873
6	Dealer Note Encoded	6.853074
7	Dealer Solution Encoded	7.299223

Fig. 10. Variance inflation factor (VIF) scores

Before clustering, the selected features are transformed into standard form using Python StandardScaler() function. This function is applied to ensure that all the input data have a balance scale where the feature's mean is zero and have a unit of standard deviation. K-Means clustering is then implemented to learn the selected features in groups or clusters. By dividing a collection of data points into multiple categories, K-Means clustering aims to generate data points within each group more similar to each other and distinct from the data points within the other groups. In essence, it is an organization of items according to their similarities and differences. The elbow method is firstly applied to determine the optimal number of clusters, k to group the features. Based on the elbow method results in Fig. 11, the elbow pattern is formed for $k = 3$, therefore the optimal number of clusters is 3 for the selected features. When evaluated against points in different clusters, the silhouette score for each point indicates the similarity between the points in that identical cluster. A high silhouette score for the majority of points indicates that the clustering is suitable. For this clustering results, the silhouette score is 0.42, considering the clusters formed are good in average.

According to Fig. 12, the clustering results for each feature are presented, while the corresponding clustering visualizations are provided in Appendix A. Analysis of the clustering outcomes based on mileage (km) against other features indicates that the most dominant quality issue across clusters is related to car starting failure caused by battery defects. The first cluster is primarily associated with battery-related problems, including items such as "12 V battery," "car starting problem," and "battery replacement," typically occurring around 8104 km of vehicle usage. The second cluster is mainly linked to steering-related issues, particularly steering rack defects, while the third cluster includes a combination of battery-related issues and tailgate faults observed at different mileage ranges, notably around 25193 km.

Further clustering analysis using part number, part description, customer notes, dealer notes, and dealer solutions reveal consistent patterns. The first and third clusters across these features are largely dominated by battery-related issues, where the recommended solution is battery replacement. In contrast, the second cluster is associated with non-battery-related issues, particularly sun visor mirror covers cracks and software defects. Similar patterns are observed in the clustering of customer notes against dealer notes and dealer solutions, where battery defects form the most frequent cluster, followed by sun visor-related defects.

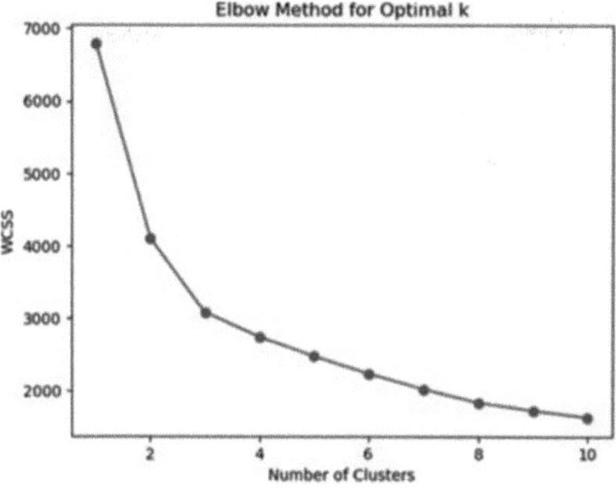

Silhouette Score: 0.419

Fig. 11. Variance inflation factor (VIF) scores

Overall, the clustering results demonstrate that battery-related problems constitute the most frequent and critical quality issue across all feature combinations, followed by sun visor defects. As battery defects dominate the clustering outcomes, they are identified as crisis-level issues requiring prioritized attention. Subsequently, feature importance analysis using the F-test is conducted to evaluate the contribution of each feature to the clustering results. Based on Fig. 15, the F-test results indicate that all features are statistically significant, with p-values less than 0.05. To enhance model reliability and accuracy, only features with F-scores greater than 1000 are retained, namely part description, dealer note, and dealer solution, which are then used as key inputs for the predictive model.

Next, classification algorithms included Decision Tree (DT), Support Vector Machines (SVM), Random Forest (RF), K-Nearest Neighbors (KNN), Naïve Bayes (NB) and Gradient Boosting Tree (GBT) are applied to train and test the predictive model using the input features and target categories. Fig. 12 represents the accuracy of each classifier based on their classification report.Among the classifiers, RF algorithms contributed the highest accuracy (97.4%) for the predictive model. RF hits 0.95 of average precision across all classes and 0.98 of average precision when considering class imbalance, meaning that there are high positive predictions for all the classes. It also able to determine most of the instances of each class with 0.92 of average recall and 0.97 of weighted average recall. For F1-score, RF classifier have the highest average score of 0.93 and weighted average score of 0.97, indicating that RF classifier has a perfect precision and recall, thus works better with the predictive model. Since the RF classifier has the highest scores based on classification report, it is used to train another set of data from year 2020 to 2023. This is to ensure the classifier are consistently work well with every set of data in the system. Based on the Fig. 13, the RF classifier accuracy shown 100%, meaning that the predictive model is able to classify every data correctly with perfect precision and recall. The predictive model is successful and ready to propose to users.

Features	Part Number	Part Description	Customer Note	Dealer Note	Dealer Solution
Km					
Part Number					
Part Description					
Customer Note					
Dealer Note					

Fig. 12. Clustering results for each selected feature

```
Feature: Km, F-Score: 3.29, P-Value: 3.75e-02
Feature: Part Number Encoded, F-Score: 755.40, P-Value: 5.88e-209
Feature: Part Description Encoded, F-Score: 1341.31, P-Value: 5.11e-299
Feature: Customer Note Encoded, F-Score: 722.79, P-Value: 7.99e-203
Feature: Dealer Note Encoded, F-Score: 1492.82, P-Value: 9.03e-318
Feature: Dealer Solution Encoded, F-Score: 1345.40, P-Value: 1.53e-299
```

Fig. 13. F-scores and P-values for selected features

3.3 Classification

According to the company quality issues management, the crisis issues usually being classified based on the number of occurrences in a specific period or based on the repair cost. In this project, the crisis issue will be classified based on the number of occurrences only due to the data confidentiality of the repair cost. Based on the F- scores, the input features selected for the predictive model are 'Part Description', 'Dealer Note' and 'Dealer Solution' while the target outputs are set as 'Crisis', 'Intermediate-crisis Level 1', 'Intermediate-crisis Level 2', 'Intermediate-crisis Level 3' based on the number of occurrences of the quality issues in the selected features to determine the level of crisis. Figure 14 represents the list of the top 10 frequent items in the selected features and the car battery will be categorized as the crisis issue as it contributed an extreme number of occurrences. The second frequent item which is the driveshaft defect will be assigned to 'Intermediate-crisis Level 1'. Next, the sun visor and fog lamp defect will be assigned to 'Intermediate-crisis Level 2' as in 'Dealer Solution', the new spare part, which refers to the fog lamp in the data system and sun visor have the same number of occurrences. While for 'Intermediate-crisis Level 3', the steering rack and shock absorber defect are being assigned to this class. Finally, the rest of the quality issues will be automatically assigned to the class 'Miscellaneous Items' as issues that are less crisis.

Top 10 frequent items in Part Description:		Top 10 frequent items in Dealer Note:		Top 10 frequent items in Dealer Solution:	
Part Description		Dealer Note		Dealer Solution	
12V BATTERY	505	Battery to be replaced	620	Replace battery	618
Unknown	181	Driveshaft defect	24	Replace driveshaft	27
NEW DRIVESHAFT	29	Steering rack defect	21	Replace new spare part	25
SUN VISOR	28	Check and follow TSB for rack noise	21	Replace sunvisor	25
NEW ELECT STEERING MECHANISM	23	AC compressor defect	18	Replace shock absorber	18
STEERING ROD BALL JOINT KIT	21	Noise from shock absorber	17	Replace steering rack	18
FRONT SUSPENSION DAMPER	18	Sunvisor mirror cover crack	17	Carried out TSB, no noise	15
GAUGE WITH PUMP	15	Software to be updated	16	Replace new part	14
AIR NON-RETURN VALVE	15	Rear foglamp defect	7	Replace fuel gauge pump	12
NEW COMPRESSOR	13	Steering rack and ball joint defect	7	Replace ball joint	10
Name: count, dtype: int64		Name: count, dtype: int64		Name: count, dtype: int64	

Fig. 14. Top 10 frequent items in selected features

Next, classification algorithms included Decision Tree (DT), Support Vector Machines (SVM), Random Forest (RF), K-Nearest Neighbors (KNN), Naïve Bayes (NB) and Gradient Boosting Tree (GBT) are applied to train and test the predictive model using the input features and target categories. Fig. 15 represents the accuracy of each classifier based on their classification report. Among the classifiers, RF algorithms contributed the highest accuracy (97.4%) for the predictive model. RF hits 0.95 of average precision across all classes and 0.98 of average precision when considering class imbalance, meaning that there are high positive predictions for all the classes. It also able to determine most of the instances of each class with 0.92 of average recall and 0.97 of weighted average recall. For F1-score, RF classifier have the highest average score of 0.93 and weighted average score of 0.97, indicating that RF classifier has a perfect precision and recall, thus works better with the predictive model. Since the RF classifier has the highest scores based on classification report, it is used to train another set of data from year 2020 to 2023. This is to ensure the classifier are consistently work well with every set of data in the system. Based on the Fig. 16, the RF classifier accuracy shown 1.00 or 100%, meaning that the predictive model is able to classify.

Before proceed to the data visualization, the quality issues are assigned to the responsible departments depending on the part description using mapping function. There is total three responsible departments which are engineering, manufacturing and supplier parts. Engineering site are separated to PRS (Propulsion, Reliability and System), BCI (Body Chassis, Interior appearance parts and all issue regarding noise, vibration and harshness), electrical parts and software while manufacturing site consists of body (welding), painting, assembly and quality control. For supplier parts, it consists of local parts and import parts, however due to the data privacy, the supplier parts do not separate into local and import. For example, the crisis issue which is the battery problem is eassigned to engineering electrical parts while driveshaft will be assigned to engineering BCI as it belongs to body chassis. For engineering PRS, it includes parts related to engine and some built-in systems. The sun visor is assigned to manufacturing assembly to replace and assemble while parts needed for mounting are assigned to manufacturing body. Finally, the fog lamp and door parts are assigned to supplier parts.

Classifier	Classification Report	Accuracy
Decision Tree (DT)	precision recall f1-score support 0 1.00 1.00 1.00 131 1 0.00 0.00 0.00 6 2 0.00 0.00 0.00 10 3 0.50 0.09 0.15 11 4 0.72 0.99 0.83 69 accuracy 0.88 227 macro avg 0.44 0.42 0.40 227 weighted avg 0.82 0.88 0.84 227	88.1%
Support Vector Machines (SVM)	precision recall f1-score support 0 0.91 1.00 0.95 131 1 0.00 0.00 0.00 6 2 0.00 0.00 0.00 10 3 0.00 0.00 0.00 11 4 0.69 0.83 0.75 69 accuracy 0.83 227 macro avg 0.32 0.37 0.34 227 weighted avg 0.73 0.83 0.78 227	82.8%
Random Forest (RF)	precision recall f1-score support 0 0.99 1.00 1.00 131 1 1.00 0.83 0.91 6 2 1.00 0.90 0.95 10 3 0.77 0.91 0.83 11 4 0.97 0.96 0.96 69 accuracy 0.97 227 macro avg 0.95 0.92 0.93 227 weighted avg 0.98 0.97 0.97 227	97.4%
K-Nearest Neighbors (KNN)	precision recall f1-score support 0 1.00 0.98 0.99 131 1 0.67 0.67 0.67 6 2 0.82 0.90 0.86 10 3 0.60 0.82 0.69 11 4 0.91 0.87 0.89 69 accuracy 0.93 227 macro avg 0.80 0.85 0.82 227 weighted avg 0.94 0.93 0.93 227	93.0%
Naïve Bayes (NB)	precision recall f1-score support 0 1.00 0.99 1.00 131 1 1.00 1.00 1.00 6 2 0.42 0.50 0.45 10 3 0.00 0.00 0.00 11 4 0.79 0.88 0.84 69 accuracy 0.89 227 macro avg 0.64 0.68 0.66 227 weighted avg 0.86 0.89 0.88 227	89.0%
Gradient Boosting Tree (GBT)		96.9%

Fig. 15. Classification report of each classifier

	precision	recall	f1-score	support
0	1.00	1.00	1.00	131
1	1.00	0.67	0.80	6
2	0.91	1.00	0.95	10
3	0.71	0.91	0.80	11
4	0.97	0.94	0.96	69
accuracy			0.97	227
macro avg	0.92	0.90	0.90	227
weighted avg	0.97	0.97	0.97	227

Fig. 15. (*continued*)

Classification Report:

	precision	recall	f1-score	support
0	1.00	1.00	1.00	215
1	1.00	1.00	1.00	23
2	1.00	1.00	1.00	16
3	1.00	1.00	1.00	42
4	1.00	1.00	1.00	629
accuracy			1.00	925
macro avg	1.00	1.00	1.00	925
weighted avg	1.00	1.00	1.00	925

Fig. 16. Classification report for training new set of data

3.4 Data Visualization

Based on the data visualization in Power BI dashboard Fig. 17, some important items are displayed especially the quality issues with the assigned categories and departments. From the upper left of the dashboard, the repair date and category slicers can be adjusted or selected to choose the range of date and category (Fig. 18). After selecting the date and category, the upper right table will present all the data in terms of 'Repair Date', 'Part Description', 'Dealer Note', 'Dealer Solution', 'Category' and 'Department' assigned. Users can easily view the quality issues whether in crisis or intermediate crisis. If the quality issue is assigned to their department, the users can quickly notice and implement actions to solve the problem. According to the lower side of the dashboard, some graphs are presented to showcase the statistics of quality issues and all are based on the range of the repair date selected. The horizontal bar chart on the lower left presents the count of the quality issues by month, users can easily do reporting by presenting this bar chart to top management. The vertical bar chart next to it displays the percentage of each category in the selected repair date while the donut chart on the lower right side presents the percentage of crisis issues by month. The total counts of the crisis quality issues will be displayed in card to highlight the total crisis amount within the selected repair date. The crisis issues by sales will be presented in visual maps to represent the counts of crisis issues in each location. For example, there is 48 crisis issues occur in the sales by Philippines from January to November 2024.

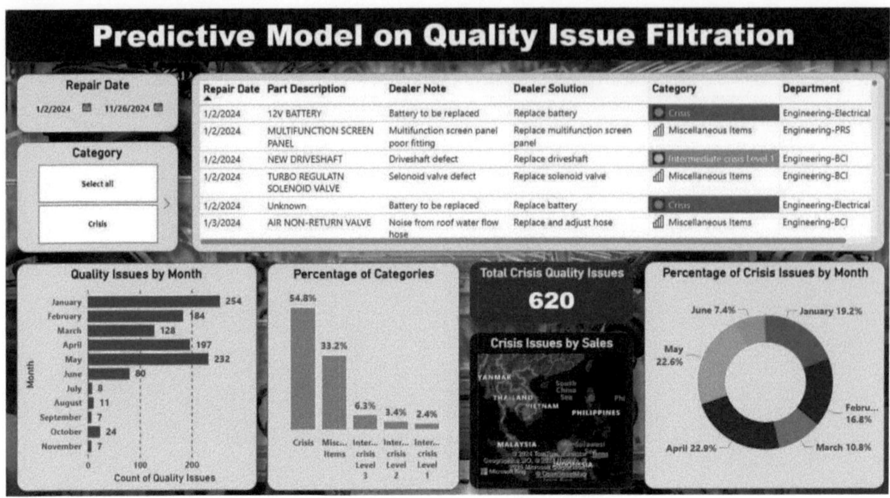

Fig. 17. Data Visualization Dashboard

Fig. 18. Select repair date and category

4 Conclusion

This study successfully achieved its research objectives by developing a machine learning-based predictive analytics framework for quality issue filtration in an automotive manufacturing context. Feature learning and selection were performed using correlation analysis, variance inflation factor, clustering, and F-test methods to identify significant inputs for the predictive model. The clustering results revealed that battery-related issues were the most frequent and critical quality problems, which were therefore classified as crisis-level issues. Based on feature importance analysis, part description, dealer note, and dealer solution were selected as key input features for model development. Several classification algorithms were evaluated, and the random forest classifier demonstrated the best performance, achieving an accuracy of 97.4%, indicating the robustness and reliability of the proposed model. The predictive results were further integrated into an interactive Power BI dashboard to support operational decision-making. The dashboard enables users to filter quality issues by repair date and category, view assigned departments and distinguish between crisis and intermediate crisis cases in a timely manner. Overall, the proposed framework enhances the timeliness and consistency of quality issue prioritization, reduces miscommunication between departments, and improves productivity through data-driven decision support aligned with Industry 4.0 practices.

Acknowledgements. The authors acknowledge funding support from Universiti Teknologi Malaysia under UTM Encouragement Research grant (UTM ER) (Q.J130000.3824.42J27).

References

1. Agag, G., et al.: Understanding the relationship between marketing analytics, customer agility, and customer satisfaction: a longitudinal perspective. J. Retail. Consum. Serv. **77**, 103663 (2024)
2. Ahad, A.A., Das, B., Khan, M.R., Saha, N., Zahid, A., Ahmad, M.: Multiclass liver disease prediction with adaptive data preprocessing and ensemble modeling. Results Eng. **22**, 102059 (2024)
3. Akhavan, F., Hassannayebi, E.: A hybrid machine learning with process analytics for predicting customer experience in online insurance services industry. Decis. Anal. J. **11**, 100452 (2024)
4. Akman, G., Yorur, B., Boyaci, A.I., Chiu, M.: Assessing innovation capabilities of manufacturing companies by combination of unsupervised and supervised machine learning approaches. Appl. Soft Comput. **147**, 110735 (2023)
5. Allenbrand, C.: Supervised and unsupervised learning models for pharmaceutical drug rating and classification using consumer generated reviews. Healthc. Analytics. **5**, 100288 (2024)
6. Bankar, V.R., Nandurkar, K.N.: Implementation of IoT technology for quality improvement in an automotive industry. Mater. Today Proc. **1–7** (2023)
7. Beinabadi, H.Z., Baradaran, V., Komijan, A.R.: Sustainable supply chain decision-making in the automotive industry: a data-driven approach. Socio Econ. Plan. Sci. **101908** (2024)
8. Benhar, H., Idri, A., Fernández-Alemán, J.L.: Data preprocessing for heart disease classification: a systematic literature review. Comput. Methods Prog. Biomed. **195**, 105635 (2020)
9. Biswas, A.K., Seethalakshmi, R., Mariappan, P., Battacharjee, D.: An ensemble learning model for predicting the intention to quit among employees using classification algorithms. Decis. Anal. J. **9**, 100335 (2023)
10. Bowen, W., Zhang, Y., Chen, J., Li, X., Wang, H., Liu, Z.: Long short-term memory deep learning model for predicting the dynamic performance of automotive PEMFC system. Energy AI. **14**, 100278 (2023)
11. Calavito, G., Lanubile, F., Novielli, N., Quaranta, L.: Impact of data quality for automatic issue classification using pre-trained language model. J. Syst. Softw. **210**, 111838 (2024)
12. Gorcun, O.F., Mishra, A.R., Aytekin, A., Simic, V., Korucuk, S.: Evaluation of industry 4.0 strategies for digital transformation in the automotive manufacturing industry using an integrated fuzzy decision-making model. J. Manuf. Syst. **74**, 922–948 (2024)
13. Gupta, U., Sharma, R.: Analysis of criminal spatial events in India using exploratory data analysis and regression. Comput. Electr. Eng. **109**, 108761 (2023)

Design of Indoor Mobile Robot with Visual SLAM Using Distributed Processing and Deep Learning

Liao Cheng[1], Sophan Wahyudi Nawawi[1(✉)], and Sulaiman Sabikan[2]

[1] Universiti Teknologi Malaysia, Johor Bahru, Malaysia
e-sophan@utm.my

[2] Universiti Teknikal Malaysia Melaka, Melaka, Malaysia

Abstract. Indoor mobile robots have a wide range of application value, such as home cleaning, indoor exploration, etc. SLAM technology is usually used to solve its localization and mapping problems, with VSLAM (visual SLAM) being a key research focus. Traditional VSLAM robots mainly rely on processors on the vehicle body. This architecture of pure on-board processor leads to problems such as low battery endurance and high vehicle body cost. In addition, traditional VSLAM frameworks, such as ORB-SLAM3, tend to exhibit degraded performance in environments with dynamic objects. This research presents a novel approach to address these challenges by designing an indoor distributed VSLAM system that offloads computationally intensive tasks to an off-board station via advanced wireless communication technology (WIFI 6) and hardware-accelerated video encoding (H.264). Additionally, the system integrates deep learning-based image segmentation to filter dynamic objects, thereby improving tracking error. The research employs ORB-SLAM3 as the foundational VSLAM framework, combined with ROS 2 for distributed communication management. The implemented system demonstrates real-time communication performance with a Sense-Move delay of 64.9 ms, network delay of 19.1 ms, and network throughput of 0.5 MBps. The integration of YOLO-based semantic segmentation for dynamic object filtering substantially improves trajectory tracking accuracy, reducing RMSE from 0.669 m to 0.015 m in dataset evaluations. The integrated system achieves 10–15 fps on Jetson Orin Nano Super hardware. This distributed architecture successfully reduces the computational burden on indoor mobile robots while maintaining real-time operation, demonstrating the effectiveness of integrating advanced wireless technologies, hardware-accelerated video encoding, and deep learning.

Keywords: Indoor Mobile Robot · Visual SLAM · Distributed Computing · Deep Learning · ORB-SLAM3 · ROS 2 · WIFI · video encoding · YOLO

1 Introduction

Robotics has always been a popular research direction, as it guides humans to develop robots that can help improve productivity, especially in labor intensive tasks. Mobile robots, or the movement of robots, are an important part of robotics, as they expand

N. A. Mohd Subha et al. (Eds.): SMS 2026, CCIS 2956, pp. 294–308, 2026.
https://doi.org/10.1007/978-981-92-0306-2_21

the robot's workspace [1, 2]. Indoor mobile robots have found widespread applications across diverse domains, including home cleaning (robotic vacuum cleaners), restaurant service (food delivery robots), factory material handling, and indoor exploration for inspection and mapping [3]. These varied applications share a fundamental requirement: robust autonomous navigation in structured indoor environments where GPS-based localization is unavailable.

For autonomous moving purposes, many technologies have been used as the basis of autonomous movement localization and mapping. SLAM (Simultaneous Localization And Mapping) technology is usually used to solve the localization and mapping problem systematically and precisely [4, 5]. Currently, two types of sensors are typically used in SLAM. One is LiDAR (Light Detection and Ranging), which uses laser light to measure the distance between the sensor and surroundings. The other is the camera, which uses optical images to obtain environmental information.

With the development of computer vision and camera manufacturing, camera sensors can acquire more information from the environment, such as colour, texture, and lighting, in a more efficient and cost-effective way compared with LiDAR. Therefore, visual SLAM (or VSLAM), which is dedicated to camera-based SLAM, has become a popular research direction and is now a practical solution, especially in indoor environments [5–7]. This paper focuses on indoor mobile robots using visual SLAM technology, which has wide-ranging applications, such as home cleaning, restaurant service, factory handling, and indoor exploration.

1.1 Project Background

Despite the growing adoption of visual SLAM technology in indoor robotics, current implementations face two fundamental challenges that constrain their practical deployment and performance in real world applications. These challenges computational resource limitations and vulnerability to dynamic environments significantly impact the viability and effectiveness of VSLAM based mobile robots.

First, traditional visual SLAM robots mainly rely on processors on the robot body, i.e., onboard processors [8, 9]. The main reason might be some of the characteristics of the VSLAM processing pipeline, i.e., large data volumes, low latency requirements, and high computational load. For current mainstream indoor mobile robots powered by batteries, the architecture of placing processors on the robot leads to problems such as poor battery endurance, high heat generation, larger body volume, and high body cost. Moreover, some of these problems are more obvious in collaborative SLAM systems involving multiple robots.

Second, traditional visual SLAM frameworks, such as ORB-SLAM, tend to exhibit degraded performance in environments with dynamic objects [10]. Dynamic objects such as moving people, pets, or other mobile entities violate the static world assumption that underlies most SLAM algorithms. When feature points are extracted from these moving objects, they introduce erroneous constraints into the SLAM optimization process, leading to degraded tracking performance and accumulated localization errors.

This section critically reviews existing technologies relevant to distributed visual SLAM systems, examining both their capabilities and constraints. The review is organized around three key technology domains that form the foundation of this research:

advanced wireless communication technologies suitable for real-time robotic applications, established VSLAM frameworks and their architectural characteristics, and the persistent challenge of dynamic objects in visual SLAM environments. By analyzing these areas systematically, this section reveals the technological gaps that motivate the distributed architecture and deep learning integration proposed in this work.

The feasibility of distributed robotic systems depends on wireless communication infrastructure supporting real-time data transmission. Several high-bandwidth wireless technologies are available for indoor applications, including WIFI (IEEE 802.11), UWB (Ultra-Wideband, IEEE 802.15.4a/z), and WiGig (IEEE 802.11ad/ay). WIFI operates in the 2.4 GHz and 5 GHz bands, with the latest WIFI 6 (IEEE 802.11ax) offering data rates up to 9.6 Gbps (Khorov et al., 2018). UWB utilizes an ultra-wide spectrum (3.1–10.6 GHz, often more than 6 GHz) with bandwidth exceeding 500 MHz, providing centimeter-level positioning accuracy and low latency, but with limited data rates (around 27 Mbps) and shorter communication range (Lee et al., 2007). WiGig operates at 60 GHz millimeter-wave frequencies, achieving extremely high data rates up to 7 Gbps (IEEE 802.11ad) with potential evolution to 20–40 Gbps (IEEE 802.11ay), but suffers from poor wall penetration and requires line-of-sight communication due to the propagation characteristics of millimeter waves (Ghasempour et al., 2017). Considering the balance between data throughput, communication range, wall penetration, device availability, and cost, WIFI remains the predominant choice for mobile robot wireless communication in indoor environments.

Sultan et al. [11] and Pandey and Parasuraman [12] characterized WIFI performance in robotic applications, measuring throughput of 30–35 Mbps and latency of 40-80 ms in non-line-of-sight indoor conditions. However, these studies exclusively utilized WIFI 5 (IEEE 802.11 ac) devices, leaving newer WIFI 6 technology unexplored despite its theoretical improvements: data rates up to 9.6 Gbps, reduced latency through OFDMA, and better dense-environment performance [13, 14].

Wei et al. [15] introduced RT-WIFI, a real-time communication protocol that modifies the MAC layer to offer deterministic, low-latency transmission in cyber-physical systems. Although highly effective, it demands a non-standard software stack and is not broadly supported in commodity platforms, limiting its practical applicability for mobile robotics.

This research gap is significant for distributed VSLAM, where WIFI 6's higher bandwidth and lower latency could enable more sophisticated architectures than previously feasible.

Visual SLAM systems have evolved substantially over the past two decades. As outlined by Gao et al. [5], VSLAM pipelines typically include Sensing, Visual Odometry, Optimization, Loop Closing, and Mapping stages. Based on the approach to processing visual information, VSLAM methods can be broadly categorized into feature-based (indirect) methods and direct methods [9, 16].

Feature-based methods extract and track distinctive keypoints from images. The ORB-SLAM series, developed by the University of Zaragoza, represents one of the most influential feature-based VSLAM systems. The system employs ORB (Oriented FAST and Rotated BRIEF) features, a binary descriptor proposed by Rublee et al. [17] that combines the FAST keypoint detector [18] with a rotation-invariant BRIEF descriptor

[19]. ORB features achieve computational efficiency approximately 100 times faster than SIFT while maintaining comparable performance, enabling real-time operation on standard CPUs without requiring GPU acceleration. The original ORB-SLAM [20] introduced a unified framework using ORB features for all SLAM tasks—tracking, mapping, relocalization, and loop closing. This work was subsequently extended to ORB-SLAM2 [21] with support for stereo and RGB-D cameras, and further evolved into ORB-SLAM3, which supports visual, visual-inertial, and multi-map SLAM with pinhole and fisheye lens models.

In contrast, direct methods operate directly on image intensities without explicit feature extraction. LSD-SLAM (Large-Scale Direct SLAM) [22] performs large-scale direct monocular SLAM by tracking and mapping high-gradient image regions, enabling semi-dense reconstruction. DSO (Direct Sparse Odometry) [23] further advances direct methods with a sparse approach that jointly optimizes photometric error and geometry, achieving high accuracy in trajectory estimation. These direct methods can be advantageous in low-texture environments where feature extraction is challenging.

Recent developments have also introduced semantic SLAM systems that integrate deep learning for scene understanding. Kimera [24] provides real-time metric-semantic localization and mapping, combining geometric and semantic information for richer environmental representations.

However, these advanced frameworks share two critical limitations. First, they require substantial computational resources for real-time operation—feature extraction, matching, and optimization are all computationally intensive, posing challenges for resource-constrained mobile platforms. Second, existing systems assume centralized on-board processing and do not natively support distributed architectures where tasks could be offloaded via wireless communication. These limitations motivate the distributed approach in this research, aiming to leverage off-board computing while maintaining real-time performance. Recent surveys [16, 25] further highlight ongoing challenges in computational efficiency and real-time performance on embedded platforms.

2 Methodology

This chapter presents the complete implementation and evaluation of the proposed distributed and deep learning-enhanced visual SLAM system.

2.1 System Architecture

The robot system is designed with the following architecture (Fig. 1):

In the station, a PC (Personal Computer) or PC-like product with a WIFI 6 module is used. Due to the need for VSLAM computation and inference of deep learning-based image segmentation, a PC with a relatively powerful CPU (x86 or ARM) and GPU is needed. Since ORB-SLAM3 is officially supported on Ubuntu systems, Ubuntu 20.04 and Ubuntu 22.04 were used and tested in this research project.

For the vehicle, a differential robot is designed with these important parts: (1) motion components, including robot chassis, battery, motors, motor drivers, etc.; (2) VSLAM (relevant) hardware, including cameras, processors, WIFI modules, etc.

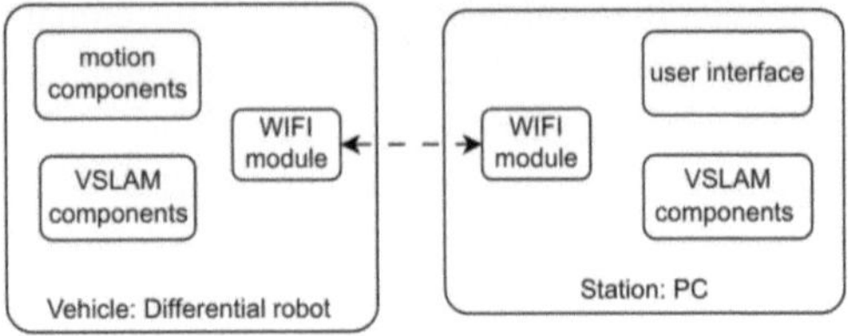

Fig. 1. System architecture

2.2 Hardware Design

Vehicle Motion Hardware. As mentioned in previous section, motion components consist of robot chassis, battery, motors, motor drivers, etc. They were selected and purchased from the market. Some of them (Fig. 2):

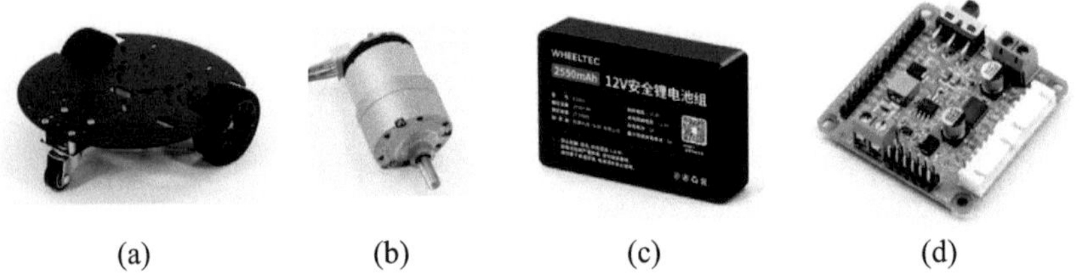

(a) (b) (c) (d)

Fig. 2. Some components of vehicle motion hardware: (a) Robot chassis (with motor), (b) 12 V encoded motors with gear, (c) 12 V Battery, (d) Motor driver

Vehicle VSLAM Hardware. Mainstream VSLAM hardware configuration, MPU system with binocular camera module, is adopted for this project, which utilizes the mature VSLAM software ecosystem and provides adequate performance. It is demonstrated as in Fig. 3:

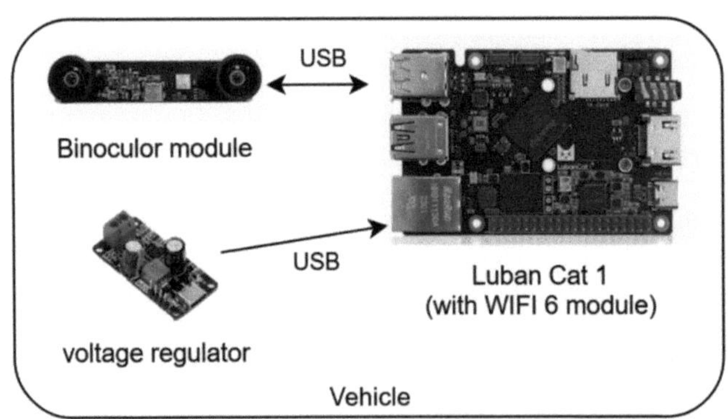

Fig. 3. Vehicle VSLAM Hardware

Vehicle Entire. As designed above, the entire vehicle was assembled as Fig. 4:

Fig. 4. The entire vehicle

2.3 Software Design

Based on the methodology described above, the software architecture is designed. The software architecture is illustrated in Fig. 5:

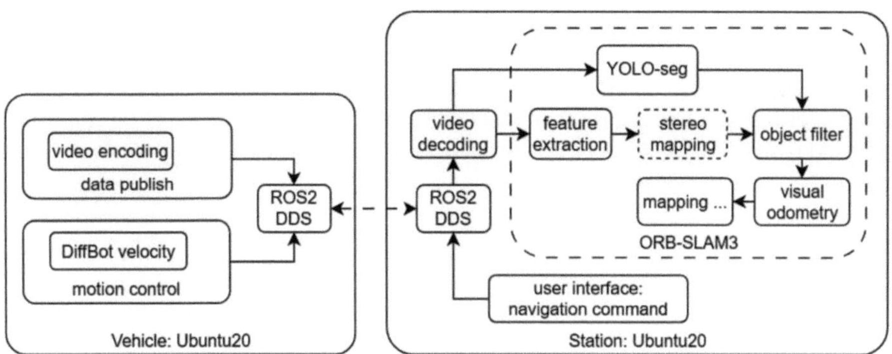

Fig. 5. Software Architecture

On the vehicle, Ubuntu 20 runs on the LubanCat 1 development board, where ROS 2 manages two concurrent threads: a motion control thread operating at 10 ms using a DiffBot model-based velocity controller, and a data publishing thread synchronized with the camera frame rate through a video encoding pipeline.

On the station, subscribed data are received via ROS 2 DDS and decoded before being processed in parallel for YOLO-based semantic segmentation and VSLAM feature extraction, which are synchronized during dynamic object filtering prior to further SLAM processing. In addition to VSLAM computation, the station also supports user interface functions such as navigation commands. In compliance with the GPLv3 license of ORB-SLAM3, the station software is released as open source.

2.4 Vehicle Motion Control

Motion is an important part of robotics. Since SLAM typically requires only velocity control, the control architecture is designed as shown in Fig. 6, taking into account the robot model and VSLAM stability.

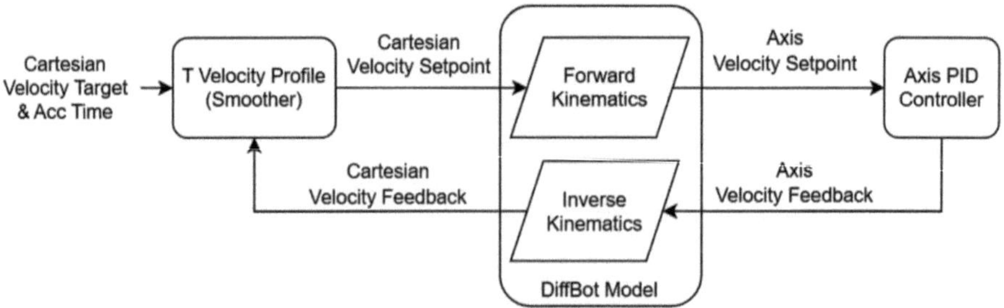

Fig. 6. Velocity control

An X-R acceleration-time-synchronized T (Trapezoidal) velocity profile is designed for smooth motion to achieve better VSLAM performance. Details are as follows.

The velocity controller receives the Cartesian velocity target from the ROS cmd_vel topic transmitted via the teleop_twist_keyboard tool, together with a user-defined Acc (acceleration) time and Cartesian velocity feedback, which are processed by a velocity profile generator. The resulting Cartesian velocity setpoint is converted into axis velocity commands through DiffBot forward kinematics and regulated by an axis PID controller, while Axis velocity feedback is transformed back into Cartesian velocity feedback. This control design achieves the required motion performance, as illustrated in Figs. 7 and 8.

2.5 Object Filter

This project evaluates the effectiveness of the deep learning-based dynamic object filtering component. The evaluation includes dataset-based quantitative analysis to assess the impact of object filtering on VSLAM tracking performance.

Dataset SLAM was performed on ORB-SLAM3 with the TUM RGB-D dataset (rgbd_dataset_freiburg3_walking_xyz), using people as the dynamic objects. The effect is demonstrated as in Fig. 9:

With the data produced by the testing mentioned above, trajectory tracking error was evaluated (Fig. 10):

Without object filtering, the RMSE and scale of the trajectory are 0.669102 m and 0.179680, respectively, while with object filtering, they are 0.015034 m and 0.978786, respectively.

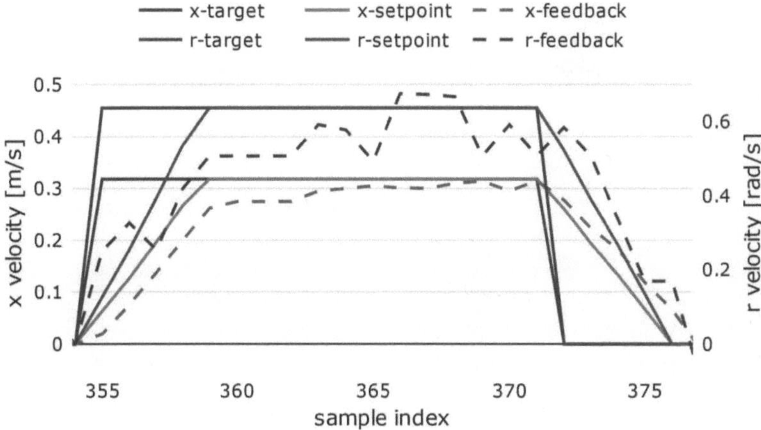

Fig. 7. Cartesian velocity control effect

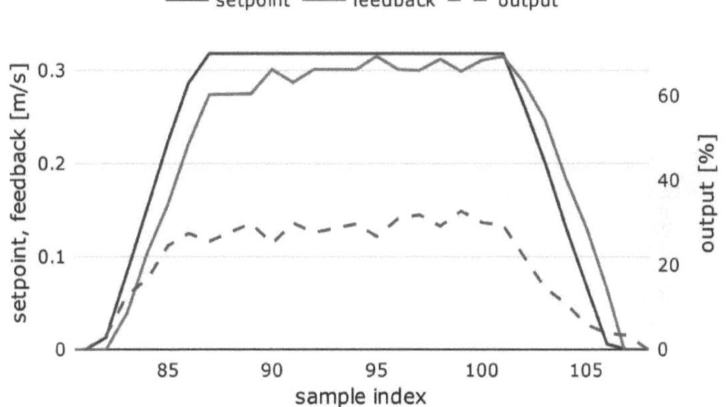

Fig. 8. Axis PID control effect (one axis)

The dataset evaluation demonstrates a significant reduction in trajectory error when object filtering is applied, validating the substantial effectiveness of deep learning integration for object filtering in dynamic environments.

2.6 Communication Evaluation

This project comprehensively evaluates the distributed system's communication performance through a series of progressive tests. The evaluation consists of four tests: measures basic network delay through round-trip testing; evaluates Sense-Move delay including codec time and network performance; validates real-time operation with integrated motion control; and examines full system performance with complete VSLAM processing. These tests collectively demonstrate the practical viability of the distributed architecture for real-time indoor mobile robotics applications. For all delay related evaluations, the data is presented such that the x-axis represents the sample index (ranging

(a) (b)

Fig. 9. Object filter effect of dataset SLAM: (a) Without object filter, (b) With object filter

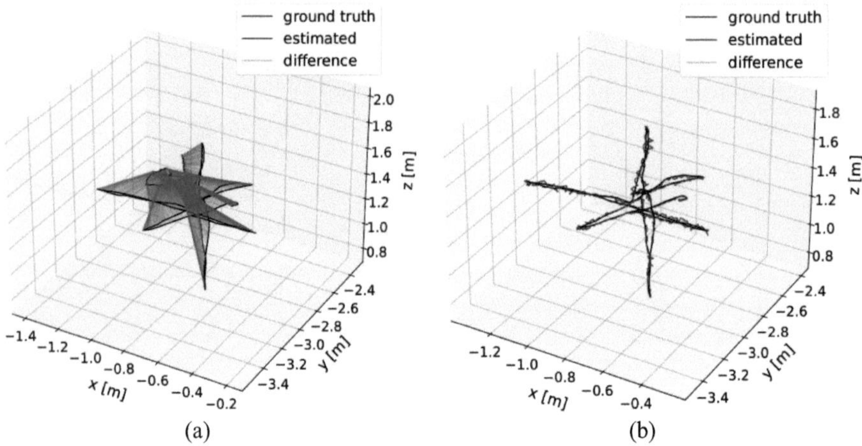

(a) (b)

Fig. 10. Dataset evaluation of object filter SLAM: (a) Without object filter, (b) With object filter

from 1 to 500 samples), while the y-axis indicates the time duration in milliseconds (ms). This format allows for the observation of latency stability across the entire testing duration

Network Delay Test. A typical network delay test is the round-trip test. The test method is as in Fig. 11:

Let p denote a timepoint, with t and r representing transmission and reception; the network delay is defined as:

$$d_n = p_r - p_t \tag{1}$$

Result with test condition—short topic message round-trip time, 500 samples, 15 s duration (30 ms/sample):

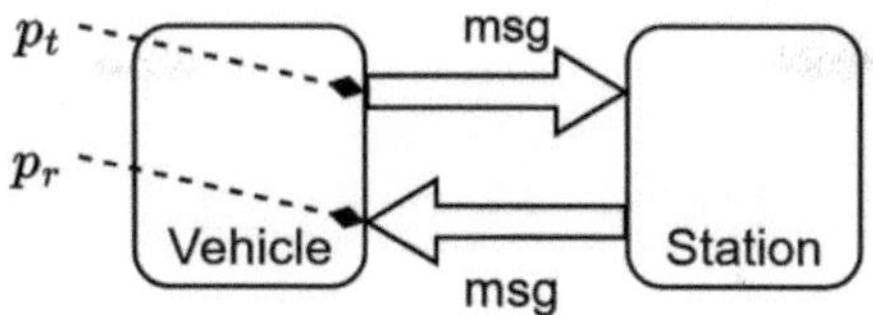

Fig. 11. Network delay (round-trip) test method

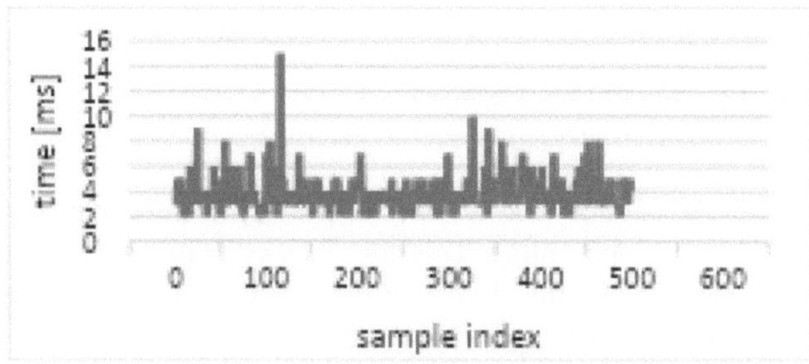

Mean	3.7
Mode	3.0
Minimum	2.0
Maximum	15.0
Range	13.0
Standard Deviation	1.3
Count	500

Fig. 12. Network delay test result

Figure 12 Network delay test result (X-axis: sample index; Y-axis: time in ms). The results show a mean round-trip delay of 3.7 ms over 500 samples, which is an expected good result for this network configuration. Most delays being below 10 ms are sufficient for motion control.

Sense-Move Delay Test. In this project, to evaluate the delay of motion control after sensing in VSLAM, Sense-Move delay is defined as in Fig. 13:

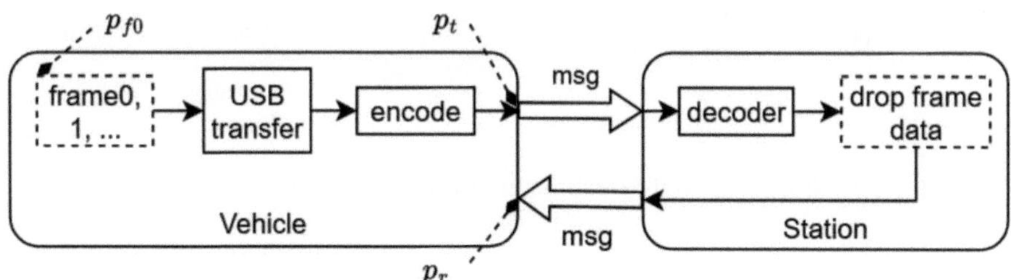

Fig. 13. Sense-Move delay test method

Let p denote a timepoint, where p_{f0} is the timepoint of the newest frame, and t_e and t_d represent encoding and decoding times; the system delay is defined as:

$$d_s = p_r - p_{f0} \tag{2}$$

Result with test condition—topic message round-trip time, 500 samples, 16.5 s duration (33 ms/sample) (Fig. 14):

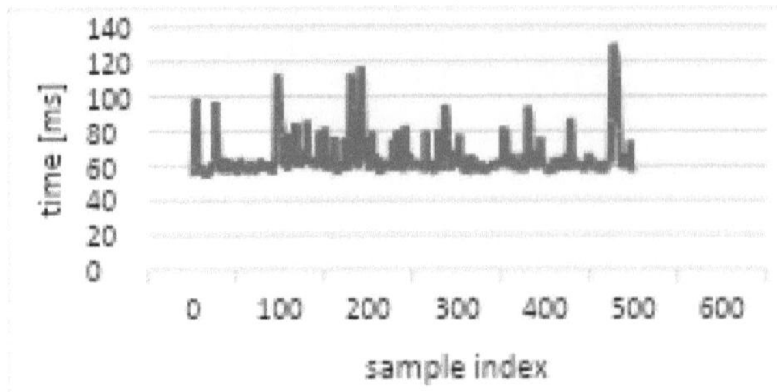

Mean	64.9
Mode	59.0
Minimum	54.0
Maximum	130.0
Range	76.0
Standard Deviation	12.7
Count	500

Fig. 14. Sense-Move delay test result

The mean delay, 64.9 ms, demonstrates that the distributed architecture maintains sufficient responsiveness for closed-loop control. This low latency confirms the system's suitability for real-time robotic control. Two key contributing factors—codec performance and network performance—are examined in detail below.

Codec Performance. The First Factor Is the Video Codec Time, Including Encoding and Decoding Time. The most Important One Is the Encoding Time with Hardware Acceleration, which Offers Key Support for the Low Latency and Low Throughput of Communication (See Next Section). The Result Is Shown in Fig. 15:

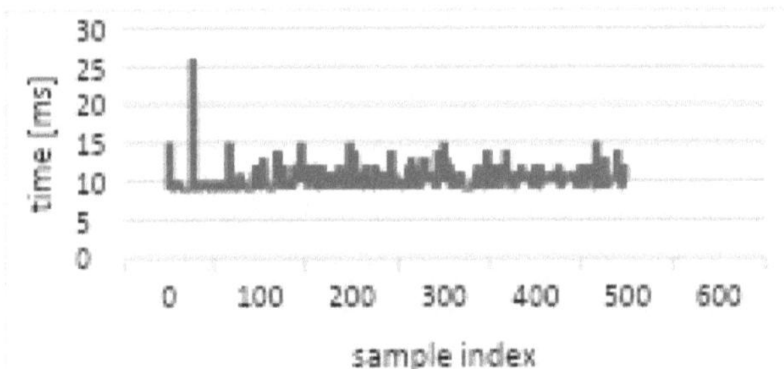

Mean	10.2
Mode	10.0
Minimum	9.0
Maximum	26.0
Range	17.0
Standard Deviation	1.4
Count	500

Fig. 15. Encoding time (hardware-accelerated) test result

Here, an extra encoding test with traditional software encoding (i.e., CPU encoding) was conducted to compare with the hardware-accelerated one (Fig. 16):

As the result, hardware-accelerated encoding time is significantly smaller and more stable than the CPU one.

Network Performance. The network performance result inside the Sense-Move delay test (Figs. 17 and 18):

With the advanced technologies, WIFI 6 and hardware-accelerated encoding, the network performance in this robot design reaches a good level.

Integrated Test with Motion. To validate the complete distributed system under realistic operating conditions, comprehensive integration testing was conducted with the

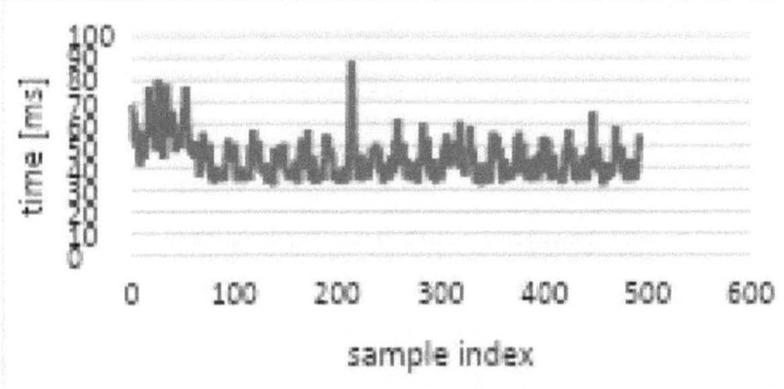

Mean	43.7
Mode	34.0
Minimum	32.0
Maximum	88.0
Range	56.0
Standard Deviation	9.1
Count	493

Fig. 16. Encoding time (CPU) test result

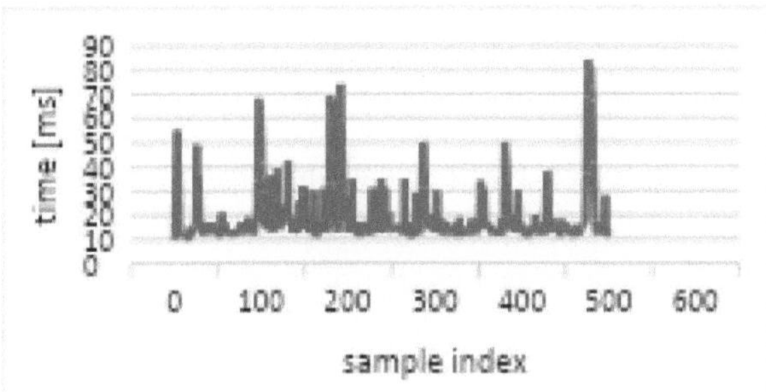

Mean	19.1
Mode	14.0
Minimum	10.0
Maximum	84.0
Range	74.0
Standard Deviation	12.4
Count	500

Fig. 17. Network delay (Sense-Move) test result

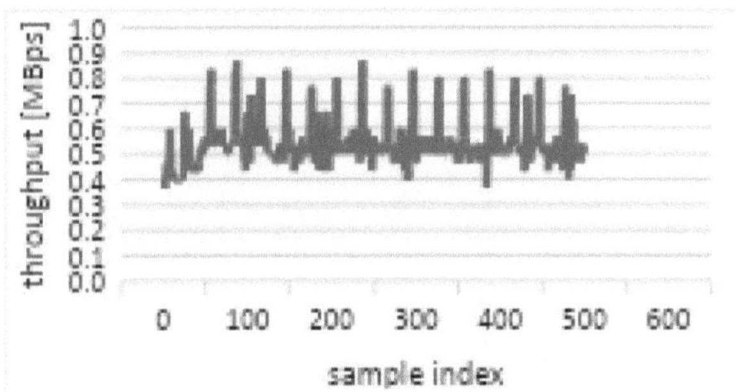

Mean	0.5
Mode	0.5
Minimum	0.4
Maximum	0.9
Range	0.5
Standard Deviation	0.1
Count	500

Fig. 18. Network throughput (Sense-Move) test result

vehicle performing continuous motion while executing the full communication pipeline. The tests employed a stereo camera configuration with frame sizes of 1600 × 800 pixels for the stereo pair (equivalent to 800 × 600 pixels for monocular mode), operating at 30 fps. During these tests, the vehicle navigated through indoor environments while maintaining continuous wireless communication with the station (Fig. 19).

The integrated test results demonstrated that the end-to-end delay remained consistent with the previously measured Sense-Move delay of 64.9 ms, confirming that the system maintains its low-latency performance. The user experience during teleoperation was smooth and responsive, with no perceptible lag between commands and robot

Fig. 19. Screenshot of integrated test with motion

actions. Most importantly, the communication latency and overall system responsiveness were sufficient to effectively support VSLAM-based motion control, enabling the robot to perform real-time localization and path planning while in motion.

Integrated Test with VSLAM. Beyond basic motion control, the system including the ORB-SLAM3 pipeline with integrated deep learning-based object filtering was evaluated under real-world operating conditions. These tests utilized frame sizes of 1280 × 480 pixels for stereo input (640 × 480 pixels for monocular mode) and employed a NVIDIA Jetson Orin Nano Super as the station (Fig. 20).

Fig. 20. Screenshot of integrated test with VSLAM

Under these conditions, the integrated system achieved frame rates of 10–15 fps, which represents an acceptable performance level for many indoor mobile robotics applications. While this frame rate is lower than the theoretical maximum of the camera sensors, it reflects the computational demands of simultaneously performing ORB-SLAM3 processing and YOLO-based semantic segmentation in real-time. Although the real-time performance of VSLAM (particularly with integrated deep learning) varies significantly across different hardware configurations, these test results still provide valuable reference data for VSLAM robot design, especially for embedded systems.

3 Conclusion

In conclusion, the distributed architecture successfully reduces computational burden on the mobile robot while maintaining real-time operation, demonstrating the viability of advanced wireless technologies and deep learning integration for enhanced visual SLAM performance in indoor environments. By addressing the research gaps in WIFI 6 exploitation for distributed VSLAM, deep learning-based dynamic object removal, and comprehensive communication performance evaluation, this work provides both theoretical insights and practical validation for continued development of distributed robotic architectures enhanced by semantic understanding.

Acknowledgements. This work was supported by the Fundamental Research Grant Scheme (FRGS), Ministry of Higher Education, Malaysia under project No. R.J130000.7851.5F586. The authors would also like to acknowledge Fakulti Teknologi dan Kejuruteraan Elektrik (FTKE), UTeM for their support throughout this research.

References

1. Huang, J., Junginger, S., Liu, H., Thurow, K.: Indoor positioning systems of mobile robots: a review. Robotics. **12**, 47 (2023)
2. Noreils, F.R.: Toward a robot architecture integrating cooperation between Mobile robots: application to indoor environment. Int. J. Robot. Res. **12**, 79–98 (1993). https://doi.org/10.1177/027836499301200106
3. Rubio, F., Valero, F., Llopis-Albert, C.: A review of mobile robots: concepts, methods, theoretical framework, and applications. Int. J. Adv. Robot. Syst. **16**, 1729881419839596 (2019). https://doi.org/10.1177/1729881419839596
4. Tzafestas, S.G.: Mobile robot control and navigation: a global overview. J. Intell. Robot. Syst. **91**, 35–58 (2018). https://doi.org/10.1007/s10846-018-0805-9
5. Gao, X., Zhang, T., Liu, Y., Yan, Q.: 14 Lectures on Visual SLAM. (2017)
6. Cadena, C., et al.: Past, present, and future of simultaneous localization and mapping: toward the robust-perception age. IEEE Trans. Robot. **32**, 1309–1332 (2016). https://doi.org/10.1109/TRO.2016.2624754
7. Liu, Y., Wang, S., Xie, Y., Xiong, T., Wu, M.: A review of sensing technologies for indoor autonomous mobile robots. Sensors. **24**, 1222 (2024)
8. Beghdadi, A., Mallem, M.: A comprehensive overview of dynamic visual SLAM and deep learning: concepts, methods and challenges. Mach. Vis. Appl. **33**, 54 (2022). https://doi.org/10.1007/s00138-022-01306-w
9. Chen, W., et al.: An overview on visual SLAM: from tradition to semantic. Remote Sens. **14**, 3010 (2022). https://doi.org/10.3390/rs14133010
10. Campos, C., Elvira, R., Rodríguez, J.J.G., Montiel, J.M.M., Tardós, J.D.: ORB-SLAM3: an accurate open-source library for visual, visual–inertial, and multimap SLAM. IEEE Trans. Robot. **37**, 1874–1890 (2021). https://doi.org/10.1109/TRO.2021.3075644

11. Sultan, J.M., Osmadi, I.A., Manap, Z.: Real-time Wi-fi network performance evaluation. Int. J. Inf. Commun. Technol. (IJ-ICT). **11**, 193 (2022). https://doi.org/10.11591/ijict.v11i3.pp193-205
12. Pandey, P., Parasuraman, R.: Empirical analysis of bi-directional Wi-fi network performance on Mobile robots in indoor environments. In: 2022 IEEE 95th Vehicular Technology Conference: (VTC2022-Spring), pp. 1–7 (2022). https://doi.org/10.1109/VTC2022-Spring54318.2022.9860438
13. Khorov, E., Kiryanov, A., Lyakhov, A., Bianchi, G.: A tutorial on IEEE 802.11 ax high efficiency WLANs. IEEE Commun Surv Tutor. **21**, 197–216 (2018)
14. Rochim, A.F., Harijadi, B., Purbanugraha, Y.P., Fuad, S., Nugroho, K.A.: Performance comparison of wireless protocol IEEE 802.11ax vs 802.11ac. In: 2020 International Conference on Smart Technology and Applications (ICoSTA), pp. 1–5 (2020). https://doi.org/10.1109/ICoSTA48221.2020.1570609404
15. Wei, Y.-H., Leng, Q., Han, S., Mok, A.K., Zhang, W., Tomizuka, M.: RT-WiFi: real-time high-speed communication protocol for wireless cyber-physical control applications. In: 2013 IEEE 34th Real-Time Systems Symposium, pp. 140–149. IEEE, Vancouver, BC, Canada (2013). https://doi.org/10.1109/RTSS.2013.22
16. Macario Barros, A., Michel, M., Moline, Y., Corre, G., Carrel, F.: A comprehensive survey of visual SLAM algorithms. Robotics. **11**, 24 (2022). https://doi.org/10.3390/robotics11010024
17. Rublee, E., Rabaud, V., Konolige, K., Bradski, G.: ORB: an efficient alternative to SIFT or SURF. In: 2011 International Conference on Computer Vision, pp. 2564–2571 (2011). https://doi.org/10.1109/ICCV.2011.6126544
18. Rosten, E., Drummond, T.: Machine learning for high-speed corner detection. In: Leonardis, A., Bischof, H., Pinz, A. (eds.) Computer Vision–ECCV 2006, pp. 430–443. Springer, Berlin, Heidelberg (2006). https://doi.org/10.1007/11744023_34
19. Calonder, M., Lepetit, V., Strecha, C., Fua, P.: BRIEF: binary robust independent elementary features. In: Daniilidis, K., Maragos, P., Paragios, N. (eds.) Computer Vision – ECCV 2010, pp. 778–792. Springer, Berlin, Heidelberg (2010). https://doi.org/10.1007/978-3-642-15561-1_56
20. Mur-Artal, R., Montiel, J.M.M., Tardós, J.D.: ORB-SLAM: A Versatile and Accurate Monocular SLAM System. IEEE Trans. Robot. **31**, 1147–1163 (2015). https://doi.org/10.1109/TRO.2015.2463671
21. Mur-Artal, R., Tardós, J.D.: ORB-SLAM2: an open-source SLAM system for monocular, stereo, and RGB-D cameras. IEEE Trans. Robot. **33**, 1255–1262 (2017). https://doi.org/10.1109/TRO.2017.2705103
22. Engel, J., Schöps, T., Cremers, D.: LSD-SLAM: Large-Scale Direct Monocular SLAM. In: Fleet, D., Pajdla, T., Schiele, B., Tuytelaars, T. (eds.) Computer Vision – ECCV 2014, pp. 834–849. Springer International Publishing, Cham (2014). https://doi.org/10.1007/978-3-319-10605-2_54
23. Engel, J., Koltun, V., Cremers, D.: Direct sparse Odometry. IEEE Trans. Pattern Anal. Mach. Intell. **40**, 611–625 (2018). https://doi.org/10.1109/TPAMI.2017.2658577
24. Rosinol, A., Abate, M., Chang, Y., Carlone, L.: Kimera: an open-source library for real-time metric-semantic localization and mapping. In: 2020 IEEE International Conference on Robotics and Automation (ICRA), pp. 1689–1696 (2020). https://doi.org/10.1109/ICRA40945.2020.9196885
25. Abaspur Kazerouni, I., Fitzgerald, L., Dooly, G., Toal, D.: A survey of state-of-the-art on visual SLAM. Expert Syst. Appl. **205**, 117734 (2022). https://doi.org/10.1016/j.eswa.2022.117734

A Comparative Study of Single-Task and Multi-task Learning for Sentiment and Emotion Classification on Reddit Text

Kai Xin Edwina Hon, Chuk Fong Ho(✉), and Chi Wee Tan

Tunku Abdul Rahman University of Management and Technology, 53300 Kuala Lumpur, Malaysia

cfho@tarc.edu.my

Abstract. Social media crises, such as the Samsung Galaxy Note 7 recall, unfold rapidly and generate intense public reactions. To respond effectively, organizations require tools that go beyond sentiment detection to capture underlying emotional patterns. However, existing sentiment classification approaches often prioritize accuracy while struggling to adapt to domain-specific crisis contexts. This study proposed a Multitask Learning (MTL) framework for jointly classifying sentiment and emotion in Reddit-based crisis communication. Transformer-based models, including BERTweet and DistilRoBERTa, were fine-tuned under both Single-Task Learning (STL) and MTL configurations using benchmark datasets and evaluated on a curated Reddit corpus from the Note 7 crisis. To ensure robustness, performance was assessed using Macro-F1 as the primary evaluation metric, with standard deviation reported across five random seeds to capture performance robustness and stability. The results show that STL models consistently outperform MTL on general-domain benchmarks for emotion classification, whereas MTL offers modest improvements for sentiment classification in crisis-specific data. Emotion classification remains particularly challenging, with most models consistently achieving lower Macro-F1 scores. Overall, the findings highlight both the potential and limitations of MTL for crisis analysis and provide a foundation for developing more reliable and transparent artificial intelligence tools to support effective crisis communication.

Keywords: Crisis communication · Sentiment classification · Emotion classification · Multitask learning · Explainable artificial intelligence · Reddit

1 Introduction

Social media has become an integral part of daily life, connecting people to information, communities, and emotions, while enabling real-time interaction with global events and local environments [1]. As of 2024, social media users reached 5.17 billion [2] and this number continues to grow. One area significantly impacted by this shift is public relations (PR), which manages the flow of information between organizations, the public, and stakeholders [3]. For example, 78% of Twitter users [4] who complained about a brand

N. A. Mohd Subha et al. (Eds.): SMS 2026, CCIS 2956, pp. 309–321, 2026.
https://doi.org/10.1007/978-981-92-0306-2_22

in 2019 expected a response within an hour, and many relied on digital platforms for timely updates during the COVID-19 pandemic [5, 6]. Such expectations have increased the pressure on brands to respond quickly and transparently to protect their reputation.

A PR crisis often involves a sudden rise in negative sentiment expressed on social media [7], which can heavily influence consumer trust and purchase intentions [8]. When negative publicity spreads rapidly online, an organization's image might be damaged within hours. A well-known example is the 2015 Volkswagen emissions scandal, where violations of the Clean Air Act sparked global outrage, causing a 30% stock decline [9], billions in penalties, and long-term brand damage. Incidents like these highlight the need for fast, strategic, and transparent crisis responses. Therefore, understanding public sentiment and emotional reactions in real time has become essential for effective crisis management [10].

Academic and industry interest in Sentiment Classification (SC) has grown in recent years, as organizations increasingly rely on these tools to better understand public opinion and emotional tone [11]. SC, a subfield of Natural Language Processing (NLP), uses Machine Learning (ML) to categorize text as *positive*, *negative*, or *neutral* [12]. However, prior studies suggest that analysing sentiment before a crisis peak is crucial for early intervention [13]. While SC techniques can measure public opinion [10], they often overlook the deeper role of emotions in shaping consumer behaviour. Without understanding emotional patterns, organizations may misinterpret public reactions and prolong reputational damage. Although transformer-based NLP models demonstrated strong performance in many language tasks [14, 15], their use in Multi-Task Learning (MTL) for jointly predicting sentiment and emotion during PR crises remains under-explored [16]. Without well-optimized models, organizations may struggle to assess emotional impact quickly enough to respond effectively.

This study addresses these gaps by evaluating transformer-based MTL models for sentiment and emotion classification in PR crisis scenarios. Specifically, it pursues two objectives: (1) to evaluate the performance of two transformer-based models, which are BERTweet (BTW) and DistilRoBERTa-base (DRB), in MTL settings for PR crisis analysis; and (2) to examine the effectiveness and limitations of domain transfer when applying models trained on general-domain benchmarks such as SST-2 and GoEmotions to a crisis-focused dataset.

2 Related Work

2.1 Crisis Management for Company Reputation

Corporate reputation is a crucial asset directly influences business performance and hence effective reputation management is essential for organizations [17]. In the digital era, social media platforms have become central channels for brand engagement and reputation management [18], offering valuable insights into shifting consumer behaviour. However, these platforms can also become breeding grounds for crises, where negative posts and customer complaints may escalate quickly, highlighting the need for proactive monitoring and timely intervention [7].

The real-time nature of social media has reshaped corporate communication by requiring organizations to detect signals of potential crises early and respond swiftly

[19]. This highlights the importance of transparency and responsiveness in crisis communication [17]. Since the way an organization handles a crisis affects public perception, crisis communication is closely tied to reputation management [20]. As a result, there is a growing need for decision-support tools that help crisis managers mitigate potential threats [21], highlighting the pivotal role of social media analytics in early crisis detection and strategic response planning.

2.2 Sentiment Classification for Social Media Applications

SC has been widely applied across domains to capture public perceptions in PR crisis contexts. One study [22] used SC to track emotional shifts among social media users and found that online sentiment trends aligned closely with traditional survey results. Another study [23] compared hotel reviews across TripAdvisor, Expedia, and Yelp, discovering notable discrepancies in how businesses were portrayed across platforms. Differences in sentiment, writing styles, ratings, and user interactions were shown to significantly influence consumer perceptions. SC has also been used to analyse customer opinions in reviews and social media interactions [24], and to monitor consumer attitudes to support brand reputation management [25], demonstrating its broad value in understanding public attitudes across industries.

Existing SC approaches can be broadly categorized into lexicon-based and ML-based methods, with the latter widely recognised for their effectiveness [26]. However, a recent study [27] analysing Reddit data with several SC tools including VADER, TextBlob, and the BERT found discrepancies between lexicon-based methods and ML models. These inconsistencies highlight the complexity of identifying sentiment and emotion, particularly during crises, and emphasize the need for more advanced approaches, such as MTL and XAI. These methods can better capture and interpret nuanced emotional reactions that shape public opinion in high-stakes PR scenarios.

2.3 MTL Models in Sentiment and Emotion Analysis

MTL is an ML framework where several related tasks are learned jointly within a single model [28]. By leveraging shared information and task interdependencies, MTL aims to improve generalization across tasks [29]. In sentiment and emotion analysis, MTL has shown clear benefits, as jointly optimizing both tasks allows models to learn richer representations, such as aspect extraction and sentiment classification leading to improved accuracy and F1 scores compared to Single-Task Learning (STL) [30]. Incorporating multilayer feature extraction can further enhance performance [28].

This capability is particularly valuable for PR crisis management, where crisis-related social media content often contains mixed or rapidly shifting sentiments and emotions that are better captured through simultaneous analysis [31]. For instance, one study [32] proposed an MTL model for multimodal sentiment and emotion classification and demonstrated its effectiveness in processing diverse inputs and modelling both tasks. Such multimodal capability is useful in PR crisis, where discussions often involve text, images and videos [31].

With the introduction of attention-based transformer models [33], attention mechanisms which enable models to selectively focus on different parts of input data during training and inference [34] have been incorporated into MTL frameworks to enhance both interpretability and task performance [33]. In PR crisis contexts, attention mechanisms can help identify specific aspects or keywords that drive sentiment and emotion, offering a more granular understanding of public perception.

Despite these advances, attention-based MTL models have not been widely applied to interpret sentiment and emotion in evolving PR crisis events. This study extends MTL for sentiment and emotion classification in crisis-related social media content by incorporating attention mechanisms to identify influential keywords and improve prediction accuracy. Supporting this direction, recent studies show that transformer-based models achieve strong performance for emotion classification. In particular, BERTweet demonstrated superior performance over strong baselines such as RoBERTa-base and XLM-R-base [35], whereas fine-tuned DistilRoBERTa can achieve strong accuracy on benchmark emotion datasets at lower computational cost [36]. Hence, in this study, BERTweet and DistilRoBERTa are adopted as backbone encoders for multi-task sentiment and emotion classification in PR crisis scenarios.

3 Research Methodology

3.1 Data Collection

This study uses Reddit data from the Samsung Galaxy Note 7 crisis–referred to as CRISIS–to evaluate the proposed MTL framework. Reddit offers rich, user-generated discussions that capture spontaneous public reactions during real-world events, making it well suited for crisis-related analysis. Data was collected using the Reddit API from the r/samsung subreddit, covering 1 August to 31 December 2016, which aligns with the timeline of the Note 7 recall. To ensure relevance, targeted search queries were designed to retrieve posts and comments specifically related to the crisis. Following the approach in [37], only highly crisis-specific terms were used to improve dataset quality and reduce noise. Non-English entries were removed to maintain linguistic consistency and align with the study's focus on English-language PR crisis communication.

Both posts and comments were included, with each treated as an independent unit of analysis. Data was stored in JSON format to preserve metadata, hierarchical structures, and conversational context. This allows the analysis to incorporate not only individual posts but also user responses, which often provide crucial insights into public reactions during crises. To further refine the dataset, a TextBlob subjectivity score (0–1) was computed for every entry, and only entries with subjectivity $\geq$0.4 were retained, ensuring the inclusion of sufficient opinionated and emotionally expressive content essential for sentiment and emotion classification. After filtering, the dataset consisted of 95 units. As shown in Fig. 1, the resulting dataset after filtering is not balanced across classes, be it emotions or sentiments. However, no balancing techniques will be applied in this study to preserve the natural class imbalance typically observed in real-world social media discussions.

3.2 Data Annotation

All retained entries were then manually annotated for both sentiment and emotion. Sentiment was labelled using a three-class scheme: *positive*, *negative*, *neutral*, while

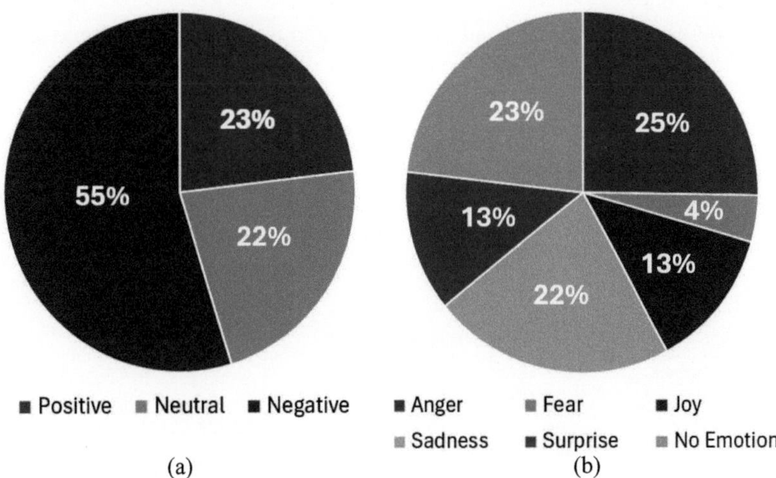

Fig. 1. Distribution of (a) sentiment classes and (b) emotion classes in the dataset.

emotion labels were based on an adapted version of Shaver's six basic categories [38]: *anger*, *fear*, *joy*, *sadness*, *surprise*, along with a *no emotion* class to capture factual or neutral content. Recent studies [39, 40] have demonstrated the robustness of Shaver's taxonomy for short-form text, supporting its use for Reddit data.

Manual annotation was selected due to the nuanced and informal nature of Reddit language. Sarcasm, idiomatic expressions, abbreviations, and fragmented conversational threads often undermine the accuracy of automated labeling methods [41], making human judgment more reliable in this context. For example, a post complaining about delayed refunds might be annotated as *anger*, whereas a message showing relief after receiving compensation may be labeled as *joy*.

3.3 Data Cleaning

Real-world social media data is often noisy, incomplete, and inconsistent. Reddit posts may include redundant metadata, informal expressions, or irrelevant content that can obscure the linguistic cues required for accurate sentiment and emotion classification. To address these issues, several text-cleaning procedures were applied, including text normalization, emoji conversion, stopword removal, and lemmatization. Non-essential fields were also removed to streamline the dataset and retain only information relevant to downstream classification tasks.

3.4 MTL Framework

The MTL framework developed in this study is designed to jointly classify sentiment and emotion in social media content by leveraging shared linguistic features across both tasks. The model uses a shared transformer encoder based on BTW and DRB to generate rich contextual embeddings from the text. The architecture then diverges into two task-specific dense layers in which the sentiment head is configured as a three-class softmax classifier, whereas the emotion head is implemented as a six-class softmax classifier. The overall architecture of the MTL framework is illustrated in Fig. 2.

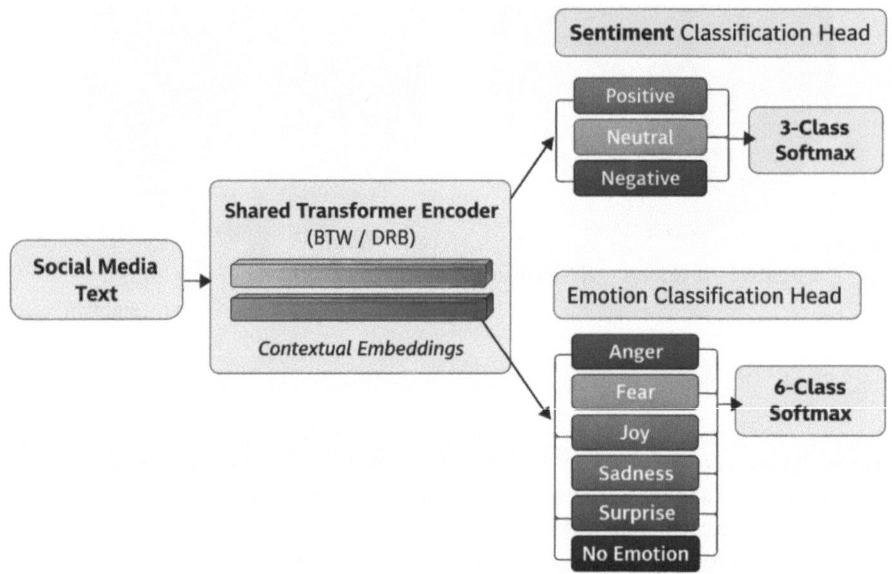

Fig. 2. MTL framework: Sentiment and emotion analysis.

All models were initialised with pretrained weights and subsequently fine-tuned on two public datasets, collectively referred to as GENERAL. For sentiment classification, SST-2, which is a widely recognised as a benchmark for binary sentiment classification, was used. For emotion classification, GoEmotions, which is a large dataset of 58,000 Reddit comments originally annotated with 27 fine-grained emotion categories, was used. These labels were mapped to six broader emotion categories to align with the emotion head of the MTL framework [42, 43]. Fine-tuning on both datasets enables models to adapt pretrained linguistic representations to task-specific and domain-relevant objectives.

The models were fine-tuned using a weighted multitask loss function, where errors from both sentiment and emotion classification contribute to overall learning:

$$L_{total} = \alpha \cdot L_{sentiment} + (1 - \alpha) \cdot L_{emotion} \tag{1}$$

Here, $L_{\text{sentiment}}$ and L_{emotion} denote the categorical cross-entropy losses for the respective tasks. For single-task experiments, which served as a standard baseline [44], α is fixed at 0.5. For multitask experiments, α is treated as a tuneable hyperparameter within [0.3, 0.7], allowing the model to balance task influence according to validation performance [45]. This design prevents one task from dominating training, encourages parameter sharing, and improves computational efficiency.

3.5 Model Fine Tuning

Two experimental configurations were evaluated: STL and MTL. In STL, BTW was fine-tuned solely for sentiment classification using three labels, while DRB was fine-tuned solely for emotion classification using six labels. In MTL, both tasks were optimized jointly within a unified framework. A shared transformer encoder produced contextualized embeddings, which were then fed into two task-specific output heads, allowing a direct comparison between STL and MTL performance using identical backbone architecture.

Model hyperparameters were selected using Optuna's RandomSampler, which explored configurations such as task loss weighting and dropout rates across multiple trials. Within each trial, model weights were updated using the AdamW optimizer to improve generalization. Overfitting was further controlled by applying dropout to both hidden and classifier layers, with rates tuned between 0.1 and 0.3. Stratified sampling was also used during data preparation to maintain consistent class distributions across training and validation sets.

Evaluation followed a consistent cross-validation protocol. Using a fixed random seed 42, pretrained models were first evaluated without fine-tuning using five-fold data partitioning, where each fold served exclusively as a test set and performance was averaged across folds. This procedure was conducted on both GENERAL and CRISIS datasets to establish baseline pretrained performance. Using the same data splits, the models were then fine-tuned under a five-fold cross-validation scheme, with training folds used for adaptation and the held-out fold for testing. Final results were obtained by averaging test performance across folds, enabling robust and comparable estimates.

3.6 Evaluation

Given the substantial class imbalance in the CRISIS dataset, Macro-F1 was adopted as the primary evaluation metric, as it better reflects performance on minority classes than accuracy or Micro-F1 [46]. Since CRISIS is limited in size, to reduce its sensitivity to random initialization, each configuration was trained using five different random seeds selected randomly, including 42, 123, 456, 789, and 999, and results are reported as mean Macro-F1 and its corresponding standard deviation.

4 Results and Discussion

Figures 3 and 4 summarize the performance of all models on sentiment and emotion classification tasks.

Overall, most models achieved stronger results on GENERAL, while their performance degraded substantially on CRISIS, highlighting the difficulty of domain adaptation. This decline can be attributed to mismatched domain signals between the datasets used for fine-tuning and evaluation. For example, SST-2 consists of relatively formal and clean text [47], whereas Reddit comments are typically informal and noisy [48]. Although GoEmotions is also sourced from Reddit and shares similar linguistic characteristics, it spans a relatively wider range of topics [49]. In contrast, the CRISIS dataset

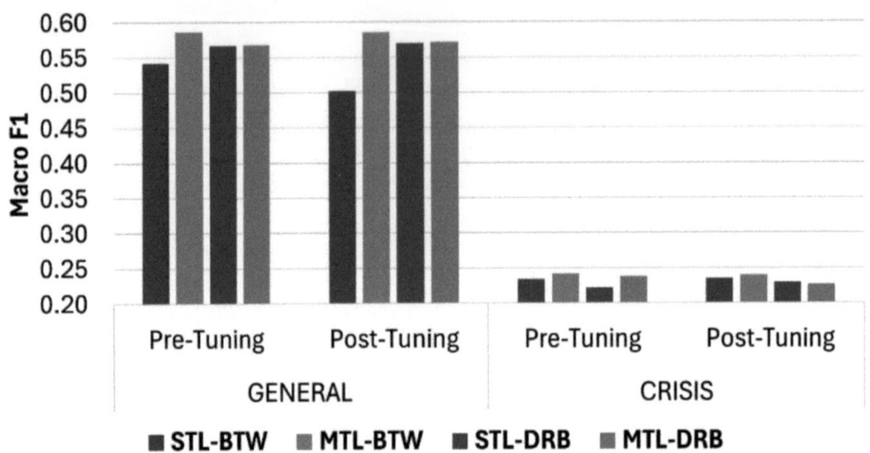

Fig. 3. Mean Macro-F1 scores for sentiment classification across the GENERAL and CRISIS datasets, before and after fine-tuning.

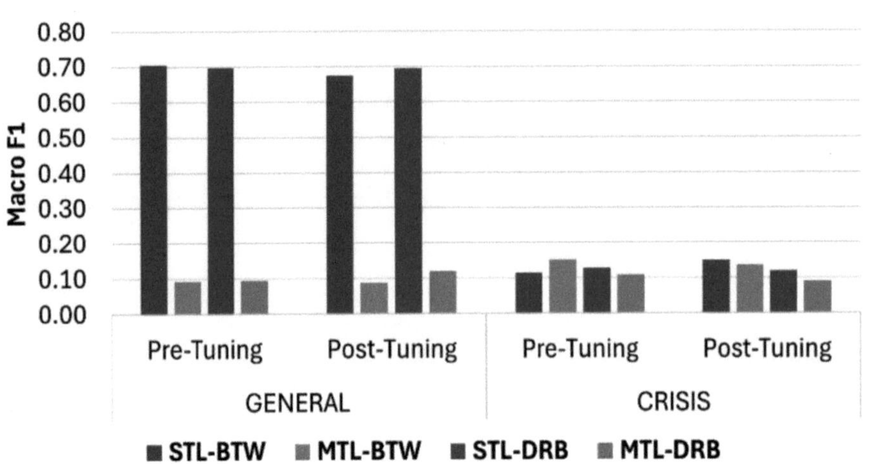

Fig. 4. Mean Macro-F1 scores for emotion classification across the GENERAL and CRISIS-datasets, before and after fine-tuning.

contains emotion expressions that are narrowly tied to a specific brand and crisis context. This topical and contextual mismatch likely explains why GoEmotions does not transfer effectively to the crisis domain.

A comparison between STL and MTL reveals that MTL models consistently underperformed in most settings. In particular, extending STL-based BTW and DRB to their MTL counterparts led to the largest performance drops on GENERAL test sets, with Macro F1 decreases ranging from 0.5745 to 0.6117. This behaviour suggests the presence of negative transfer, where learning one task impairs performance on another [50]. One possible explanation is that finer emotion categories can create confusion or ambiguity in labeling [51], making emotion classification with seven labels inherently more challenging than the three-class sentiment classification. For instance, the emotion *surprise* can convey either *positive* or *negative* sentiment depending on context [52], leading to sentiment-ambiguous emotion labels and conflicting optimization signals, ultimately harming emotion classification performance. Moreover, the common practice of aggregating GoEmotions into broader emotion categories may further intensify this ambiguity. In contrast, sentiment labels typically have a fixed polarity, resulting in clearer decision boundaries. This difference likely explains why MTL models consistently outperformed STL models on sentiment classification.

Fine-tuning aims to adapt pretrained models to CRISIS to improve task-specific performance and generalization. However, its effectiveness was not consistently observed in this study. While fine-tuning provided marginal gains for sentiment classification, with Macro-F1 improvements of only 0.0009 to 0.0033, it had a more pronounced negative impact on emotion classification. In most cases, fine-tuning reduced emotion Macro-F1 by 0.0038 to 0.0316, and MTL models in particular often deteriorated after fine-tuning. The greater complexity and label granularity of emotion classification likely contributes to this outcome, leading to negative task interference [53], where the model prioritizes the easier sentiment task at the expense of learning more nuanced emotional patterns [54].

In summary, when evaluated on GENERAL test sets, STL-DRB appears to be the most robust model, consistently demonstrating strong and balanced performance across both tasks, despite not always being the top performer. Conversely, MTL-BTW achieved the best results on CRISIS test sets for both sentiment and emotion classification, reflecting its suitability for short, informal social media text. This outcome aligns with the pretraining characteristics of the models: BTW is trained on a large corpus of Twitter data that closely resembles noisy social media language [55], whereas DRB, distilled from RoBERTa, is pretrained primarily on cleaner, more formal text [56].

Overall, Tables 1 and 2 show that several models exhibit low standard deviations, indicating stable predictions, particularly for sentiment classification on general-domain datasets. In contrast, emotion classification on Reddit-Crisis data show higher variability, with most models exceeding 0.01. This suggests that predictive stability varies by task, dataset, and model, highlighting the importance of fine-tuning and dataset-specific evaluation. More importantly, the achievement of MTL-BTW demonstrates the potential of MTL to leverage shared linguistic features across related tasks, enabling more efficient and context-aware understanding of sentiment and emotion in complex, real-world social media data.

Table 1. Standard deviation for sentiment classification across GENERAL and CRISIS datasets, before and after fine-tuning.

Models	GENERAL		CRISIS	
	Pre-Trained	Fine-Tuned	Pre-Trained	Fine-Tuned
STL-BTW	0.0659	0.0668	0.0045	0.0055
MTL-BTW	0.0011	0.0021	0.0150	0.0230
STL-DRB	0.0023	0.0045	0.0079	0.0117

(continued)

Table 1. (*continued*)

Models	GENERAL		CRISIS	
	Pre-Trained	Fine-Tuned	Pre-Trained	Fine-Tuned
MTL-DRB	0.0028	0.0041	0.0098	0.0159

Table 2. Standard deviation for emotion classification across GENERAL and CRISIS datasets, before and after fine-tuning.

Models	GENERAL		CRISIS	
	Pre-Trained	Fine-Tuned	Pre-Trained	Fine-Tuned
STL-BTW	0.0054	0.0537	0.0221	0.0176
MTL-BTW	0.0019	0.0018	0.0203	0.0190
STL-DRB	0.0051	0.0066	0.0186	0.0364
MTL-DRB	0.0027	0.0051	0.0211	0.0222

5 Conclusion

This study introduced a MTL framework for jointly classifying sentiment and emotion in Reddit-based crisis communication, using BTW and DRB under both STL and MTL settings. To ensure robustness, results averaged over five random seeds were prioritised, providing a more conservative and reproducible basis for analysis. The findings show that STL models achieved the strongest performance for sentiment classification on GENERAL, while MTL offered only limited gains for sentiment at the expense of emotion classification. Emotion recognition remained consistently challenging, reflecting the difficulty of capturing nuanced emotional signals in crisis discourse, particularly in small, noisy, and domain-specific CRISIS datasets. Overall, the results indicate that although MTL has potential to model multiple aspects of public reactions simultaneously, careful consideration of task complexity and dataset characteristics is essential for practical deployment in real-time crisis monitoring and decision-making.

Future work should more explicitly discuss practical implications for real-world PR crisis monitoring and explore how the proposed MTL framework could be adapted and evaluated in actual deployment scenarios to support timely decision-making. Future work should also expand datasets across platforms, incorporate multimodal information such as images and videos, and integrate explainability methods to improve usability and reduce the "black-box" nature of the models. Developing a user-facing prototype could further translate these findings into practical tools for public relations teams. In summary, this study shows that MTL frameworks hold promise for crisis communication but require effective domain adaptation and robust interpretability to support trustworthy AI systems during real-world crises.

Acknowledgments. The authors would like to sincerely extend their deepest appreciation to Tunku Abdul Rahman University of Management and Technology (TAR UMT), Malaysia, for their generous financial support and the provision of essential resources, facilities, and technical assistance that have played a pivotal role in enabling the smooth execution of this study.

References

1. Naja, I., Valdez, M., Antonini, A.: Social media and cities: a case-study based analysis of how digital life influences the urban physical environment. In: HT '24: Proceedings of the 35th ACM Conference on Hypertext and Social Media, pp. 239–245 (2024)
2. Statista: (2025). Number of worldwide social network users
3. Fatema, S., Li, Y., Dong, F.: Impact of social media on politician/citizens relationship. In: ICEME '20: Proceedings of the 2020 11th International Conference on E-Business, Management and Economics, pp. 109–114 (2020)
4. Standberry, S.: 32 social media marketing statistics that will blow your mind. LYFE Marketing Blog. (2025)
5. Sharma, K., Seo, S., Meng, C., Rambhatla, S., & Liu, Y.: COVID-19 on Social Media: Analyzing Misinformation in Twitter Conversations. (2020). https://arxiv.org/abs/2003.12309.
6. World Health Organization: Coronavirus disease (COVID-19) pandemic, WHO
7. Lambret, C.V., Barki, E.: Social media crisis management: aligning corporate response strategies with stakeholders' emotions online. J. Contingencies Crisis Manage. **26**(8) (2017)
8. Baek, H., Ahn, J., Oh, S.: Impact of tweets on box office revenue: focusing on when tweets are written. ETRI J. **36**(4), 581–590 (2014)
9. Thompson, M.: Volkswagen stock has plunged 50%. Will it ever recover? CNNMoney. (2015)
10. Che, S., Nan, D., Kamphuis, P., Zhang, S., Kim, J.H.: Examining crisis communication using semantic network and sentiment analysis: a case study on NetEase games. Front. Psychol. **13**(823415) (2022)
11. Nohh, N.H.M., Zainuddin, N.M.M., Anuar, S., Azmi, N.F.M.: Sentiment analysis towards hotel reviews. Open Int. J. Inform. (OIJI). **7**(2) (2019)
12. Aftab, F., et al.: A comprehensive survey on sentiment analysis techniques. Int. J. Technol. **14**(6), 1288–1298 (2024)
13. Watimin, N. H., Zanuddin, H., Rahamad, M. S., & Yadegaridehkordi, E.: Content framing role on public sentiment formation for pre-crisis detection on sensitive issue via sentiment analysis and content analysis. (2023)
14. Gowda, K.P., et al.: Transformers in sentiment analysis: a paradigm shift in deep learning research. J. Inf. Syst. Eng. Manage. **10**(5) (2025)
15. Win Myint, P.Y., Lo, S.L., Zhang, Y.: Unveiling the dynamics of crisis events: sentiment and emotion analysis via multi-task learning with attention mechanism and subject-based intent prediction. Inf. Process. Manag. **61**(4) (2024)
16. De Goma, J.C., Mercado, R.M.C., Saporna, T.J.B., Vallo, S.P.D.: Sentiment analysis of tweets towards vaccines via bidirectional encoder representations from transformers (BERT). In: ICCAI '24: Proceedings of the 2024 10th International Conference on Computing and Artificial Intelligence, pp. 419–424 (2024)
17. Armutlu, I.I.: Corporate reputation and crisis Management in Social Media: the Patiswiss case study. Int. J. Res. Innovation Soc. Sci. **8**(10), 682–692 (2024)
18. Bell, C., Olukemi, A., & Broklyn, P. (2024). Social Media Sentiment Analysis for Brand Reputation Management. ResearchGate

19. James, E.H., Wooten, L.P.: Leadership as (un)usual: how to display competence in times of crisis. Organ. Dyn. **34**(2), 141–152 (2005)
20. Gasana, K.: Crisis communication and reputation Management in the age of fake news. J. Public Relat. **3**(1), 28–39 (2024)
21. Stieglitz, S., Mirbabaie, M., Fromm, J., Melzer, S.: The adoption of social media analytics for crisis management - challenges and opportunities. Res. Pap. **4** (2018)
22. O'Connor, B., Balasubramanyan, R., Routledge, B., & Smith, N.: From tweets to polls: linking text sentiment to public opinion time series. Proceedings of the International AAAI Conference on Web and Social Media, 4(1), 122–129 (2010)
23. Xiang, Z., Du, Q., Ma, Y., Fan, W.: A comparative analysis of major online review platforms: implications for social media analytics in hospitality and tourism. Tour. Manag. **58**, 51–65 (2017)
24. Manjappa, R. S., & Kumar, A.: Twitter Sentiment Analysis. SSRN Electron. J. (2020)
25. Barunaha, A., Prakash, M.R., Naresh, R.: Real-time sentiment analysis of social media content for brand improvement and topic tracking. In: Advances in Computer Science Research, Proceedings of the 6th International Conference on Intelligent Computing (ICIC-6 2023) (2023)
26. Medhat, W., Hassan, A., Korashy, H.: Sentiment analysis algorithms and applications: a survey. Ain Shams Eng. J. **5**(4) (2014)
27. Pham, K., Kathala, K.C.R., Palakurthi, S.: Reddit sentiment analysis on the impact of AI using VADER, TextBlob, and BERT. Procedia Comput. Sci. **258**, 886–892 (2025) ScienceDirect
28. Cai, Y., Li, X., Zhang, Y., Li, J., Zhu, F., & Rao, L.: Multimodal sentiment analysis based on multi-layer feature fusion and multi-task learning. (2025)
29. Akhtar, M.S., Chauhan, D., Ghosal, D., Poria, S., Ekbal, A., Bhattacharyya, P.: Multi-task learning for multi-modal emotion recognition and sentiment analysis. In: Proceedings of the 2019 Conference of the North American Chapter of the Association for Computational Linguistics: Human Language Technologies, vol. 1. (Long and Short Papers), pp. 370–379 (2019)
30. Garg, S., Chaturvedi, D., Jain, T., Mishra, A., & Kapoor, A.: Sentiment Analysis of Twitter Data using Machine Learning: A Case Study of SVM Algorithm. (2023)
31. Du, Q., Hoste, V.: A multi-task framework with enhanced hierarchical attention for sentiment analysis on classical Chinese poetry: utilizing information from short lines. In: Proceedings of the 4th International Conference on Natural Language Processing for Digital Humanities, pp. 113–122 (2024)
32. Chauhan, D.S., Dhanush, S.R., Ekbal, A., Bhattacharyya, P.: Sentiment and emotion help sarcasm? A multi-task learning framework for multi-modal sarcasm, sentiment and emotion analysis. In: Conference: Proceedings of the 58th Annual Meeting of the Association for Computational Linguistics (2020)
33. Vaswani, A., et al.: Attention is all you need. Adv. Neural Inf. Proc. Syst. 30, 5998–6008 (2017). arXiv:1706.03762v5.
34. Kotei, E., Thirunavukarasu, R.: A systematic review of transformer-based pre-trained language models through self-supervised learning. Information. **14**(3) (2023)
35. Nguyen, D.Q., Vu, T., Nguyen, A.T.: BERTweet: a pre-trained language model for English tweets. In: Proceedings of the 2020 Conference on Empirical Methods in Natural Language Processing: System Demonstrations (EMNLP-Demos), pp. 9–14 (2020)
36. Vu, B., Keshri, N., Chandna, S., Jalali, M., Mehraeen, S.: A systematic approach to fine-tuning transformers for emotion detection on the empathetic dialogues benchmark. Int. J. Inf. Technol. **17**(7), 3895–3912 (2025)
37. Kryvasheyeu, Y., Chen, H., Moro, E., Hentenryck, P. V., & Cebrian, M.: Performance of Social Network Sensors during Hurricane Sandy. (2015)

38. Shaver, P., Schwartz, J., Kirson, D., O'Connor, C.: Emotion knowledge: further exploration of a prototype approach. J. Pers. Soc. Psychol. **52**(6), 1061–1086 (1987)
39. Bruyne, L.D., Clercq, O.D., Hoste, V.: An emotional mess! Deciding on a framework for building a Dutch emotion-annotated corpus. In: Conference on Language Resources and Evaluation (2020)
40. Williams, G., Mahmoud, A.: Analyzing, classifying, and interpreting emotions in software users' tweets. In: IEEE/ACM International Workshop on Emotional Awareness in Software Engineering (SEmotion) (2017)
41. Najafabadi, M.K., Ko, T.Z.C., Chaeikar, S.S., Shabani, N.: A multi-level embedding framework for decoding sarcasm using context, emotion, and sentiment feature. Electronics. **13**(22), 4429 (2024)
42. Wang, K., Jing, Z., Su, Y., & Han, Y.: Large Language Models on Fine-grained Emotion Detection Dataset with Data Augmentation and Transfer Learning. (2024). https://arxiv.org/abs/2403.06108.
43. Praveena, S.: Emotion classification using BERT: a comprehensive study. Tuijin Jishu/Journal Propul. Technol. **45**(4) (2024)
44. Kendall, A., Gal, Y., Cipolla, R.: Multi-task learning using uncertainty to weigh losses for scene geometry and semantics. In: Proceedings of the IEEE/CVF Conference on Computer Vision and Pattern Recognition (CVPR), pp. 7482–7491 (2018)
45. Little, A.: Emotion classification in text using multi-task learning with transformers. In: CS224N Natural Language Processing with Deep Learning Final Project. Stanford University (2024)
46. Zhang, T., Irsan, I.C., Thung, F., Lo, D.: Revisiting sentiment analysis for software engineering in the era of large language models. ACM Trans. Softw. Eng. Methodol. **34**(3), 1–30 (2025)
47. Socher, R., et al.: Recursive deep models for semantic compositionality over a sentiment treebank. In: Proceedings of the Conference on Empirical Methods in Natural Language Processing (EMNLP) (2013)
48. Michel, P., & Neubig, G.: MTNT: A testbed for machine translation of noisy text. (2018). arXiv preprint https://arxiv.org/abs/1809.00388.
49. Demszky, D., Movshovitz-Attias, D., Ko, J., Cowen, A., Nemade, G., Ravi, S.: GoEmotions: a dataset of fine-grained emotions. In: Proceedings of the 58th Annual Meeting of the Association for Computational Linguistics (ACL), pp. 4040–4054 (2020)
50. Zhang, W., Deng, L.F., Zhang, L., Wu, D.R.: A survey on negative transfer. IEEE/CAA J. Automatica Sinica. **10**(2), 305–329 (2023)
51. Wu, W., et al.: Handling ambiguity in emotion: From out-of-domain detection to distribution estimation. (2024). https://arxiv.org/abs/2402.12862.
52. Hinojosa Lee, M.C., Braet, J., Springael, J.: Performance metrics for multilabel emotion classification: comparing micro, macro, and weighted F1-scores. Appl. Sci. **14**(21), 9863 (2024)
53. Ding, D., Lu, Z., Wang, S., Cheng, R., & Boddeti, V. N.: Mitigating task interference in multi-task learning via explicit task routing with non-learnable primitives. (2023). https://arxiv.org/abs/2308.02066.
54. Son, S., Saurez, A., & Har, D.: Continuous adversarial text representation learning for affective recognition. (2025). https://arxiv.org/abs/2502.20613.
55. Nguyen, D.Q., Vu, T., Nguyen, A.T.: BERTweet: a pre-trained language model for English tweets. In: Proceedings of the 2020 Conference on Empirical Methods in Natural Language Processing: System Demonstrations, pp. 9–14 (2020)
56. Hugging Face: DistilRoBERTa-base model card. HuggingFace.co. (2023)

Adaptive Transformer-Based Tri-Modal Fusion of RGB, Thermal, and LiDAR Data for Robust Object Detection in UAV Surveillance

Mohamad Tamim Bakrou, Hazlina Selamat(✉), and Anas Aburaya

Faculty of Electrical Engineering, Universiti Teknologi Malaysia (UTM), 81310 Johor Bahru, Malaysia
hazlina@utm.my

Abstract. Robust object detection in Unmanned Aerial Vehicle (UAV) surveillance is frequently compromised by dynamic environmental degradations such as low illumination, atmospheric obscurants (fog, smoke), and sensor failures. While multi-modal sensor fusion offers a potential solution by leveraging complementary data, existing approaches often rely on static integration rules that fail to adapt when specific sensors become unreliable. This rigidity can lead to "negative transfer," where the inclusion of noisy or degraded data reduces performance below that of the single best sensor. To address this, this paper presents FuseFormerNet, an adaptive tri-modal fusion framework that robustly integrates RGB, thermal, and LiDAR data. The proposed architecture introduces a novel mid-level fusion mechanism comprising two key components: (1) a context-conditioned reliability gating module that dynamically regresses modality importance weights to suppress degraded sensors before fusion, and (2) a transformer-based fusion block that utilizes multi-head self-attention to facilitate content-aware feature refinement across modalities. The approach is benchmarked on the challenging R-LiViT dataset using a rigorous "detector-fair" protocol to isolate fusion benefits. Experimental results demonstrate that FuseFormerNet achieves 67.14% mAP@0.5, outperforming static tri-modal fusion by 1.41% and the RGB-only baseline by 4.80%. Crucially, the system exhibits superior robustness, maintaining high recall in darkness and synthetic smoke where single-modality detectors fail.

Keywords: Adaptive sensor fusion · Object detection · RGB · thermal · LiDAR · Transformer attention · UAV surveillance

1 Introduction

Reliable object detection forms the foundation of intelligent perception systems across diverse platforms, particularly Unmanned Aerial Vehicles (UAVs)

The authors thank Universiti Teknologi Malaysia and the Ministry of Higher Education Malaysia for their supports via research grants 00Q18 and 00Q58.

N. A. Mohd Subha et al. (Eds.): SMS 2026, CCIS 2956, pp. 322–336, 2026.
https://doi.org/10.1007/978-981-92-0306-2_23

deployed for surveillance and search-and-rescue. Real-world deployments face dynamic environmental challenges, ranging from abrupt illumination changes and atmospheric interference (fog, smoke) to partial occlusions, that can degrade perception performance unpredictably [1–5]. While single-modality approaches based on RGB cameras remain predominant due to their low cost and rich texture information, they exhibit fundamental limitations under adversarial conditions. For instance, standard RGB detectors fail catastrophically under low illumination (e.g., nighttime) or when visual features are obscured by haze.

To overcome these limitations, multi-sensor fusion has emerged as a promising direction. The three sensing modalities considered in this work, RGB, thermal, and LiDAR, offer fundamentally complementary strengths. RGB cameras capture rich texture and color essential for classification; thermal sensors detect heat signatures independent of ambient light, enabling visibility in total darkness or through smoke; and LiDAR provides precise 3D geometric structure unaffected by lighting or thermal crossover. This complementarity implies that an intelligent fusion of all three should robustly outperform any single sensor.

However, realizing this potential is non-trivial because each sensor also has unique failure modes: thermal cameras suffer from saturation and texture loss, while LiDAR becomes sparse at extended ranges or degrades in heavy precipitation [6]. Therefore, a robust fusion system must not only combine modalities but also dynamically adapt to the varying reliability of each sensor to prevent "negative transfer," where a degraded sensor corrupts the fused representation.

2 Related Work

Modern object detection has evolved from two-stage methods to efficient single-stage architectures like YOLOv8 [7,8], achieving real-time performance critical for UAVs [9–11]. Recent edge-efficient designs further emphasize lightweight architectures for constrained platforms [12–14]. Despite these architectural advances, single-modality detectors remain bound by sensor physics. RGB blind spots in darkness, thermal texture loss, and LiDAR sparsity at range cannot be solved by deeper networks alone.

2.1 Multi-modal Fusion Strategies

Sensor fusion typically occurs at one of three stages: early (pixel-level), mid-level (feature-level), or late (decision-level). Early fusion requires precise spatial alignment often difficult on vibrating UAVs. Late fusion is simple but discards rich cross-modal correlations [15]. Mid-level fusion offers a principled balance, allowing encoders to extract specialized features before integrating them [16, 17]. Extensive surveys confirm its efficacy [18,19]. Bi-modal approaches (e.g., RGB-Thermal [20–22] or LiDAR-Thermal [23,24]) have shown success but still struggle in complex conditions like fog [25].

2.2 Attention Schemes

Transformer-based attention has revolutionized feature refinement [26,27]. Cross-attention [28] and feedback loops [29] help resolve occlusions [30] and memory constraints [31], motivating our design.

2.3 Existing Tri-Modal Approaches

While RGB-thermal fusion is well-studied [32], tri-modal fusion (RGB, thermal, and LiDAR) remains underexplored. Recent works like Zhang et al. [33] utilize concatenation-based fusion for foggy conditions but rely on static weights (e.g., 1/3 each). This rigidity is problematic in dynamic flights where conditions shift rapidly. Other approaches [34] modify detection heads, confounding validation of the fusion mechanism itself. Critically, few existing methods employ reliability-aware gating that can prevent negative transfer from a failed sensor without requiring explicit supervision. This gap motivates our adaptive, transformer-based approach.

2.4 Challenges and Research Gap

Table 1 reveals critical gaps in existing tri-modal integration. Three fundamental challenges remain:

1. *Heterogeneous failure modes:* RGB, thermal, and LiDAR sensors degrade differently under environmental stress (e.g., low light, thermal crossover, or geometric sparsity). Static fusion rules cannot adapt to these dynamic shifts.
2. *Negative transfer risk:* When a degraded sensor is given equal weight, it injects noise that can drag overall performance below single-modality baselines. This is critical in UAV scenarios where conditions change rapidly [17,35].
3. *Representation heterogeneity:* Combining dense 2D feature maps (RGB/Thermal) with sparse 3D point clouds (LiDAR) requires learning complex cross-modal correspondences rather than simple alignment.

No existing approach simultaneously addresses these with adaptive fusion and rigorous detector-fair evaluation. Prior methods either use static weights [33] or confounding architectural changes [34].

2.5 Contributions

This paper presents FuseFormerNet, a mid-level fusion framework for robust tri-modal object detection. Our key contributions are:

1. *Performance-driven reliability gating:* A mechanism that dynamically weights modalities based on their detection utility without requiring ground-truth reliability labels.
2. *Transformer-based fusion:* A cross-modal attention block that enables content-aware feature interaction, allowing sensors to refine each other.

Table 1. Comparison with Representative Fusion Approaches

Work	Mod.	Stage	Adapt.	Trans.	Det. Fair	UAV	Interp.
Chen et al. [36]	R+T	Early	×	×	×	×	Ltd.
Zhang et al. [16]	Bi	Mid	Part.	✓	×	×	Mod.
Zhang et al. [33]	R+T+L	Hybrid	×	×	×	×	Ltd.
Mehrotra et al. [34]	R+T+L	Late	×	Part.	×	×	Ltd.
Kim et al. [37]	Multi	Hybrid	×	✓	×	✓	Ltd.
Proposed	**R+T+L**	**Mid**	✓	✓	✓	✓	**Comp.**

3. *Detector-fair protocol:* We maintain an identical YOLOv8 detection head across all experiments to isolate fusion benefits from architectural confounds.
4. *Robustness and Interpretability:* The framework features graceful degradation via modality dropout and provides feature-level analysis validating the learned fusion behavior.

3 Proposed Method

3.1 Overall Architecture

Figure 1 illustrates the FuseFormerNet pipeline:

1. Modality Preparation: RGB I_{rgb}, thermal I_{th}, LiDAR BEV I_{ld} resized to 640×640.
2. Feature Extraction: Modality-specific YOLOv8 encoders extract P4 features $F_m \in \mathbb{R}^{512 \times 40 \times 40}$.
3. Adaptive Fusion: Context-conditioned gating produces weights $\mathbf{w}$; transformer processes weighted tokens to produce F_{fused}.
4. Detection: Unchanged YOLOv8 head produces predictions.

FuseFormerNet follows a mid-level, detector-fair design: all modalities are fused at the same semantic depth (P4) and the detection head is kept identical across all configurations. The fusion stage combines two complementary mechanisms. First, a reliability gate performs per-sample global reweighting to suppress degraded modalities and reduce negative transfer. Second, a transformer fusion block performs content-aware cross-modal interaction after reweighting, enabling each modality to refine others via attention.

3.2 Feature Extraction

Each modality is processed by an encoder E_m with the same YOLOv8 architecture but independent weights:

$$F_m = E_m(I_m) \in \mathbb{R}^{C \times H \times W}, \tag{1}$$

where $C = 512$ and $H = W = 40$ at P4 (stride 16) for 640×640 inputs. Independent weights allow specialization (texture for RGB, heat for thermal, occupancy for LiDAR) while maintaining consistent shapes for P4 fusion.

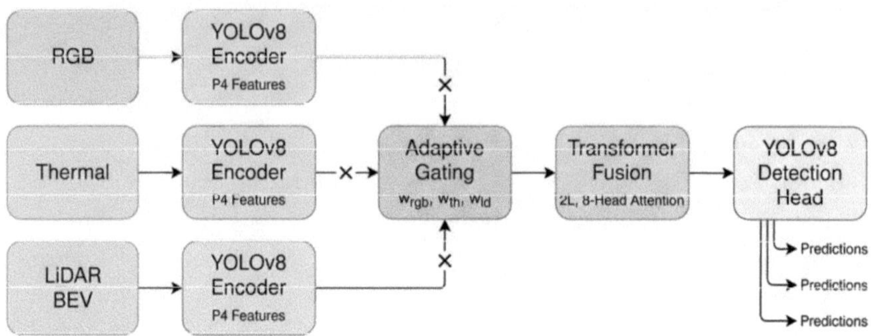

Fig. 1. FuseFormerNet Architecture: Tri-modal inputs are encoded, reliability-weighted, fused via transformer, and processed by YOLOv8.

3.3 LiDAR BEV Representation

Point clouds are projected into a BEV grid ($x, y \in [-50, 50]$m, 0.25m res) to create a dense 2D tensor (z_{max}, density ρ, intensity $\bar{I}$). This is resized to 640×640 to align with other modalities.

3.4 Context-Conditioned Reliability Gating

Global average pooling produces a descriptor $c_m = \text{GAP}(F_m)$. Concatenated descriptors $\mathbf{c}$ feed a lightweight MLP:

$$\mathbf{z} = W_2\,\sigma(W_1\mathbf{c} + b_1) + b_2, \quad \mathbf{w} = \text{softmax}(\mathbf{z}). \tag{2}$$

$$\mathbf{z} = W_2\,\sigma(W_1\mathbf{c} + b_1) + b_2, \quad \mathbf{w} = \text{softmax}(\mathbf{z}). \tag{3}$$

The gate structure, illustrated in Fig. 2, processes the global context vectors through a two-layer perceptron to regress modality-specific importance weights $\mathbf{w}$. This allows the network to learn to upweight reliable modalities and suppress those contributing noise, effectively implementing a soft-switching mechanism driven entirely by detection supervision.

3.5 Transformer-Based Fusion

We weight features $\tilde{F}_m = w_m \cdot F_m$ and flatten them to tokens X_m. Concatenated tokens X enter a 2-layer transformer:

$$Z = \mathcal{T}(X), \tag{4}$$

$$Z = \mathcal{T}(X), \tag{5}$$

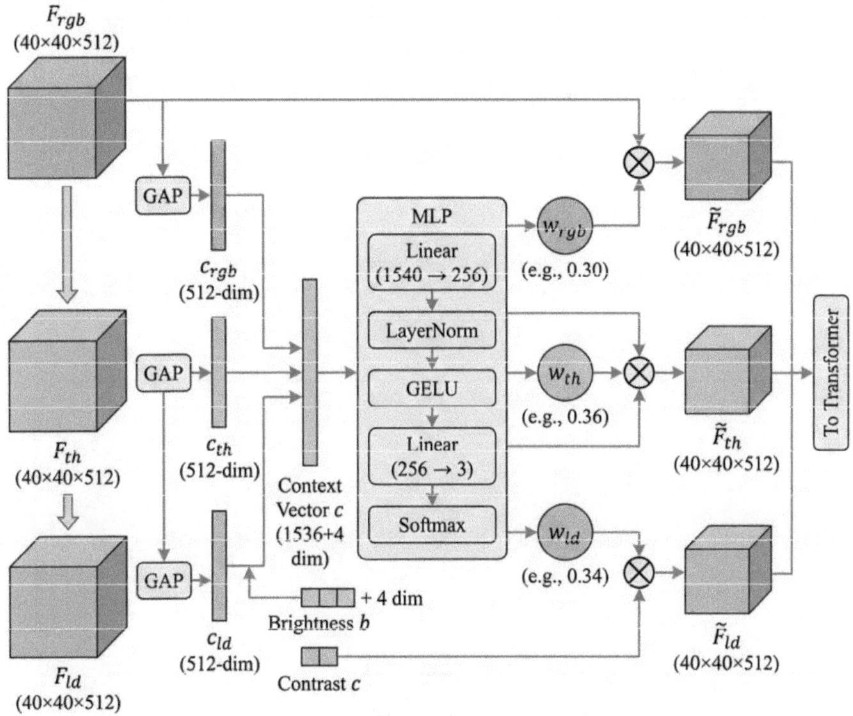

Fig. 2. Reliability Gating: Pooled P4 features drive MLP weighting.

enabling content-aware cross-modal interaction. As depicted in Fig. 3, the transformer block consists of multi-head self-attention layers that allow features from one modality to attend to complementary features in others (e.g., thermal texture clarifying an ambiguous LiDAR shape). Fused P4 features are recovered by averaging and reshaping modality segments. This local refinement complements the global gating mechanism shown in Fig. 1, providing a hierarchical approach to fusion: gating handles sensor reliability, while the transformer handles feature integration.

3.6 Detector Fairness and Robustness

We use an identical 11.2M parameter YOLOv8 head for all experiments. Training uses modality dropout ($p = 0.1$); at inference, weights are renormalized if sensors are missing $\mathbf{w}' = (\mathbf{w} \odot \mathbf{m})/(\sum w_k m_k)$.

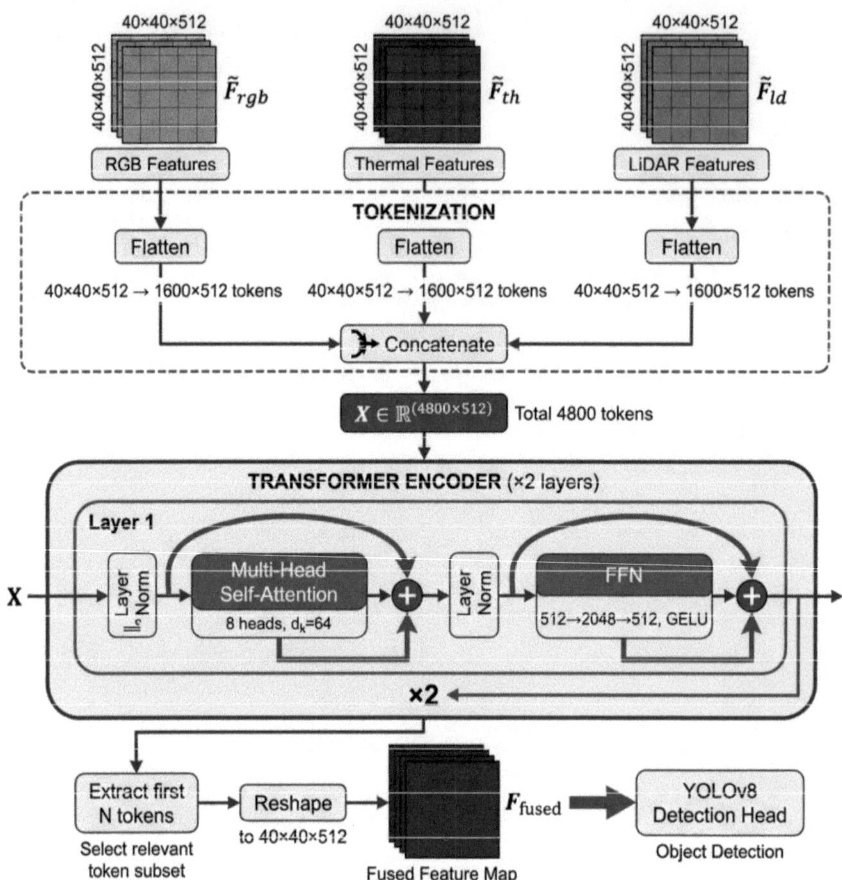

Fig. 3. Transformer Fusion Block: Enables cross-modal attention.

3.7 Training Objective

The model is trained end-to-end with standard YOLOv8 losses: $\mathcal{L} = \lambda_{box}\mathcal{L}_{box} + \lambda_{cls}\mathcal{L}_{cls} + \lambda_{dfl}\mathcal{L}_{dfl}$.

4 Experimental Setup

4.1 Dataset and Implementation

We use the R-LiViT benchmark [38] (2,400 RGB-thermal-LiDAR frames, 8 classes), extending earlier multispectral efforts [39]. Implementation uses YOLOv8s (small version), resizing inputs to 640^2. Training: batch 8, 20 epochs (10 frozen backbone), AdamW [40] (lr=0.01). The gating MLP reduces channels to $C/4$, and the transformer has 2 layers (8 heads, dim 512).

4.2 Implementation Details

To ensure reproducibility, the experiments are implemented as follows:

1. **Data Preprocessing**: All input images (RGB, Thermal) are resized to 640 × 640 pixels. LiDAR point clouds are voxelized and projected into a 3-channel Bird's Eye View (BEV) map encoding density, maximum height, and intensity, also resized to 640 × 640. Augmentations including mosaic and mixup are applied during training.
2. **Model Initialization**: The YOLOv8 backbones are initialized with COCO-pretrained weights to leverage general visual features. The fusion modules (gating MLP and transformer) are initialized randomly using Xavier initialization.
3. **Training Protocol**: Training proceeds in two stages. First, the backbones are frozen for 10 epochs to allow the fusion layers to adapt to the multi-modal feature space. Second, the entire network is un-frozen and fine-tuned end-to-end for another 10 epochs.
4. **Inference**: During testing, the reliability gate predicts weights $\mathbf{w}$ for each test sample. If a sensor is simulated as "missing" (for robustness testing), its corresponding input is zeroed-out, and the remaining weights are renormalized to sum to one before fusion.

5 Results and Discussion

Table 2 shows independently trained single-modality baselines. RGB leads in favorable lighting (62.34% mAP), while thermal (58.91%) and LiDAR (50.43%) provide critical backup when visual data fails.

Table 3 presents a comprehensive evaluation of the proposed framework. FuseFormerNet achieves 67.14% mAP@0.5, surpassing the static baseline (65.73%) by +1.41% and the RGB baseline (62.34%) by +4.80%.

Table 2. Single-Modality Baseline Performance

Mod.	mAP@.5	mAP@.5:.95	P	R
RGB	62.34	44.12	85.12	51.02
Thermal	58.91	41.56	82.34	48.76
LiDAR	50.43	35.21	78.92	42.34

The Component Ablation section of Table 3 (formerly Tables 4 and 5 in preliminary versions) isolates the contribution of each module. Adding reliability gating alone improves mAP by +0.48%, showing the value of downweighting noisy sensors. Adding transformer fusion alone yields +0.75%, validating cross-modal attention. The combination achieves +1.41%, demonstrating that

Table 3. Comprehensive Performance Analysis: Component Ablation and Modality Robustness

Configuration	mAP@.5(%)	mAP@.5:.95(%)	R(%)
Component Ablation			
Static Baseline	65.73	47.06	54.69
+ Reliability Gating	66.21	–	56.12
+ Transformer Fusion	66.48	–	56.87
FuseFormerNet (Full)	**67.14**	**47.90**	**58.23**
Modality Robustness (Inference)			
RGB+Thermal	65.23	–	–
RGB+LiDAR	63.87	–	–
Thermal+LiDAR	61.02	–	–

Fig. 4. Dark Scene Comparison: (a) RGB-only detection struggling with low illumination, (b) thermal-only detection identifying heat signatures, and (c) tri-modal fusion successfully detecting all pedestrians and vehicles.

gating and attention are complementary: gating globally suppresses unreliable modalities, while the transformer locally refines features.

Figure 4 provides a visual comparison in low-light conditions, demonstrating the capability of the fusion model to recover objects completely missed by the RGB baseline. While the RGB image is dominated by noise and poor contrast, the fusion model leverages thermal signatures to accurately localize the vehicle and pedestrians, highlighting the critical role of non-visual modalities in darkness.

The benefit of fusion is significant in challenging conditions. In the night scene shown in Fig. 5, RGB detection fails completely (TP = 0) due to low illumination. However, the fused model successfully detects all 7 objects (TP = 7) with high confidence by leveraging thermal heat signatures and LiDAR geometry. This confirms that the system effectively prevents negative transfer from the dark RGB input.

Figure 6 evaluates robustness under synthetic smoke. While RGB performance typically degrades significantly as smoke density increases, the proposed fusion framework maintains robustness by relying on thermal and LiDAR data. Figure 6 visually juxtaposes the failure of the RGB detector against the stable

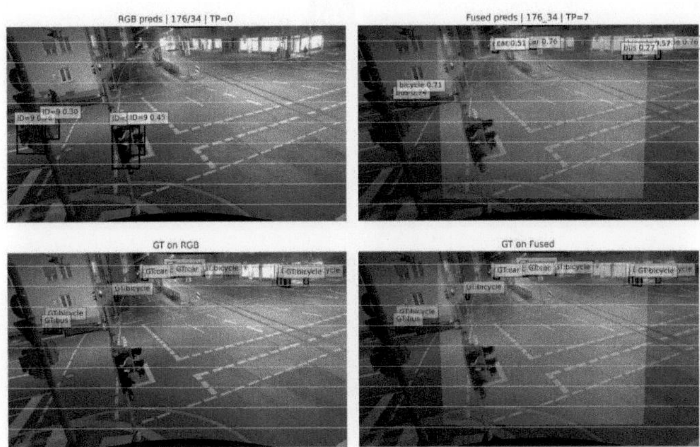

Fig. 5. Night Scene Comparison: (a) RGB detections failing (TP=0) in low light, (b) FuseFormerNet detections with high recall (TP=7), (c) ground truth on RGB, and (d) ground truth on fused output.

detections of the Fusion model, illustrating the system's ability to switch effective modalities under potential visual obstruction.

To validate that the model learns meaningful features, Fig. 7 visualizes the fused P4 representations. Figure 7(b) shows high activation intensity concen-

Fig. 6. Smoke Robustness Comparison: (a) performance in clear conditions, (b) light smoke, (c) medium smoke, and (d) heavy smoke, demonstrating FuseFormerNet's stability as visibility degrades.

Fig. 7. P4 Feature Analysis: (a) fused feature map, (b) intensity, (c) entropy, (d) variance.

trated on objects. Figure 7(c) reveals low entropy on object regions (indicating high confidence) and high entropy on backgrounds. Figure 7(d) shows controlled variance. This confirms the transformer fusion is refining features to reduce ambiguity rather than just mixing noise.

A key contribution is the learned scene-adaptive weighting (Fig. 8). Under optimal lighting, weights are balanced (33% each). In darkness, reliance shifts to thermal (40%) and LiDAR (35%) while RGB drops (25%). Under smoke, RGB is further suppressed to 20%. This automatic reweighting aligns with physical sensor reliability without requiring explicit supervision.

This reweighting reflects the physical reality that heat signatures remain discriminative when visible-spectrum information degrades, while geometric cues from LiDAR gain relative importance. Under smoke conditions representing severe degradation, thermal receives the highest weight (42%), LiDAR 38%, and RGB only 20%, demonstrating automatic suppression of the most compromised modality. Critically, these weights emerge purely from end-to-end detection supervision without ground-truth reliability labels, validating the performance-driven adaptivity approach. The learned weighting patterns align with physical sensor characteristics and demonstrate that the network has internalized modality-specific failure modes, enabling robust operation across diverse environmental conditions.

The Modality Robustness section of Table 3 demonstrates remarkable resilience to sensor failure, reporting mAP averaged across the entire R-LiViT test set. Even when reduced to thermal+LiDAR (missing the texture-rich RGB), the system retains 61.02% mAP, validating the effectiveness of the modality-dropout training strategy.

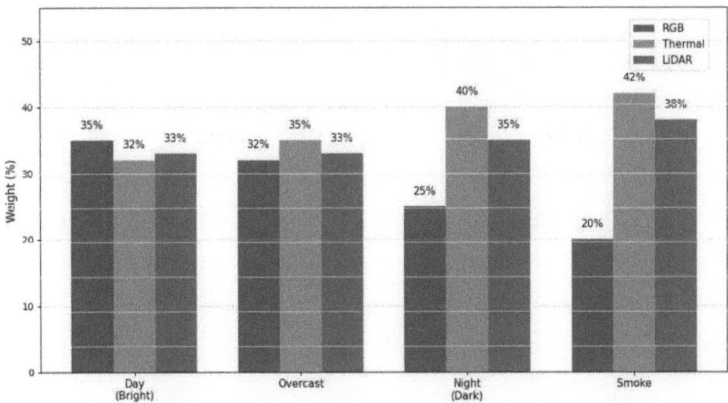

Fig. 8. Learned Adaptive Weights Across Environments: (a) balance in bright day, (b) shift in overcast conditions, (c) thermal/LiDAR dominance in night (dark), and (d) maximum RGB suppression in heavy smoke.

A comprehensive qualitative comparison of detection performance across all modalities in a challenging night scene is presented in Fig. 9, organized in a two-row layout. The top row displays raw input modalities and reference: (a) RGB input showing the visible-spectrum image with severely degraded visibility due to low illumination, (b) thermal input displaying heat signatures as bright regions, (c) LiDAR BEV representation showing geometric structure, and (d) ground truth bounding boxes. The bottom row presents the corresponding detection results: (e) RGB-only detection struggling significantly due to poor illumination, (f) thermal-only detection performing better by leveraging heat signatures, (g) LiDAR-only detection providing geometric cues, and (h) FuseFormerNet (ours) that successfully combines the complementary strengths of all three modalities, achieving the most complete and accurate detection coverage with higher confidence scores. The fusion approach effectively compensates for RGB's failure under low light by incorporating thermal's illumination-invariant features and LiDAR's geometric cues.

6 Limitations

While promising, our approach has limitations. First, it assumes perfect sensor calibration; real-world misalignments from vibration or different frame rates remain a challenge. Second, fusion occurs only at the P4 scale, potentially reducing precision for very small or large objects compared to multi-scale approaches, though this trade-off favors efficiency.

Third, validaton utilizes the R-LiViT benchmark; testing on broader datasets with extreme weather (e.g., heavy rain or sandstorms) is needed for full generalization. Finally, the transformer block increases inference latency (from 15ms to

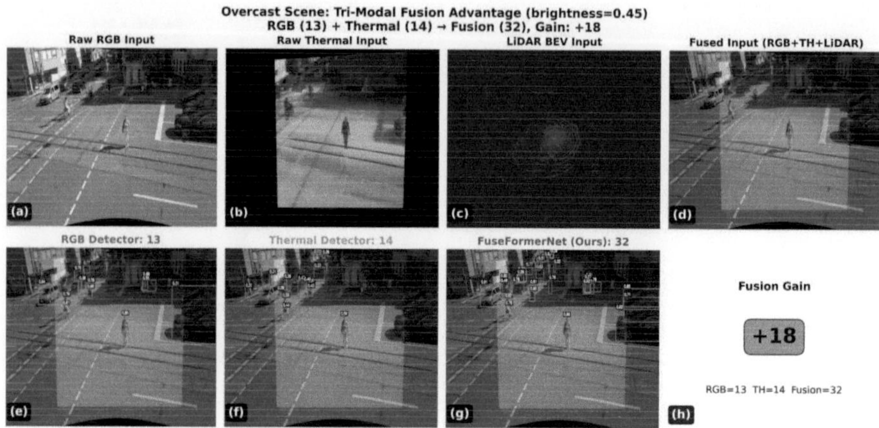

Fig. 9. Qualitative Comparison: (a) RGB image, (b) thermal, (c) LiDAR BEV, (d) RGB detection, (e) thermal detection, (f) LiDAR detection, (g) ground truth, and (h) FuseFormerNet (ours).

28ms), which may require optimization (e.g., pruning or distillation) for deployment on highly constrained edge platforms like the Jetson Orin. Learned weights currently rely on implicit supervision, where explicit reliability labels could further enhance robustness.

7 Conclusion

This paper presented FuseFormerNet, an adaptive transformer-based mid-level fusion framework for tri-modal object detection integrating RGB, thermal, and LiDAR modalities. The core contribution is a performance-driven adaptive fusion mechanism that learns to weight modalities based on their relative informativeness for detection, combined with transformer-based cross-modal attention and a detector fairness protocol for rigorous evaluation. Experiments on the R-LiViT benchmark demonstrate 67.14% mAP@0.5, outperforming static tri-modal fusion by 1.41% and the RGB-only baseline by 4.80%. While evaluated on UAV surveillance data, the framework is designed as a generalizable approach applicable to other multi-modal perception platforms. Future work includes extending to multi-scale and temporal fusion, developing efficient edge-deployable variants, and incorporating explicit reliability supervision for improved robustness to novel degradation patterns.

Acknowledgements. *AI Tool Disclosure.* AI-assisted tools were used for language polishing and formatting; all technical content and experiments were conducted by the authors.

References

1. Kumar, A., et al.: Multi-sensor uav platforms for environmental monitoring. IEEE Sens. J. **25**(8), 5432–5445 (2024)
2. Zhang, H., et al.: Uav-detr: Efficient end-to-end object detection for uav imagery. IEEE Access **12**, 4567–4580 (2024a)
3. Smith, J., Zhao, L.: Multi-sensor data fusion for autonomous flight of unmanned aerial vehicles in complex flight environments. J. Aerospace Inf. Syst. **21**(3), 456–470 (2024)
4. Dong, Y., et al.: Securing the skies: a comprehensive survey on anti-uav methods. arXiv preprint arXiv:2504.11967 (2025)
5. Patel, S., Roy, D.: Adaptive sensor fusion for uav navigation in dynamic environments. IEEE Sens. J. **24**(5), 7890–7901 (2024)
6. Lin, D., et al.: Adverse weather robust lidar object detection. IEEE Trans. Intell. Transp. Syst. **24**(5), 4532–4544 (2023)
7. Li, M., Zhao, Q., Liu, T.: Real-time object detection in uav imagery using enhanced yolov5. IEEE Access **11**, 45678–45689 (2023)
8. Wang, P., et al.: Yolov8-based object detection for uav systems in urban environments. IEEE Access **13**, 45678–45690 (2025)
9. Ali, M., Park, K.: Yolov3-drone: Fast object detection for aerial vehicles. Sensors **22**(11), 4102 (2022)
10. Silva, R., et al.: Yolov8 for uav-based small object detection: dataset challenges and model improvements. IEEE Trans. Image Process. **33**, 9021–9034 (2024)
11. Yang, S.-Y., Cheng, H.-Y., Yu, C.-C.: Real-time object detection and tracking for uavs using cnns. Electronics **12**(24), 4928 (2023)
12. Choi, K., Moon, J., Jung, H.G., Suhr, J.K.: Real-time object detection and tracking based on embedded edge devices. Electronics **13**(5), 811 (2024)
13. Roy, A., et al.: Edge-efficient multi-modal fusion for robotics: Lightweight design and deployment. IEEE Trans. Rob. **40**(2), 245–259 (2025)
14. Zhu, J., et al.: Aerialvision: a survey on object detection in uav imagery. IEEE Access **11**, 88123–88145 (2023)
15. Kim, Y., et al.: Late fusion strategies for multi-modal perception in autonomous driving. IEEE Trans. Intell. Transp. Syst. **24**(2), 1499–1509 (2024)
16. Zhang, L., et al.: Transformer-based multi-modal fusion for uav surveillance. IEEE Trans. Geosci. Remote Sens. **61**, 1–12 (2024b)
17. Han, S., Liu, L., Zhang, Y.: A survey on multi-modal fusion strategies for robotics: early, mid, and late fusion tradeoffs. IEEE Transactions on Cybernetics **54**(2), 1234–1249 (2024)
18. Stein, A.N., et al.: A survey of multi-sensor fusion methods in object detection. IEEE Trans. Cybern. **53**(1), 221–233 (2023)
19. Brenner, M., Smith, A., Wang, L.: Rgb-d and thermal sensor fusion: a systematic review. IEEE Access **11**, 82411–82430 (2023)
20. Wang, H., et al.: Illumination-aware object detection using rgb and thermal images. IEEE Trans. Instrum. Meas. **72**, 1–13 (2023)
21. Kim, J., et al.: Low-light object detection via thermal-rgb fusion networks. Sensors **21**(7), 2456 (2021)
22. Wei, N., et al.: Thermal-rgb fusion networks for uav night-time detection. Sensors **21**(15), 4987 (2021)
23. Lee, D., Kim, M.: Fusion of lidar and thermal imaging for robotic navigation. IEEE Access **13**, 23456–23469 (2024)

24. Tan, D., et al.: Feature-level fusion of lidar and thermal for occlusion-robust detection. IEEE Access **11**, 56789–56799 (2023)
25. Elshamy, M., et al.: Challenges in uav vision under fog and low light: an rgb-based study. Remote Sensing **15**(1), 112 (2023)
26. Nguyen, T., Tran, P., Vo, L.: Efficient transformer architectures for uav-based object detection. IEEE Trans. Image Process. **33**, 987–998 (2024)
27. Luo, H., Zhang, K., Chen, X.: Cross-attention-based transformer fusion for multi-sensor robotic perception. IEEE Trans. Rob. **40**(1), 88–101 (2025)
28. Sun, X., et al.: Target-aware bidirectional fusion transformer for aerial object tracking. arXiv preprint arXiv:2503.09951 (2025)
29. Gao, N., et al.: Feedback-aware transformer architectures for occlusion-resilient object detection. IEEE Trans. Image Process. **34**(6), 11523–11535 (2025)
30. Zhao, Z., et al.: Occluawarenet: Occlusion-robust fusion for aerial object tracking. IEEE Trans. Image Process. **32**(8), 10123–10136 (2024)
31. Fu, M., et al.: Attention-based memory for improving uav navigation. IEEE Access **10**, 76598–76609 (2022)
32. Wang, Y., Zheng, Z., Chai, L.: Llvip: a visible-infrared paired dataset for low-light vision. In: Proc. IEEE/CVF Conf. Computer Vision and Pattern Recognition Workshops (CVPRW), pp. 3496–3504 (2021)
33. Zhang, M., et al.: Tri-modal fusion for robust object detection in foggy and night-time environments. IEEE Access **11**, 8874–8886 (2023)
34. Mehrotra, A., Huang, P.: Adaptive feature fusion with attention for rgb-thermal-lidar object detection. IEEE Sens. J. **25**(10), 11345–11359 (2025)
35. Liu, B., et al.: Safefusion: Sensor fusion for adversarial robustness in uav surveillance. IEEE Access **11**, 12345–12357 (2023)
36. Chen, R., et al.: Pixel-level alignment for cross-modal early fusion. Sensors **22**(6), 3132 (2022)
37. Kim, D., Park, J.: Aerofusionnet: occlusion-aware multi-scale detection for uav surveillance. IEEE Trans. Intell. Transp. Syst. **25**(1), 234–246 (2025)
38. Wang, X., et al.: R-livit: Rgb, lidar, and thermal dataset for uav surveillance. IEEE Trans. Intell. Transp. Syst. **25**(2), 1123–1135 (2024)
39. Hwang, S., Park, J., Kim, N., Choi, Y., Kweon, I.S.: Multispectral pedestrian detection: Benchmark dataset and baseline. In: Proc. IEEE Conf. Computer Vision and Pattern Recognition (CVPR), pp. 1037–1045 (2015)
40. Loshchilov, I., Hutter, F.: Decoupled weight decay regularization. In: Proc. Int. Conf. Learning Representations (ICLR), pp. 1–11 (2019)

Author Index

N. A. Mohd Subha et al. (Eds.): SMS 2026, CCIS 2956, pp. 337–338, 2026.
https://doi.org/10.1007/978-981-92-0306-2

MIX
Papier aus verantwortungsvollen Quellen
Paper from responsible sources
FSC® C105338

If you have any concerns about our products,
you can contact us on
ProductSafety@springernature.com

In case Publisher is established outside the EU,
the EU authorized representative is:
Springer Nature Customer Service Center GmbH
Europaplatz 3, 69115 Heidelberg, Germany

Printed by Libri Plureos GmbH
in Hamburg, Germany